AF389389

Transitioning to
Affordable and Clean Energy

Transitioning to Sustainability Series: Volume 7

Series Editor: Manfred Max Bergman

Volumes in the series:

Volume 1: Transitioning to No Poverty
ISBN 978-3-03897-860-2 (Hbk);
ISBN 978-3-03897-861-9 (PDF)

Volume 2: Transitioning to Zero Hunger
ISBN 978-3-03897-862-6 (Hbk);
ISBN 978-3-03897-863-3 (PDF)

Volume 3: Transitioning to Good Health
and Well-Being
ISBN 978-3-03897-864-0 (Hbk);
ISBN 978-3-03897-865-7 (PDF)

Volume 4: Transitioning to Quality
Education
ISBN 978-3-03897-892-3 (Hbk);
ISBN 978-3-03897-893-0 (PDF)

Volume 5: Transitioning to Gender Equality
ISBN 978-3-03897-866-4 (Hbk);
ISBN 978-3-03897-867-1 (PDF)

Volume 6: Transitioning to Clean Water
and Sanitation
ISBN 978-3-03897-774-2 (Hbk);
ISBN 978-3-03897-775-9 (PDF)

Volume 7: Transitioning to Affordable and
Clean Energy
ISBN 978-3-03897-776-6 (Hbk);
ISBN 978-3-03897-777-3 (PDF)

Volume 8: Transitioning to Decent Work
and Economic Growth
ISBN 978-3-03897-778-0 (Hbk);
ISBN 978-3-03897-779-7 (PDF)

Volume 9: Transitioning to Sustainable
Industry, Innovation and Infrastructure
ISBN 978-3-03897-868-8 (Hbk);
ISBN 978-3-03897-869-5 (PDF)

Volume 10: Transitioning to Reduced
Inequalities
ISBN 978-3-03921-160-9 (Hbk);
ISBN 978-3-03921-161-6 (PDF)

Volume 11: Transitioning to Sustainable
Cities and Communities
ISBN 978-3-03897-870-1 (Hbk);
ISBN 978-3-03897-871-8 (PDF)

Volume 12: Transitioning to Responsible
Consumption and Production
ISBN 978-3-03897-872-5 (Hbk);
ISBN 978-3-03897-873-2 (PDF)

Volume 13: Transitioning to Climate Action
ISBN 978-3-03897-874-9 (Hbk);
ISBN 978-3-03897-875-6 (PDF)

Volume 14: Transitioning to Sustainable
Life below Water
ISBN 978-3-03897-876-3 (Hbk);
ISBN 978-3-03897-877-0 (PDF)

Volume 15: Transitioning to Sustainable
Life on Land
ISBN 978-3-03897-878-7 (Hbk);
ISBN 978-3-03897-879-4 (PDF)

Volume 16: Transitioning to Peace, Justice
and Strong Institutions
ISBN 978-3-03897-880-0 (Hbk);
ISBN 978-3-03897-881-7 (PDF)

Volume 17: Transitioning to Strong
Partnerships for the Sustainable
Development Goals
ISBN 978-3-03897-882-4 (Hbk);
ISBN 978-3-03897-883-1 (PDF)

Edwin C. Constable (Ed.)

Transitioning to
Affordable and Clean Energy

Transitioning to Sustainability Series

MDPI • Basel • Beijing • Wuhan • Barcelona • Belgrade • Manchester • Tianjin • Tokyo • Cluj

EDITOR
Edwin C. Constable
University of Basel
Switzerland

EDITORIAL OFFICE
MDPI
St. Alban-Anlage 66
4052 Basel, Switzerland

For citation purposes, cite each article independently as indicated below:

Author 1, and Author 2. 2022. Chapter Title. In *Transitioning to Affordable and Clean Energy*. Edited by Edwin C. Constable. Transitioning to Sustainability Series 7. Basel: MDPI, Page Range.

Published with the generous support of the Swiss National Science Foundation.

ISBN 978-3-03897-776-6 (Hbk)
ISBN 978-3-03897-777-3 (PDF)
ISSN: 2624-9324 (Print)
ISSN: 2624-9332 (Online)
doi.org/10.3390/books978-3-03897-777-3

Contents

About the Editor

Edwin C. Constable was born in Edinburgh but grew up in Hastings. He studied chemistry at Oxford, where he gained a B.A. and D.Phil. He moved to the University of Cambridge before being appointed to a University Lectureship and Fellowship of Robinson College in 1984. He remained at Cambridge until 1993 when he was appointed Professor of Inorganic Chemistry at the University of Basel. In 2000 he took up a position as Professor of Chemistry at the University of Birmingham (England) where he was appointed Head of School. He returned to the University of Basel as Professor of Chemistry in 2002. He publishes in all areas of chemistry with over 650 peer-reviewed publications and is highly cited (22,000 citations, h-index 74). His interests cover all aspects of chemistry, chemical history and the communication of science. He is Editor-in-Chief for a new open access journal Chemistry. He has held an ERC Advanced Grant and has been Research Dean and Vice-President of the University of Basel. He is currently Chairman of the Swiss Academies Expert Group on Research Integrity and President of Euresearch. When he is not being a chemist, he is to be found chasing insects and pursuing his love of photography.

Contributors

ALBERTO BEZAMA
Prof., Department of Bioenergy, Helmholtz
Centre for Environmental Research, UFZ,
Leipzig, Germany

ARIF AHMED
Dr., Energy and Power Systems
Department, TUMCREATE, Singapore

BIHTER AVŞAR
Dr., Faculty of Engineering and Natural
Sciences, Sabanci University, Istanbul,
Turkey

BILAL A. BHATTI
Dr., Pacific Northwest National Laboratory,
Richland, WA, USA

CRISTINA RODRIGUES
Dr., UFP Energy, Environment and Health
Research Unit (FP-ENAS), Fernando Pessoa
University, Porto, Portugal

DANIAL ESMAEILI ALIABADI
Dr., Department of Bioenergy, Helmholtz
Centre for Environmental Research, UFZ,
Leipzig, Germany
Faculty of Engineering and Natural
Sciences, Sabanci University, Istanbul,
Turkey

DANIELA THRÄN
Prof., Department of Bioenergy, Helmholtz
Centre for Environmental Research, UFZ,
Leipzig, Germany
Department of Bioenergy Systems, German
Biomass Research Center, DBFZ, Leipzig,
Germany

EKATERINA SERMYAGINA
Dr., LUT School of Energy Systems,
Lappeenranta-Lahti University of
Technology, Lappeenranta, Finland

FABIANO PALLONETTO
Dr., School of Business, Maynooth
University, Mariavilla, Maynooth, Co.
Kildare, Ireland

GERARDO FASCIA
M.Sc., Innovation Manager, Social
Enterprise "Prometeus", Lucera, Italy

HENRIQUE PINHEIRO
Dr., UFP Energy, Environment and Health
Research Unit (FP-ENAS), Fernando Pessoa
University, Porto, Portugal

JUVET NCHE FRU
Dr., Department of Physics, University of
Pretoria, Pretoria, South Africa

KAI ZHANG
Dr., Institute for High Performance
Computing, A*STAR, Singapore

KATJA KUPARINEN
Dr., LUT School of Energy Systems,
Lappeenranta-Lahti University of
Technology, Lappeenranta, Finland

MANUEL LEMOS DE SOUSA
Dr., UFP Energy, Environment and Health
Research Unit (FP-ENAS), Fernando Pessoa
University, Porto, Portugal

MARIAROSARIA LOMBARDI
Asst. Prof., Department of Economics,
University of Foggia, Foggia, Italy

MAURIZIO PROSPERI
Asst. Prof., Department of Agriculture,
Food, Natural Resource and Engineering
(DAFNE), University of Foggia, Foggia,
Italy

MD JAN ALAM
Dr., Pacific Northwest National Laboratory,
Richland, WA, USA

MMANTSAE DIALE
Prof., Department of Physics, University of
Pretoria, Pretoria, South Africa

PANNAN I. KYESMEN
Dr., Department of Physics, University of
Pretoria, Pretoria, South Africa

SARMAD HANIF
Dr., Pacific Northwest National Laboratory,
Richland, WA, USA

SATU LIPIÄINEN
M.Sc., LUT School of Energy Systems,
Lappeenranta-Lahti University of
Technology, Lappeenranta, Finland

SEBASTIAN TROITZSCH
M.Sc., Institute for High Performance
Computing, A*STAR, Singapore

TOBIAS MASSIER
Dr., Energy and Power Systems
Department, TUMCREATE, Singapore

VICKI STEVENSON
Dr., Architecture, Cardiff University,
Cardiff, Wales, UK

WARREN S. VAZ
Asst. Prof., Department of Engineering
and Engineering Technology, University of
Wisconsin-Oshkosh, Fox Cities Campus,
Menasha, WI, United States

Abstracts

The Transition towards Affordable Electricity: Tools and Methods
by Sebastian Troitzsch, Sarmad Hanif, Tobias Massier, Kai Zhang,
Bilal A. Bhatti, Arif Ahmed and Md Jan Alam

To decarbonize the electric energy sector, renewable energies are increasingly integrated into the generation mix. The main challenge, apart from the efficiency of renewable energy conversion, is maintaining the reliability of the electric grid, which is responsible for linking electric generators and consumers. To this end, the whole electric power system, covering generation, transmission, distribution, and consumption needs to be planned and operated in a cost-effective as well as reliable manner. The research in this domain has led to the development of tailor-made open-source software tools for electric grid modeling, simulation, and optimization. This chapter discusses the tools MATPOWER, GridLAB-D, MESMO, and URBS, which cater to the integration of renewable energies and other distributed energy resources (DERs) in the electric grid. The key features and applications for each tool are highlighted and compared, with a focus on district-scale electric grids, i.e., electric distribution systems. Furthermore, exemplary results are presented to emphasize suitable applications for each tool, based on a synthetic distribution system test case for Singapore.

Clean Energy Transition Challenge: The Contributions of Geology
by Cristina Rodrigues, Henrique Pinheiro and Manuel Lemos de Sousa

The transition from fossil fuel-dominant energy production to so-called carbon-neutral sources has been identified as an important new challenge seeking to address climate change. Climate change, specifically global warming, is presently considered as being intimately related to carbon dioxide (CO_2) emissions, especially those of an anthropogenic origin. The issue of CO_2 emissions of an anthropogenic origin from the combustion of fossil fuels remains rather controversial, due to the following main reasons: other greenhouse gases (GHGs) such as methane (CH_4) produce a more negative environmental effect than CO_2, and natural causes such as the sun and volcanic activities also play an important role. In addition, an important part of CO_2 emissions is unrelated to energy production, but concerns other industries such as chemical and cement production. Furthermore, it should be stated that there still exists considerable disagreement in climate models and scenarios used by the UN Framework Convention on Climate Change (UNFCCC). A workable and viable strategy towards the production of clean energy must include the capture and storage of CO_2 as one of the main targets in the energy and climate binomial strategy, despite facing criticism from some environmental organizations. The contribution of geology is not only related to the need of carbon capture and storage technologies, as already admitted in the Paris Agreement, but also to the exploitation of mineral raw materials essential to build renewable energy equipments, and, ultimately, to the underground energy storage associated to hydrogen energy production.

Use of Storage and Renewable Electricity Generation to Reduce Domestic and Transport Carbon Emissions—Whole Life Energy, Carbon and Cost Analysis of Single Dwelling Case Study (UK)
by Vicki Stevenson

This case study is a detached dwelling situated in South Wales, UK. It had a 3.6 kWp vertical photovoltaic (PV) system installed in 2014 and a 6.12 kWp roof-mounted PV system installed in 2020, along with a 13.5 kWh electricity storage device, closely followed by an electric vehicle and charger. The impact of these interventions on the reduction in domestic and transport carbon emissions is considered in relation to energy tariffs which encourage the user to shift consumption from high-carbon intensity generation times (generally matching peak consumption in the evening) to low-carbon intensity generation times (overnight). Based on the initial monitored data, the combination of renewable generation, energy storage and swapping to an electric vehicle is likely to avoid 1655.6 kg CO_2 per year operational emissions based on the UK electricity grid carbon intensity in 2020.

Sustainable Energy Future with Materials for Solar Energy Collection, Conversion, and Storage
by Juvet Nche Fru, Pannan I. Kyesmen and Mmantsae Diale

The transition to a sustainable energy future is dependent on a clean and efficient power supply. Solar power is the most attractive source of clean energy because of its abundance and numerous ways of harnessing it. Harvesting solar energy involves the use of a wide range of materials including metal oxides and halide perovskites (HaP) for conversion into hydrogen and electricity via photoelectrochemical (PEC) water splitting and photovoltaic technologies, respectively. Hematite has emerged as one of the most suitable metal oxide photocatalysts for solar hydrogen production due to its small bandgap (~2.0 eV) and stability in solution. However, the major challenges limiting the use of hematite in PEC water splitting include its low conductivity, poor charge separation, and short charge carrier lifetime. Additionally, HaP solar cells are the fastest emerging photovoltaic technology in terms of power conversion efficiency. However, their instability and toxicity of lead and solvents are major bottlenecks blocking the commercialization of this technology. This chapter reviews the strategies that have been engaged towards overcoming the limitations of using hematite and HaP for direct conversion of solar energy into hydrogen fuels and electricity, respectively. The simultaneous engagement of strategies such as nanostructuring, doping, formation of heterostructures, use of co-catalysts, and plasmonic enhancement effects has shown great promise in improving the photocatalytic water splitting capabilities of hematite. Vapor methods for preparing HaP have the potential for improving their stability and eliminate the use of toxic solvents during fabrication. More research will be required for the eventual commercialization of solar hydrogen production and photovoltaic technologies using hematite and halide perovskites, respectively.

Advanced Energy Management Systems and Demand-Side Measures for Buildings towards the Decarbonisation of Our Society
by Fabiano Pallonetto

Electricity supply/demand balancing measures have traditionally been achieved by controlling conventional generation in response to energy demand variations. However, increasing renewable generation can lead to more significant changes on the supply side, requiring a faster balancing response from grid operators. Standard generation units may not have sufficient ramping capabilities to counter high volatility in renewable energy generation. Modern forecasting techniques and advanced control systems can mitigate such challenges and flexible demand resources and demand-side management measures. Among demand-side management measures, demand response has been promoted as a critical mechanism to increase the percentage of renewable energies in the system. The widespread adoption of demand response programs leads to a paradigm shift in the way operators manage the grid. Such changes require a bi-directional communication link and advanced energy management systems that monitor building consumption and operations. Innovations such as building home automation, diffusion of intelligent appliances and energy management system integration are necessary prerequisites to boost the power system's efficiency while increasing the renewable penetration towards an affordable and clean energy supply. Combining these measures with energy management systems equipped with advanced artificial intelligence algorithms enables electricity end-users to modulate their electricity consumption by dynamically responding to fluctuations in the power generation caused by renewable. The increased capacity of the controllable load through these devices actively contributes to the higher penetration of renewable energy and the decarbonisation of our society.

Social Innovation for Energy Transition: Activation of Community Entrepreneurship in Inner Areas of Southern Italy
by Mariarosaria Lombardi, Maurizio Prosperi and Gerardo Fascia

Sustainable Development Goal no. 7 of the UN 2030 Agenda refers to changing the route of energy production and consumption to contrast climate change. One way to reach this objective is to foster the development of the renewable energy sector by promoting community entrepreneurship in rural and remote areas endowed with relevant environmental resources and containing important cultural assets. The European Union (EU) legislative framework already formally acknowledges and defines specific types of community energy initiatives that can reinforce positive social norms and support the energy transition. However, there are some concerns regarding the economic feasibility and sustainability of these initiatives in difficult contexts, such as the case of inner areas of Southern Italy, which are affected by progressive abandonment and desertification, and where the recent widespread implementation of large-scale renewable energy plants has occurred without the engagement of the local community. This has led to limited social acceptance of new investment projects in renewable energy. In light of these premises, the aim of this chapter is to propose an operational approach for developing community entrepreneurship where the renewable energy sector will provide the financial flow needed to activate initiatives for the valorization of cultural assets, tourism initiatives and civic revitalization. In this way, the energy transition may represent a unique opportunity to spur economic growth in less developed regions across the EU, capable of exerting a multiplier effect on local development and the social revitalization of local communities.

Finnish Forest Industry and Its Role in Mitigating Global Environmental Changes
by Ekaterina Sermyagina, Satu Lipiäinen and Katja Kuparinen

The forest industry is an energy-intensive sector that emits approximately 2% of industrial fossil CO_2 emissions worldwide. In Finland, the forest industry is a major contributor to wellbeing and has constantly worked on sustainability issues for several decades. The intensity of fossil fuel use has been continuously decreasing within the sector; however, there is still a lot of potential to contribute to the mitigation of environmental change. Considering the ambitious Finnish climate target to reach carbon-neutrality by 2035, the forest industry is aiming for net zero emissions by switching fossil fuels to bio-based alternatives and reducing energy demand by improving energy efficiency. Modern pulp mills are expanding the traditional concept of pulp mills by introducing the effective combination of multifunctional biorefineries and energy plants. Sustainably sourced wood resources are used to produce not only pulp and paper products but also electricity and heat as well as different types of novel high-value products, such as biofuels, textile fibres, biocomposites, fertilizers, and various cellulose and lignin derivatives. Thus, the forest industry provides a platform to tackle global challenges and substitute greenhouse-gas-intensive materials and fossil fuels with renewable alternatives.

Public Transit Challenges in Sparsely Populated Countries: Case Study of the United States
by Warren S. Vaz

Countries like Canada and the United States have a relatively low population density. Their population centers are located much further away, making nationwide public transit particularly challenging. As such, individuals travel predominantly via airplane and passenger cars. This results in an inefficient use of resources and pollution. These countries have some of the highest numbers of vehicles per person and per capita emissions in the world. Even public transit within cities is a challenge due to suburbs and urban sprawl. These countries have cities with relatively large areas and higher commute distances and times. These factors have historically been an impediment to widespread adoption of efficient public transit and clean transportation. Electric and hydrogen vehicles have yet to see significant market penetration in large part due to lagging infrastructure. These issues are explored in greater detail, including some of their socioeconomic impacts. Potential solutions and recent developments are also presented.

A Systematic Analysis of Bioenergy Potentials for Fuels and Electricity in Turkey: A Bottom-Up Modeling

by Danial Esmaeili Aliabadi, Daniela Thrän, Alberto Bezama and Bihter Avşar

Turkey is a member of the Organization for Economic Co-operation and Development (OECD) that enjoys suitable geography for renewable resources and, simultaneously, suffers from modest domestic fossil fuel reserves. The combination of mentioned factors supports devising new strategies through which renewable resources are not only being used in the power sector but also in industries and transportation. Unfortunately, bioenergy is underplayed in the past despite the country's potential; nonetheless, this view is transforming rapidly. In this chapter, using a bottom-up optimization model, we analyze bioenergy sources in Turkey and offer a pathway in which renewable resources are utilized to lower future greenhouse gas emissions. Although Turkey is chosen as the case study, the outcomes and recommendations are generic; thus, they can be used by policymakers in other developing countries.

Preface to Transitioning to Affordable and Clean Energy

Edwin C. Constable

In 2015, the United Nations Member States adopted 'The 2030 Agenda for Sustainable Development', which provided a vision and a roadmap for a sustainable future, bringing peace and prosperity for people and our planet. This document is not just a utopian vision of the future but has at its core 17 Sustainable Development Goals (SDGs), which were a call for all countries to act in a global partnership to end poverty, improve health and education, reduce inequality, encourage economic growth, tackle climate change, and preserve the forests and oceans.

This book is dedicated to SDG 7, which is tasked to 'Ensure access to affordable, reliable, sustainable and modern energy for all'. The importance of the goal is emphasised by a few statistics identified by the United Nations. Over three-quarters of a billion people lack access to electricity, and one-third of the World's population relies on dangerous and inefficient cooking systems. In particular, there is a need to improve the efficiency of energy generation and usage and to accelerate the development of modern renewable energy technologies and infrastructure. Of all of the SDGs, SDG 7 is the most likely to have an immediate impact on people's daily lives, whether through changes in technology, transport systems, or energy tariffs.

For me, SDG 7 is also one of the most interesting goals, as it involves dialogue and interactions between scientists, technologists, the private sector, general society, as well as decision-, policy-, and lawmakers. I have had a long research involvement with the development of new compounds and materials for use in photonic applications, in particular, solar energy photoconversion using dye-sensitised solar cells, and efficient lighting devices incorporating ionic transition metal complexes in light-emitting diodes and light-emitting electrochemical cells. A number of years ago, I realised that our search for proof-of-concept materials or improved efficiency was resulting in the development of technologies that were themselves non-sustainable and were based upon rare and expensive resources such as the platinum group metals. This resulted in a change in emphasis within the research group to concentrate upon the use of Earth-abundant elements with minimal ecological or biomedical impact.

This volume comprises nine chapters which emerge from many of the sectors involved in the implementation of SDG 7. The first chapter begins in the science and technology area; in 'Materials Development towards a Sustainable Energy Future',

Diale, Fru, and Kyesmen survey the emerging materials which are attracting attention in photovoltaic and related devices. In particular, they discuss the use of haematite, an Earth-abundant iron-based material which does not have optimal electronic properties, and the halide perovskites, which exhibit excellent performance but are based on the use of environmentally hazardous materials such as lead.

This is then followed by three chapters relating to the deployment of novel technologies and their societal impact. The first, by Aliabadi, Thrän, and Bezama, 'A Systematic Analysis of Bioenergy Potentials for Fuels and Electricity in Turkey: A Bottom-Up Modelling' examines the under-exploitation of bioenergy in the mosaic models for energy production. This is followed by a study of the use of storage and renewable electricity generation for reducing domestic and transport carbon emissions by Stevenson. This real-world study is subtitled 'Energy and Cost Analysis of Single Dwelling Case Study (UK)'. Rodrigues and Lemos de Sousa discuss the contribution of geology for carbon capture and storage, as well as underground energy storage, in their chapter entitled 'Clean Energy Transition Challenge: The Contributions of Geology'.

One of the very important ways in which the broader society will be impacted by, or at least made aware of, the implementation of SDG 7 is in changes to transport systems, and this is examined by Vaz in the chapter 'Clean Transportation and Public Transit Challenges in Sparsely Populated Countries'.

Sermyagina, Lipiäinen, and Kuparinen examine measures currently being considered and evaluated by the wood-related industry in Finland, which is responsible for 2% of industrial fossil CO_2 emissions worldwide, in their contribution 'Finnish Forest Industry and Its Role in Mitigating Global Environmental Changes'.

The chapters in the last section of the book comprise the final group of chapters, which are related to societal aspects of the energy turn. An operational approach for developing community entrepreneurship, driven by financial flow from the renewable energy sector, is examined by Lombardi, Prosperi, and Fascia, in their chapter 'Social Innovation for Energy Transition: Activation of Community Entrepreneurship in Inner Areas of Southern Italy'. In particular, they analyse cultural assets, tourism initiatives, and the potential for civic revitalization and conclude that the energy transition could represent an opportunity to spur economic growth with a multiplier effect on local communities. The final two chapters are concerned with the systems requirements of the energy transition. Pallonetto identifies the consequences of decarbonization strategies on the energy supply systems as intelligent buildings become more common and recognises the need for system-wide integration, in the chapter 'Advanced Energy Management Systems and Demand-Side Measures for the Full Decarbonisation of Our Society'. In their chapter 'Transition towards Affordable Electricity: Tools and Methods', Hanif, Bhatti, Alam, Massier, Ramachandran, Ahmad, and Zhang analyse the tools available for modelling and evaluating district-scale electric grids and suitable applications for specific tools are identified based upon a distribution system test case for Singapore.

Overall, this volume brings together many of the aspects with which society will be confronted in the energy transition required by the implementation of SDG 7. I hope that this heterogeneous collection will encourage the reader to delve into areas outside her or his area of expertise.

Basel, November 2021

Part 1: Methodology and Analysis

The Transition towards Affordable Electricity: Tools and Methods

Sebastian Troitzsch, Sarmad Hanif, Tobias Massier, Kai Zhang,
Bilal Ahmad Bhatti, Arif Ahmed, and Md Jan Alam

1. Introduction

In an effort to counteract climate change, the electric grid is subjected to significant transformations, such as the integration of renewable generators, demand-side flexibility, and new electrified transportation. These changing paradigms highlight the importance of research into new methods and tools, which can equip the power system stakeholders to plan and operate the electric grid of the future. The electric grid has been divided into three major subsystems: (1) generation, (2) transmission, and (3) distribution. For the analysis of the electric grid, modeling each subsystem in full detail may not be feasible in terms of complexity management and computational requirements. Instead, different modeling methodologies and software tools exist for the analysis of each subsystem. At the same time, the advancement of computational technology and the increasing interdependency between different subsystems is pushing the boundaries of these software frameworks. For example, (1) demand response techniques can activate demand-side control of the load to match generation, (2) battery charge/discharge can mimic both load/generator, (3) microgrids can act as small independent grid operators, and (4) renewable energies are considered as zero-variable-cost resources, which are highly variable and uncertain in nature.

To manage the diverse modeling requirements in this context, the research community is continuously developing open-source software tools for electric grid analysis. The open-source aspect allows greater control for researchers to extend and customize modeling workflows to match the requirements of particular studies, which ensures a broad adoption of such tools alongside the more conventional commercial power system tools. An exemplary set of open-source tools for electric grid analysis includes (1) OpenDSS and GridLAB-D for distribution grid analysis, (2) GRIDAPPS-D as a platform to standardize distribution grid interoperability with respect to modeling and data exchange, (3) TESP as a co-simulation platform that integrates multiple open-source tools, (4) MESMO for operational optimization of DER dispatch in the distribution grid, (5) MATPOWER for large-scale system integration studies, and (6) URBS for planning optimization of renewable energy deployment in the generation mix.

This chapter examines the landscape of open-source tools for electric grid analysis to identify the suitability and applicability of various tools for specific problem types in this domain. As a representative set of software frameworks, the following tools are considered: (1) MATPOWER, (2) GridLAB-D, (3) MESMO, and (4) URBS. The first part of this chapter begins with an introduction to the requirements of different problem types and a feature comparison of the software tools in order to differentiate the purpose for each of the different tools. This is complemented with a brief introduction to each of the four frameworks. Furthermore, to characterize the requirements for test case preparation, as well as for results post-processing, input/output specifications are compared for the four tools. In this context, the model conversion and co-simulation platforms GridAPPS-D and TESP are introduced to highlight possible workflows for test case preparation. In the second part of the chapter, the key capabilities of each tool are demonstrated for a district-scale test case based in Singapore. The test case considers a synthetic electric grid model, thermal building demand models, EV charging models, and photovoltaic (PV) generation potentials. The key results are discussed to highlight the core analyses that the different software tools can support. Eventually, the discussion section serves as a guideline for the choice of open-source tools for different electric grid modeling and analysis tasks.

Existing reviews for open-source tools energy system modeling and optimization, e.g., Després et al. (2015); Kriechbaum et al. (2018); Ringkjøb et al. (2018); van Beuzekom et al. (2015), focus on providing a classification of the tools by mathematical model types, as well as temporal, geographical, and sector coverage. Studies in Ringkjøb et al. (2018); van Beuzekom et al. (2015) provide a detailed comparison for a large number of tools across these dimensions. Another study (Després et al. 2015) compares a smaller number of tools in a similar fashion. Lastly, Kriechbaum et al. (2018) seeks to identify current challenges associated with the available tools. In contrast to these methical reviews, the core objective of this chapter is to differentiate key use cases for the presented software tools and to enable the reader to pick the best tool for their problem. The chapter points out specific features and application examples for each tool, such that choosing the right tool for a specific study is made easy. Since the presented tools only represent a small fraction of the available open-source frameworks, possible alternatives for each tool are indicated in Section 2.1.

2. Software Frameworks

2.1. Overview and Features

Software tools for distribution system analysis typically cater to specific problem types and stakeholders of the electric grid. Therefore, the features of these software frameworks are driven by the requirements arising from different problem types. To begin with, the following problem types for energy system analysis can be generalized based on Ringkjøb et al. (2018) (Section 2.2.1) and Klemm and Vennemann (2021) (Section 3.1):

- Operational problems, which describe the analysis of the system at an operational timescale with the purpose of providing operation decision support. Examples for this category are unit commitment problems, optimal control/model predictive control problems, and market-clearing problems. Operational problems can be cast into simulation problems and optimization problems depending on the application. For example, market-clearing problems would be expressed as optimization problems, whereas simulation is more suited for studying the nominal behavior of the distribution system with regard to a known set of control variables.

- Planning problems, which characterize design decisions for the energy system, i.e., at a planning timescale, with the goal of providing investment decision support. These studies can be addressed in terms of simulation-based scenario analysis or optimal planning problems. The simulation-based approach captures a conventional method for district-scale energy system design, whereas optimal planning seeks to determine optimal values for the design decision variables, e.g., component sizing and placement.

Essentially, problem type governs the temporal scale and resolution of the mathematical model as well as the selection of decision variables. Independent of the problem type, the solution method can be categorized into (1) simulation and (2) optimization. Simulation or forecasting tools calculate the state variables of the energy system based on fixed inputs for control and disturbance variables, whereas optimization tools determine state and control variables that optimize some objective subject to operational constraints.

The different problem types rely on casting a mathematical model for the electric grid into the particular solution logic. Mathematical models for the electric grid are essentially obtained by aggregating the models of its subsystems, i.e., generators, transmission systems, distribution systems, and DERs. To this end, complexity management is an important aspect of electric grid modeling in managing model formulation effort, model parameter data requirements, and computational limitations. In line with this, different software tools for electric analysis typically focus on a limited subset of features. To compare the capabilities of the selected software tools, the following features are considered in Table 1:

- Power flow simulation describes the ability to solve the nonlinear steady-state electric power flow;
- Power flow optimization refers to the ability to solve an optimization problem based on the electric power flow;
- Balanced AC model highlights whether steady-state properties, i.e., voltage, branch flow, and losses, can be represented for single-phase electric grids;
- Multi-phase AC model highlights whether steady-state properties, i.e., voltage, branch flow, and losses, can be modeled for multi-phase unbalanced electric grids;
- Transient dynamics model describes the ability to model transient properties of the electric grid, in addition to steady-state properties;
- Convex electric grid model denotes whether the electric grid model can be obtained in a convex form;
- DER simulation describes the ability to simulate the system dynamics and behavior of DERs assuming fixed control inputs;
- DER optimization describes the ability to solve an optimization problem considering system dynamics and behavior of DERs;
- Convex DER model notes whether the DER model can be obtained in a convex form;
- Operational problems indicate the ability to express operational problems as outlined above;
- Planning problems denote the capability to model planning problems as described above;
- Simulation-based solution highlights whether the tool is suited for simulation-based analysis, in which control or decision variables are provided as an input;
- Optimization-based solution describes the inclusion of interfaces to numerical optimization solvers, such that optimization problems can be modeled and solved, where decision variables are obtained as outputs from the optimal solution;

- Multi-period modeling refers to the ability to consider multiple time steps and capture inter-temporal linkages during simulation/optimization.

Table 1 reviews these features for a selected set of open-source tools. For the sake of brevity, only the following four representative software frameworks are included in the discussions:

- MATPOWER is an open-source software tool for power system analysis in MATLAB or GNU Octave (Zimmerman et al. 2011). It originated as a tool for balanced AC power flow solutions but has since been extended for optimal power flow (OPF) and optimal scheduling applications (Murillo-Sanchez et al. 2013). Similar tools are available for other language platforms, e.g., pandapower (Thurner et al. 2018), PYPOWER (Lincoln 2021), and PowerModels.jl (Coffrin et al. 2018).
- GridLAB-D is a software tool that connects distribution system simulation and DER simulation (Chassin et al. 2008). Its core capability is to coordinate the simulation of various subsystems in an agent-based fashion, where each subsystem model can be implemented independently. Through a modular approach, GridLAB-D supports studies ranging from classical power flow analysis to integrated energy market simulation with detailed models for the behavior of individual DERs. A similar tool in this category is OpenDSS (Dugan and McDermott 2011).
- MESMO is a Python-based framework for Multi-Energy System Modeling and Optimization, which enables the convex optimization of district-scale energy system operational problems. A similar feature set is provided by the software platform OPEN (Morstyn et al. 2020).
- URBS is an open-source software tool for energy system optimization (Dorfner et al. 2019), with a focus on capacity expansion planning and unit commitment of DERs. The sibling project FICUS provides an extension for modeling multi-commodity energy systems in factories (Atabay 2017). Similar open-source frameworks are Calliope (Pfenninger and Pickering 2018), oemof (Hilpert et al. 2018), and Temoa (Hunter et al. 2013).

Table 1. Differentiating components under each scenario.

Feature	MATPOWER	GridLAB-D	MESMO	URBS
Electric grid modeling				
Power flow simulation	✓	✓	✓	
Power flow optimization	✓		✓	✓ [a]
Balanced AC model	✓	✓	✓	
Multi-phase AC model		✓	✓	
Transient dynamics model		✓		
Convex electric grid model			✓	✓ [a]
DER modeling				
DER simulation		✓	✓	
DER optimization			✓	✓
Convex DER model			✓	✓
Problem types and solution methods				
Operational problems	✓	✓	✓	✓
Planning problems				✓
Simulation-based solution	✓	✓	✓	
Optimization-based solution	✓		✓	✓
Multi-period modeling	✓ [b]	✓	✓	✓

[a] URBS implements a simplified nodal flow balance model for the electric grid. [b] Requires using the MATPOWER Optimal Scheduling Tool (MOST) extension. Source: Table by authors.

2.2. MATPOWER

MATPOWER (Zimmerman et al. 2011) is a MATLAB package that solves the nonlinear power flow, as well as OPF. Among other tools that are able to solve power flow and OPF problems, MATPOWER stands out due to its computational efficiency and extension capability, particularly in dealing with large-scale system operation and optimization problems. Researchers across the globe have relied on the MATPOWER extension capability to solve a broad spectrum of power system operation and planning problems.

For OPF problems, MATPOWER implements multiple state-of-the-art methods including the primal-dual interior-point method (Wang et al. 2007), the trust-region-based augmented Lagrangian method and relaxation-based convexified OPF models. Furthermore, MATPOWER and its underlying modules are suitable for electricity market applications. For example, MATPOWER Optimal Scheduling Tool (MOST) (Murillo-Sanchez et al. 2013) is able to solve problems as simple as a deterministic, single-period economic dispatch problem or as complex as a stochastic, security-constrained, combined unit-commitment and multi-period OPF problem with locational contingency and load-following reserve, e.g., in Cho et al. (2019); Murillo-Sanchez et al. (2013). Further studies have adopted MATPOWER in distribution network analysis and market applications (Hanif et al. 2019).

The basic features of MATPOWER are summarized in the following:

1. Modeling capabilities

    - AC and DC single-phase electric grid models;
    - Nonlinear OPF models;
    - Relaxation-based convexified OPF models.

2. OPF problems types

    - AC- and DC-OPFs;
    - Co-optimize energy and reserves;
    - Unit commitment problems;
    - Stochastic and contingency-constrained OPF problems;
    - Parallelizable OPF formulations.

2.3. GridLAB-D

As a mature open-source simulation tool, GridLAB-D (Chassin et al. 2008, 2014) combines traditional power flow simulation capabilities with advanced DER modeling and control. Its event-driven solution logic is able to simulate various interacting DERs of the electric grid, e.g., the room temperature evolution within buildings is simulated, along with the resulting load flow in the electric grid. Apart from its traditional simulation features, the recent new capabilities of GridLAB-D can be summarized as follows:

- End-use models, including thermostatically coupled and non-coupled appliances and equipment models (Pratt and Taylor 1994; Taylor et al. 2008);
- Event-driven agent-based simulation environment to allow for behavioral decision modeling;

- Module to simulate market-based control, e.g., retail market modeling tools, including contract selection, business and operations simulation tools, models of SCADA controls, and metering technologies;
- Extension to high-level languages such as MATLAB and Python through programming interfaces;
- Possibility to run parallel power flows for large-scale system simulation.

GridLAB-D's thermal end-use models consist of commercial and residential end uses, implemented using the equivalent thermal parameters model (Pratt and Taylor 1994). The innovation in these models is that they solve differential equations of their end uses such that state changes can trigger an event for the power flow base simulator to stop and sync with the end-use models. Currently, advanced models such as heat pumps, resistance heating, electric hot water heaters, washer and dryers, cooking appliances (range and microwave), electronic plugs, and lights are captured in the model.

2.4. Multi-Energy System Modeling and Optimization (MESMO)

The Multi-Energy System Modeling and Optimization (MESMO) is a Python-based software tool for optimal operation problems of electric and thermal distribution grids along with distributed energy resources (DERs), such as flexible building loads, electric vehicle (EV) chargers, distributed generators (DGs), and energy storage systems (ESS). It implements convex modeling techniques for electric grids, thermal grids, and DERs, along with a set of optimization-focused utilities. Essentially, MESMO is a software framework for defining and solving numerical optimization problems in the electric/thermal grid context, such as OPF, distributed market clearing with network constraints, strategic offering, or multi-energy system dispatch problems. Additionally, the tool also includes classical steady-state nonlinear power flow models for electric and thermal grids.

The need for its development stems from the observation that numerical-optimization-based studies of district-level energy systems often require significant upfront implementation effort, due to domain-specific models being implemented in different software tools. Additionally, applications such as distribution locational marginal pricing and distributed/decentralized market clearing further require the underlying mathematical models to be implemented in a convex manner, which often necessitates the implementation of custom mathematical formulations. To improve this workflow, MESMO implements interoperational convex modeling techniques for electric grids, thermal grids, and DERs, along with a set of optimization-focused utilities.

MESMO is intended to complement the existing software in the domain of district-level energy system simulation and optimization. Therefore, it combines (1) convex multi-energy system modeling for (2) optimization-focused studies on the operational timescale with a focus on (3) market-clearing and distribution locational marginal price (DLMP) mechanisms. Essentially, MESMO is developed as a software framework for defining and solving numerical optimization problems for multi-energy system operation. It implements convex models for electric grids, thermal grids, and DERs, along with a set of optimization-focused utilities. To this end, the feature set of MESMO can be summarized as follows:

1. Electric grid modeling

   - Obtain nodal/branch admittance matrices and incidence matrices for the electric grid;
   - Obtain steady-state power flow solution for nodal voltage/branch flows/losses via fixed-point algorithm;
   - Obtain sensitivity matrices of global/local linear approximate grid model;
   - All electric grid modeling is fully enabled for unbalanced/multi-phase grid configuration.

2. Thermal grid modeling

   - Obtain nodal/branch incidence matrices and friction factors;
   - Obtain thermal power flow solution for nodal head/branch flows/pumping losses;
   - Obtain sensitivity matrices of global linear approximate grid model.

3. Distributed energy resource (DER) modeling

   - Obtain time series models for fixed DERs;
   - Obtain state-space models for flexible DERs;
   - Enable detailed flexible building modeling with the Control-Oriented Thermal Building Model (CoBMo) (Troitzsch and Hamacher 2020).

4. Multi-energy system operation

   - Obtain and solve nominal operation problems, i.e., steady-state simulations, of electric/thermal grids with DERs, i.e., multi-energy systems;
   - Define and solve numerical optimization problems for combined optimal operation for electric/thermal grids with DERs, i.e., multi-energy systems;

- Obtain DLMPs for the electric/thermal grids.

Figure 1 depicts a contextual view for the software architecture of MESMO, i.e., a high-level overview of the most important components and the interaction between the software system and its stakeholders, based on the C4 model (Brown 2015) for representing software architecture.

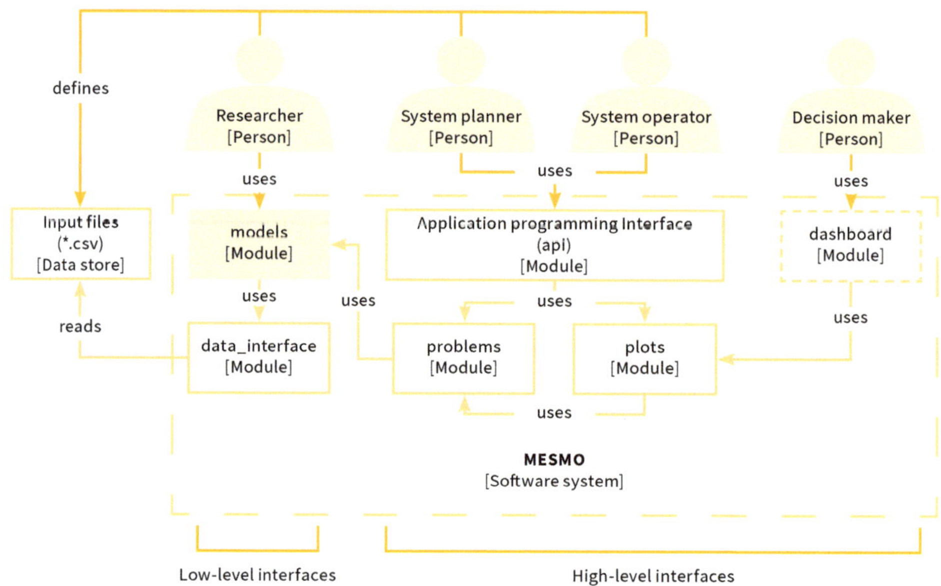

Figure 1. Software architecture of MESMO depicted as a contextual view based on the C4 model (Brown 2015). Source: Graphic by authors.

The user interfaces can be distinguished into high-level interfaces, i.e., the api module and the dashboard module, as well as low-level interfaces, i.e., the models modules. To this end, the api module and models modules describe programming interfaces, whereas the dashboard refers to a graphical user interface (GUI). Researchers primarily interface MESMO directly through the models modules, because they require highly granular access and modifiability of the modeled objects for custom workflows. System planners and system operators interface MESMO through the api module, which provides convenient access to the most common workflows, i.e., running of planning/operation problems and producing results plots. Decision makers interface MESMO through the GUI of the dashboard module. Note that the dashboard module has not yet been implemented at the time of writing.

Scenario and model data definitions are enabled through a standardized CSV-based input file format, which is referred to in Figure 1 as "Input files (*.csv)". The input files are expected to be defined by researchers, system planners, and system operators, where decision makers would rely on these actors to define appropriate scenarios for their review.

Internally, the `api` module implements API functions that rely on the `problem` module and `plots` module. The `dashboard` module implements the GUI framework but relies on the `plots` module to generate individual plots. The `problems` module implements the main workflows for setup and solution of different problem types, for which it uses the mathematical models defined in the `models` modules. The `problems` module also implements a standardized results object across all problem types, which is used by the `plots` module. The `models` modules further rely on the `data_interface` module to obtain the model data definitions from the input files.

MESMO has been utilized for a small number of studies focusing on multi-energy systems modeling and operation in Kleinschmidt et al. (2021); Schelo et al. (2021) as well as the design of market mechanisms for distribution-level energy systems in Troitzsch et al. (2020, 2021). The initial software architecture iteration of MESMO was developed as the Flexible Distribution Grid Demonstrator (FLEDGE) in Troitzsch et al. (2019).

2.5. URBS

URBS is an open-source linear energy system model (Dorfner et al. 2019). It is time-step based, with the default time-step size being 1 h. URBS sets up an optimization problem in which the objective is the minimization of costs or emissions in scenarios specified by the user. It is implemented in Python using Pyomo for the formulation of the optimization problem. Various numerical optimization solvers can be connected to URBS. The user can define various sites (e.g., countries or districts) and specify the following input data for each site:

- Sites (e.g., countries or districts);
- Commodities (e.g., gas, coal, electricity) and their market prices;
- Processes (i.e., power generators) and their characteristics such as installed capacity, minimum load factor, efficiency, and costs;
- Transmission and storage capacities, and costs;
- Time series of demand and intermittent generation;
- Demand-side management capacities.

Moreover, buying and selling prices for electricity and limits for costs and emissions can be specified, among others. Within the boundaries specified by the user, URBS determines which generators to use and to what capacity in order to satisfy the demand in each time step. URBS also decides whether to change the installed capacity of the given generators within the set boundaries. The user can also specify which outputs to be generated in the form of spreadsheets and plots. For each scenario, the output of URBS comprises emissions; prices and costs; installed, added, and retired capacities; transmission and storage for each site.

URBS has mostly been used for studies on transmission networks, e.g., in Europe (Schaber et al. 2012) or the Asia-Pacific region (Huber et al. 2015; Ramachandran et al. 2021; Stich et al. 2014; Stich and Massier 2015). However, it has also been used for smaller networks (Fleischhacker et al. 2019; Zwickl-Bernhard and Auer 2021) or specific applications such as managing the integration of intermittent sources of energy or electric vehicles (Massier et al. 2018), with some modifications. Due to its open-source availability, URBS can easily be modified and extended. Recently, uncertainty modeling has been integrated (Stüber and Odersky 2020), and first efforts to combine it with life cycle assessment have been made (Ramachandran et al. 2021).

3. Workflows for Test Case Preparation

3.1. Input/Output Specification

Inputs and outputs for different electric grid analysis software are typically governed by the underlying mathematical model specifications that are implemented within each tool. Therefore, each tool usually defines custom input and output formats in line with its internal data models. In this context, inputs refer to the technical system and problem parameters that define the test case, i.e., the subject of the study. Outputs are the results that are obtained after a successful solution of the simulation or optimization within the software tool. Tables 2 and 3 outline the different input and output data items for the presented software tools.

Input definitions can be provided either as (1) file-based input in text-based and table-based format or (2) script-based input. The file-based input is often the default avenue, as it allows encoding the complete test case into a single data container. At the same time, the script-based input allows for a more flexible way of defining and modifying models during runtime. This is an important capability for studies in which custom problem definitions or model coupling is desired. For example, MATPOWER can be utilized to iteratively obtain power flow (PF) solutions through continuous modification of model parameters, thereby extending beyond

MATPOWER's base functionality. In order to document their functionality, to allow for benchmarking, and to serve as tutorials, most software tools provide a bundled set of input data definitions for selected test cases. For the presented tools, the available test cases are summarized in Table 4.

Table 2. Input specifications.

	MATPOWER	GridLAB-D	MESMO	URBS
Input	MATLAB-based format	Text-based format [b]	CSV-based format [a]	XLS-based format
Electric grid parameters	• Nodes: Nominal voltage. • Lines: Node connections, resistance/reactance/capacitance, rated current limit. • Transformers: Node connections, phases, ratio, angles, rated power, resistance/reactance parameters.	• Nodes: Nominal voltage, phases. • Lines: Node connections, phases, resistance/ reactance/ capacitance matrices, rated current limit. • Transformers: Node connections, phases, connection scheme (wye/delta), rated power, resistance/ reactance parameters. • Line and transformer parameters are encapsulated into line type and transformer type definitions.	• Nodes: Nominal voltage, phases. • Lines: Node connections, phases, resistance/reactance/ capacitance matrices, rated current limit. • Transformers: Node connections, phases, connection scheme (wye/delta), rated power, resistance/ reactance parameters. • Line and transformer parameters are encapsulated into line type and transformer type definitions.	• Lines: Node connections, efficiency, reactance, voltage angle, base voltage, installed capacity, and minimum and maximum permitted capacity for expansion planning.
DER parameters	• N.A.	• Connection: Node, phases, connection scheme (wye/delta), nominal active/reactive power. • Fixed DERs: Dispatch time series. • Flexible DERs: Equivalent thermal parameters inputs such as thermal resistance and thermal capacitance of thermal electric loads; Battery model parameters such as inverter ratings, operation strategy.	• Connection: Node, phases, connection scheme (wye/delta), nominal active/reactive power. • Fixed DERs: Dispatch time series. • Flexible DERs: Detailed state-space model parameters, e.g., thermal building parameters, battery model parameters, EV charger efficiencies, generator model parameters.	• Connection: DERs are aggregated in the sites they are located in. • Time series: Fixed for each site (demand and intermittent supply). • Demand-side management: Can be specified for each commodity.
Cost parameters	• Operation costs: Price value. • Customizable cost functions.	• Tariff type based on customer class.	• Operation costs: Energy price time series, price sensitivity.	• Investment, fixed, variable, and fuel costs of processes, storage, transmission, weighted average cost of capital, depreciation periods, CO_2 abatement costs.

[a] MESMO input data reference: https://purl.org/mesmo/docs/0.5.0/data_reference.html (accessed on 25 August 2021). [b] Base script format is .glm, whereas additional input files could be .txt, .csv, etc. For an introduction to input/output of GridLAB-D refer to: http://gridlab-d.shoutwiki.com/wiki/GridLAB-D_Wiki:GridLAB-D_Tutorial_Chapter_4_-_Data_Input_and_Output (accessed on 25 August 2021). [c] For more information on DER parameters refer to: http://gridlab-d.shoutwiki.com/wiki/Residential_module_user%27s_guide (accessed on 25 August 2021). Source: Table by authors.

Table 3. Output specifications.

	MATPOWER	GridLAB-D	MESMO	URBS
Input			CSV-based format	XLS-based format
Electric grid results	• State variables: Nodal voltage magnitude, nodal voltage angles, branch power flow, total losses, single-branch losses.	• State variables: Per-phase voltage magnitude, nodal voltage angles, branch power flow, total losses, single-branch losses, and reactive power flows.	• State variables time series: Per-phase nodal voltage, branch power flow, total losses.	• Capacities: Initial and newly installed (processes, transmission, storage, etc.). • Time series of import and export (i.e., transmission between sites) of electricity.
DER results	• N.A.	• Temperature evolution and dispatch time series of thermostatically controlled loads and active/reactive power injection by batteries.	• Fixed DERs: Dispatch time series. • Flexible DERs: Dispatch time series, detailed state/control variable time series.	• Time series of electricity generation by process per time step, emissions by generator per time step, storage utilization, demand-side management, etc.)
Cost results	• System-level costs. • DLMP values.	• Billing results of customers with different classes, energy costs etc.	• System-level / DER-level operation costs. • DLMP time series, decomposed into congestion, voltage, loss, and energy components.	• Costs: Total investment, fixed and variable costs of processes, storage, transmission, fuel costs. • Emissions: Total emissions (CO_2 and others if specified).

Source: Table by authors.

Table 4. Included test cases.

	MATPOWER [a]	GridLAB-D [b]	MESMO	URBS
Available test cases	• Small Transmission System Test Cases • Small Distribution System Test Cases • Synthetic Grid Test Cases • European System Test Cases • French System Test Cases	• Small-scale distribution system test cases (e.g., IEEE 4-Node and IEEE 37-Node systems) • A medium-scale test system with residential models to test DER modeling capabilities • IEEE 8500-Node test system	• Small-scale distribution test cases (e.g., IEEE 4-Node, Singapore 6-Node) • Medium-scale distribution test case (Singapore Geylang, see Section 4) • Multi-energy system test case (Singapore Tanjong Pagar)	• Fictitious three-region example (Dorfner et al. 2019) • The 16 states of Germany connected to a few other regions (Dorfner et al. 2019) • HVDC connection between Australia and Singapore (Siala 2021b) • Mekong region (Siala 2021a) • Southeast Asia (Siala et al. 2021)

[a] Refer to Zimmerman and Murillo-Sánchez (2019) for more information for individual test cases. [b] MESMO test cases are currently located in the `data` directory of the MESMO repository https://purl.org/mesmo/repository (accessed on 25 August 2021). Source: Table by authors.

Outputs can similarly be obtained either as (1) file-based format or (2) runtime objects. The former can include basic, text-based and table-based outputs, as well as more sophisticated data visualization in image-based formats, e.g., through the integrated plotting capabilities of MESMO and URBS. Runtime objects are data containers that cater to custom post-processing workflows of the user and enable coupling with other software tools. Such workflows for software coupling are further discussed in the context of TESP in Section 3.3.

Since each software typically defines custom data formats, the user eventually ends up preparing dedicated pre-processing and post-processing workflows for each tool. In order to reduce the upfront effort for test case input data preparation, conversion between common data formats may be possible through third-party translation tool chains, e.g., by means of GridAPPS-D, according to Section 3.2. In the long term, input/output data are foreseen to converge towards the Common Interface Model (CIM), i.e., the standardized format for electric grid models according to IEC 61970/61968.

3.2. Model Conversion via GridAPPS-D

GridAPPS-D (Melton et al. 2017; Pacific Northwest National Laboratory 2021a) is a software tool that handles model input and output conversions, which enables utilizing multiple models to build new application workflows; see Figure 2.

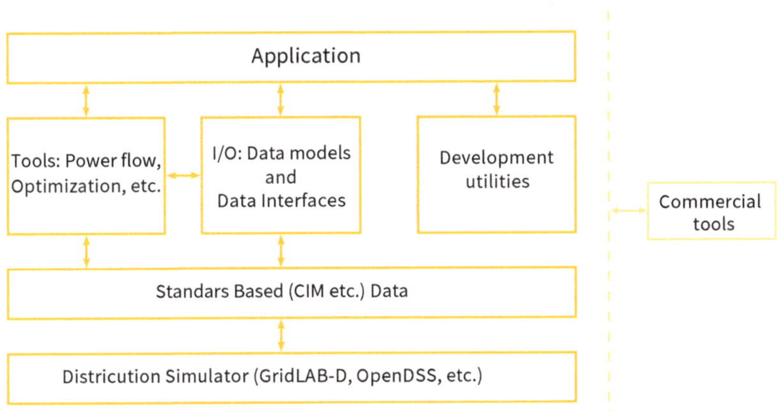

Figure 2. Overview of the software architecture of GridAPPS-D. Source: Graphic by authors, information adapted from (Melton et al. 2019).

For the test case preparation of this chapter, we utilized the Common Information Model Hub (CIMHub) to demonstrate one of the key capabilities of GridAPPS-D, i.e., the transformation of grid models across various data formats. CIMHub, as shown in Figure 3, is a module of GridAPPS-D that translates power distribution network models between different tools using the IEC 61970/61968 Common Interface Model (CIM) as a hub. CIMHub can convert models from commercial tools such as CYMDist to open-source research-grade tools such as OpenDSS and GridLAB-D. The supported inputs are CYMDist, CIM XML, OpenDSS, and Synergi Electric for distribution networks. The supported output formats are OpenDSS, GridLAB-D, and comma-separated value (CSV) for distribution systems. CIMHub can also be used to develop and propose extensions to the CIM standard.

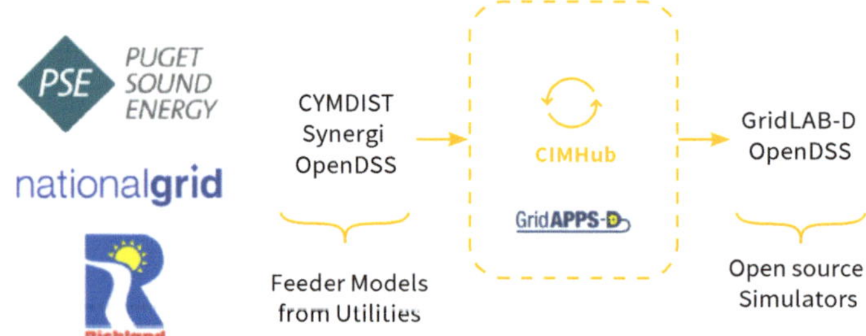

Figure 3. Power distribution network model conversion workflow via CIMHub. Source: Graphic by authors.

One goal of the GridAPPS-D program is to encourage CIM adoption by many tool vendors, lowering the burden of model conversion and other costs of integration. Details describing the overall project and the CIM transformer model can be found in Melton et al. (2017), whereas CIM unbalanced line model and database are explained in McDermott et al. (2018). CIMHub is open source under the Berkeley Software Distribution (BSD) license.

3.3. Co-Simulation via TESP

With the utilities developed to bring models from CIM to common distribution grid analysis software, a co-simulation platform can be utilized to run legacy software in an integrated fashion (Huang et al. 2018). An example of such a co-simulation platform is given in Figure 4, called the Transactive Energy Simulation Platform (TESP). Summarizing the functionality of TESP briefly, various utilities are utilized to translate the passive distribution grid into an active one by adding interactive DERs in the GridLAB-D distribution grid model, e.g., using the `feederGenerator.py` (Pacific Northwest National Laboratory 2021b). The `feederGenerator.py` script provides a systematic way of changing the distribution grid passive loads to responsive buildings, modeled using an equivalent thermal parameter approach from Taylor et al. (2008). This is important, as this makes the distribution grid load responsive to the events occurring external to it, e.g., weather impacts, outages, and wholesale market price changes.

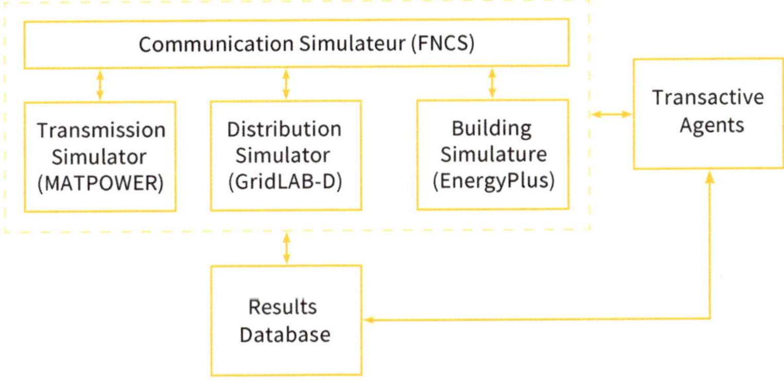

Figure 4. Co-Simulation via the Transactive Energy Simulation Platform (TESP). Source: Graphic by authors, information adapted from Huang et al. (2018).

4. Test Case

4.1. Electric Grid

The synthetic electric grid model from Trpovski (2021) was used in this test case to demonstrate the district-scale modeling capabilities of the selected tools. The following serves as a brief overview of the methodology that was applied for the preparation of the grid model, but the interested reader is referred to Trpovski (2021) for more detail. An overview of the synthetic grid layout for the Geylang District is provided in Figure 5, where 66/22 kV substations are depicted with larger nodes and 22/0.4 kV substations with smaller nodes. Note that, although depicted as direct connections between nodes, the grid lines are assumed to follow the street layout, i.e., the layout laid as underground cables.

Figure 5. Synthetic grid layout for the Geylang District in Singapore. Source: Graphic by authors.

The synthetic grid was derived based on information for (1) postal-code-clustered demand estimates and (2) 66/22 kV substation locations. Since every building block in Singapore is assigned an individual postal code, this served as a relatively detailed input for generating the 22 kV load clusters. A power system planning approach was devised to obtain the mapping and line layout between 66/22 kV substations and 22/0.4 kV substations, i.e., transformers at 22 kV load clusters. For the presented test case, the substation rating was assumed to be in 100 MVA units for 66/22 kV transformers and 1 MVA units for 22/0.4 kV transformers. This means that the minimum transformer rating for 22/0.4 kV was

1 MVA, and an appropriate integer value of transformers was deployed depending on the aggregate peak load at each 22/0.4 kV substation, where maximum utilization of 0.9, i.e., a safety factor of 1.11, was assumed for the transformer rating. The baseload time series was homogeneously defined for all 22 kV load clusters based on a representative load shape from the aggregate demand data for Singapore, which is published, along with price data, by the EMC at Energy Market Company (2021).

The final test case for the Singapore Geylang District comprised 4 subnetworks of the 22 kV distribution grid, where each subnetwork was connected to exactly one 66/22 kV substation. The total network consisted of 391 nodes and 387 lines. The lines of the synthetic grid were characterized by two line types, which defined electric parameters and current-carrying capacity, as documented in Table 5. Both line types represented underground cables, and their parameter values were based on cable supplier information, as outlined in Trpovski (2021).

Table 5. Electric line types.

Line Type	Max. Current	Resistance	Reactance
Type A	585 A	$0.23\,\Omega\,\mathrm{km}^{-1}$	$0.325\,\Omega\,\mathrm{km}^{-1}$
Type B	455 A	$0.39\,\Omega\,\mathrm{km}^{-1}$	$0.325\,\Omega\,\mathrm{km}^{-1}$

Source: Table by authors.

4.2. Building Models

The overview of the inputs, methods, and outputs for processing the feeder to attach flexible DERs to the passive loads is given in Figure 6. From the provided inputs, the stages to change the passive loads to an active one, i.e., loads that can dynamically change response based on external (weather) and internal (temperature) variables are shown. The workflow is as follows:

1. First, back-bone feeders with information on substations, lines, and loads were inputted in GLM format.
2. The module used the `networkx` Python package to perform graph-based capacity analysis and upgrades relevant protection equipment such as fuses, transformers, and lines to serve the expected load. For example, transformers can be oversized with a margin of 20%, and the circuit breaker to the air conditioner can be rated 2 times the nominal electrical load.
3. Each load was then changed to contain ZIP load, plug loads schedules, and thermostatically controlled load, as intended with the percentage population.

For example, 40% thermostatically controlled load penetration would result in convergence of only 40% load to thermodynamic models, and the rest would be left as fixed loads with a time series, which could be modified to change their load shape.

4.   For each thermostatically controlled load, the equivalent thermal properties parameter (Taylor et al. 2008) were randomized to represent a certain population of devices.

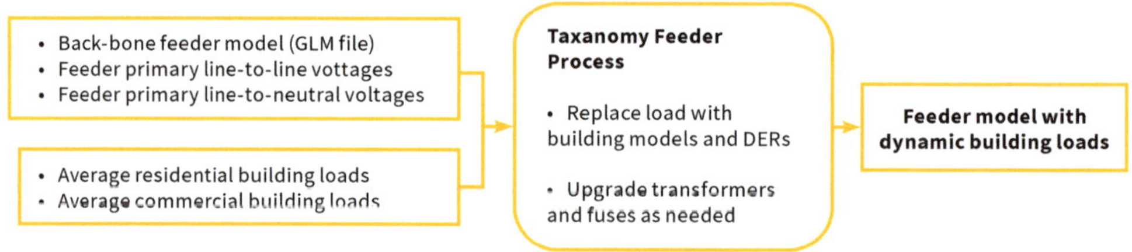

Figure 6. Building model generation workflow with `feederGenerator.py`. Source: Graphic by authors, information adapted from Pacific Northwest National Laboratory (2021b).

4.3. EV Chargers

Figure 7 highlights the main steps for the derivation of the EV charger models for private EV charging. First, historical car park availability data was used to derive representative vehicle inflow and outflow time series for existing car parks in the study area through probabilistic modeling. Second, a car park charging simulation was computed based on the representative vehicle inflow, outflow, EV penetration, and EV/charger parameters. This served to obtain the required input time series for the definition of fixed EV chargers and flexible EV chargers in MESMO.

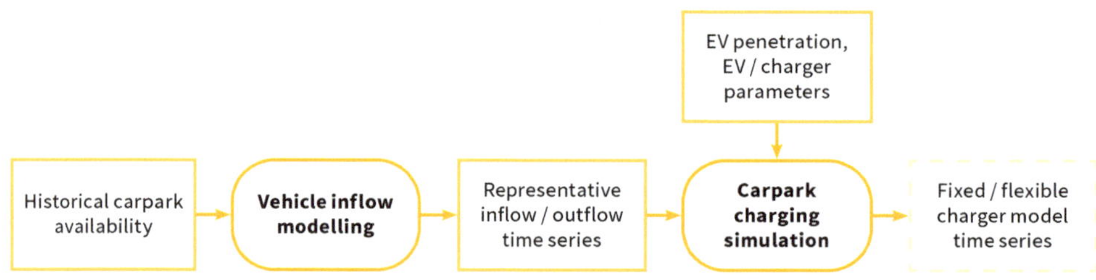

Figure 7. Workflow for EV charger demand modeling. Source: Graphic by authors.

The main input data items for the synthetic EV charger demand modeling were (1) the historical car park availability and (2) car park capacity. Both inputs were obtained from the LTA Datamall API (Land Transport Authority 2021b), which contains a selection of public residential, commercial, and mixed-use car parks in Singapore, particularly at public housing developments, government-operated general public car parks, and large-scale mixed-use developments, e.g., malls with attached office blocks. The input data was recorded at 10 min intervals between September 2018 and March 2020. Additional technical model parameters for EVs and chargers are defined in Table 6. In this test case, four EV penetration scenarios were considered: 0% (baseline scenario), 25%, 75%, and 100%.

Table 6. Electric line types.

Parameter	Value	Source
Private car population (Singapore)	520,000	(Land Transport Authority 2020a, 2020b)
Vehicle driving distance (Singapore), mean value	$48 \, \mathrm{km \, d^{-1}}$	(Land Transport Authority 2021a)
Vehicle driving distance (Singapore), standard deviation	$16 \, \mathrm{km \, d^{-1}}$	Assumed
EV energy consumption	$170 \, \mathrm{W \, h \, km^{-1}}$	(EV Database 2021)
Charger efficiency	95%	Assumed
Charger power factor	0.95	Assumed
Slow charger active power	7.4 kW	Assumed
Slow charger share	75%	Assumed
Fast charger active power	50 kW	Assumed
Fast charger share	25%	Assumed

Source: Table by authors.

4.4. Photovoltaic Generators

Photovoltaic (PV) generation potentials were estimated for this test case to demonstrate the ability of URBS for determining the cost-optimal deployment of renewable generators. To this end, PV generators were not directly modeled, but instead, the generation potential was estimated for each node of the electric grid. Importantly, PV deployment was assumed only at building surfaces, as the considered test case was based in an urban environment. Therefore, in the first step, the horizontal building surfaces were obtained from geographical information system (GIS) data for the building polygons in the test case area. In the second step, the PV generation potential at individual buildings was estimated from the available surface area and historical solar irradiation data for Singapore. Third and last, the PV generation potential of each building was mapped to the corresponding node of the synthetic electric grid. Input data items for this workflow were (1) the GIS data for building polygons and (2) the historical solar irradiation time series for Singapore. The former was obtained from Open Street Map through the Overpass API, based on the data in August 2021. The total installable PV capacity in the study area amounts to 2023 MW. Half-hourly time series of solar irradiance in Singapore were used.

Costs for rooftop PV system installations in Singapore were taken from Solar Energy Research Institute of Singapore (SERIS) (2020). For a high-efficiency system of more than 1 MW_p, a cost of ca. SGD 0.92 per W (USD 0.68 per W) was reported for 2021. For smaller systems (below 600 kW_p), the reported cost was USD 0.74 per W, and for below 300 kW_p, it was USD 0.95 per W.

4.5. Other Generators and Demand

Information on electricity generation capacity by generator type in Singapore was taken from Energy Market Authority (2020a). The installed capacity of the different generator types is listed in Table 7. These do not include the potential PV capacity in Geylang. Cost efficiencies of power plants in Singapore were obtained from Lacal Arantegui et al. (2014). They are given in Table 8.

The half-hourly electricity demand of Singapore was taken from Energy Market Authority (2020b), which is made available under the terms of the Singapore Open Data Licence version 1.0[1]

Table 7. Installed capacity of existing generators in Singapore.

Generator	Installed Capacity (GW)	Efficiency (%)
Gas (CCGT)	10.50	59
Gas (OCGT)	0.18	40
Gas (steam turbine)	2.06	38
Oil	0.49	38
Waste-to-energy	0.26	28
PV	0.17	16

Source: Table by authors, values based on Energy Market Authority (2020a).

---

[1]   https://www.ema.gov.sg/Terms_of_use.aspx (accessed on 25 August 2021).

Fuel	Price (USD/MWh)
Gas	31
Oil	40
Waste	12
Biomass	6

Source: Table by authors, values based on Lacal Arantegui et al. (2014).

5. Results

5.1. MATPOWER

This part of the study utilizes the load flow analysis from MATPOWER to study the impact on the electrical grid from different EV penetration levels. The result is shown in Figure 8, where the indicators are the branch losses and voltage profile at 22 kV distribution lines in the Geylang test case. It is worth noting that the voltage drop increases linearly to the additional peak demand from EV charging power, whereas losses increase exponentially to the additional EV charging demand.

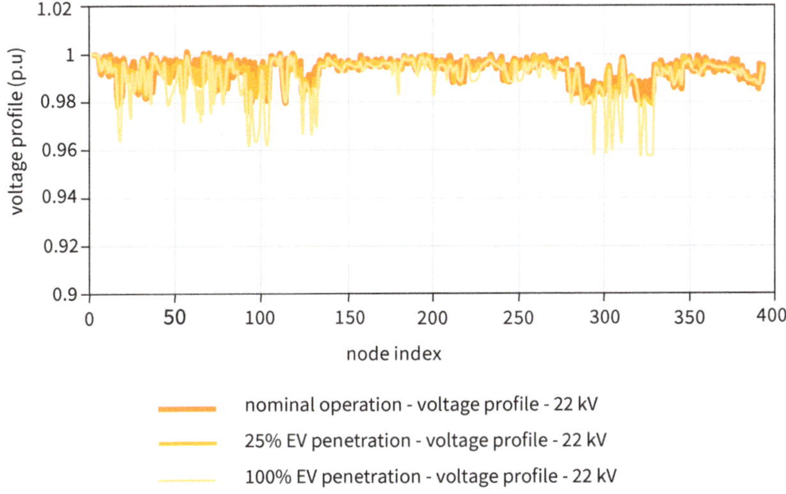

Figure 8. *Cont.*

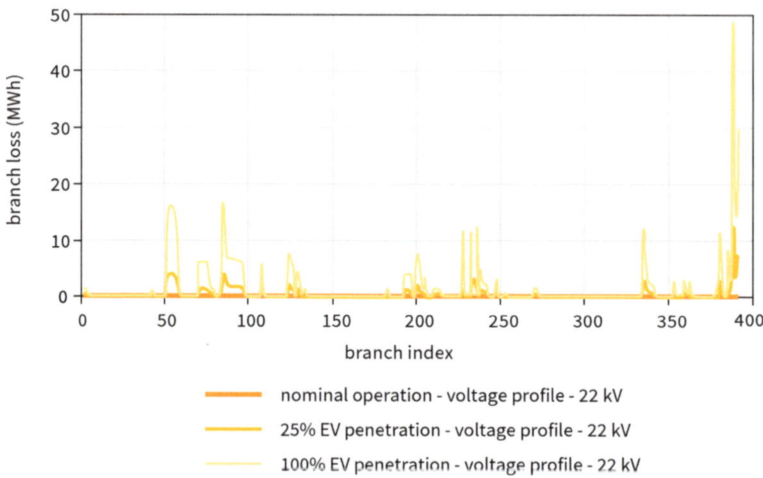

Figure 8. Voltage profile and branch losses for different scenarios from MATPOWER. Source: Graphic by authors.

5.2. GridLAB-D

Figure 9 shows the feeder load profile of the grid, which has been shown to contain thermostatically controlled loads. For the sake of simulation, we populated 525 houses with an average 3 kW load. We can observe the capability of the disaggregating contribution of thermostatically controlled load from the total load of the feeder.

Furthermore, one can select one of the thermostatically controlled loads from the population and plot its internal variables, such as temperature evolution, setpoints, and load; see Figure 10. Note that in Figure 10, the cooling of two buildings is shown by plotting the aggregated temperature of the room. For both buildings, note that aggregated temperature decreases with an increase in consumption, showing the powering of the air-conditioner and staying close to its set point.

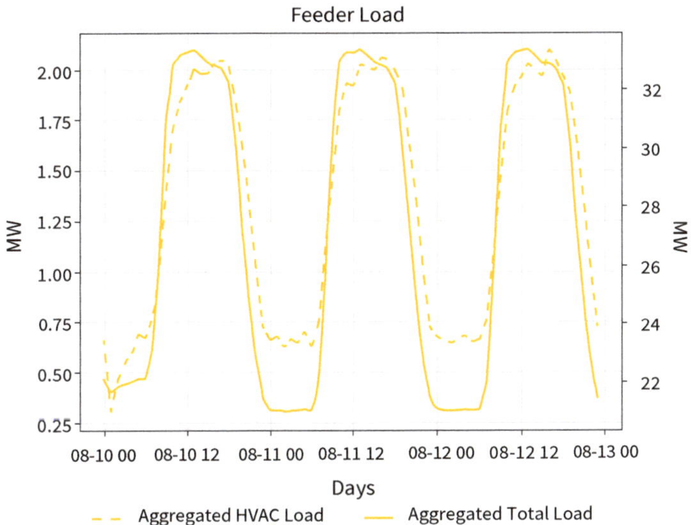

Figure 9. Feeder load profile—averaged over 5 min. Source: Graphic by authors.

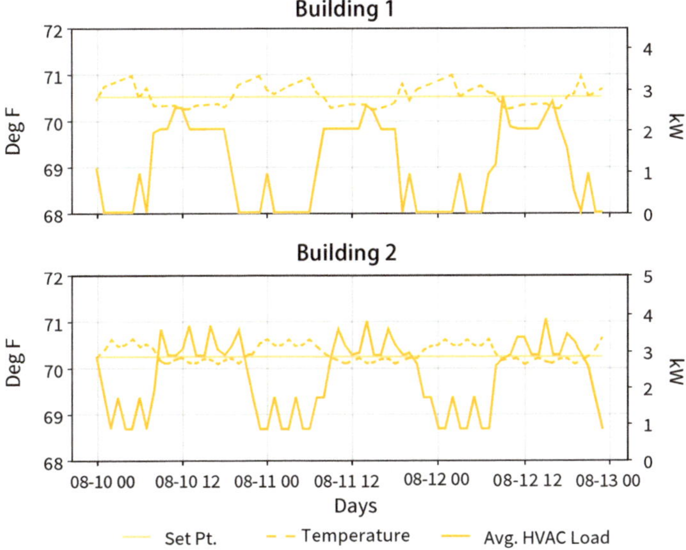

Figure 10. Building temperature and consumption—averaged over 5 min. Source: Graphic by authors.

5.3. MESMO

For the uncontrolled charging scenario, Figure 11, in its lower portion, depicts the cumulative distribution of substation transformer utilization in the test case area. The upper portion of Figure 11 describes the distribution of the transformer utilization with a box plot. The utilization level is calculated as the ratio of peak loading to the rated loading of the transformers. Recall from Section 4.1 that 22/0.4 kV transformers are assumed to occur in 1 MVA units. To this end, a large proportion, i.e., as much as 85% of transformers experience between 0.1 and 0.3 utilization in the baseline scenario (0% EVs) because load clusters can be significantly smaller than 1 MVA in the synthetic grid. In the baseline scenario, nearly 100% of substations are loaded below 0.9, where the median utilization occurs at approx. 0.11 and the mean utilization at approx. 0.21. With increasing EV penetration, the share of substation transformers loaded below 0.9 falls to approx. 94% for 25% EVs and below 90% for both 75% and 100% EVs. The median substation utilization increases to approx. 0.15 and remains constant across the higher penetration levels, since the substations with allocated charging demand occur consistently above the median level. The mean utilization increases proportionally to the EV penetration level, i.e., up to approx. 0.5. Note that the plot is truncated at the utilization level 1.0 for consistency, although higher-level EV penetration scenarios can cause significant overloading of selected transformers due to the highly localized nature of these loads.

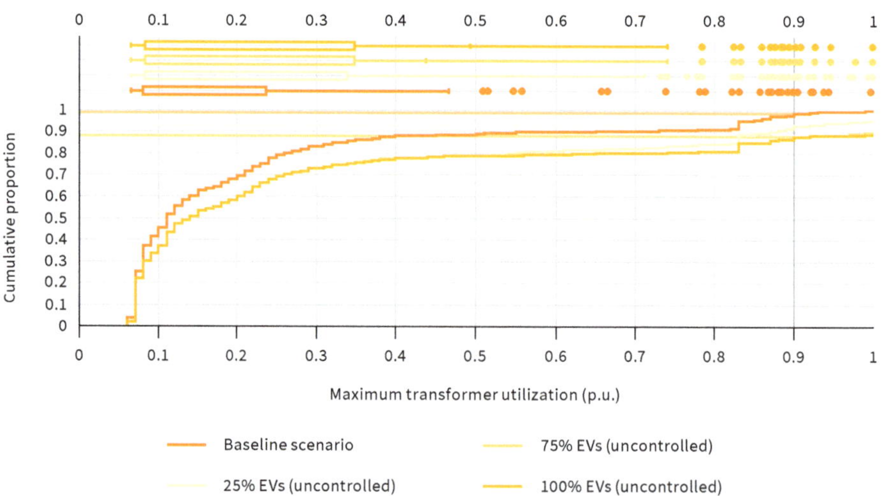

Figure 11. Substation transformer utilization for uncontrolled charging. Source: Graphic by authors.

Figure 12 depicts a comparison of the substation utilization for smart charging and uncontrolled charging across the different EV penetration levels. For 25% EVs, the smart charging increases the share of transformers loaded below 0.9 from approx. 94% to approx. 99%, and the mean utilization is decreased by approx. 0.08. For higher penetration levels, the benefit of smart charging reduces proportionally. At 100% EVs, the share of transformers loaded below 0.9 only increases by approx. 1%, although mean utilization decreases by approx. 0.08, i.e., similar to 25% EVs and 75% EVs. This behavior is due to the very high peak load at local substations, i.e., even a significant flattening of demand peaks still leads to highly overloaded substations in the 100% EV penetration scenario.

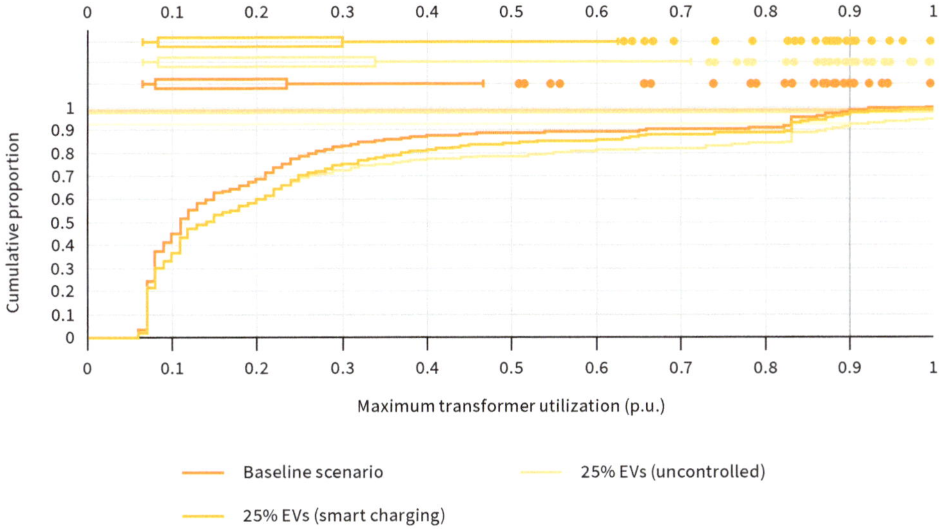

Figure 12. *Cont.*

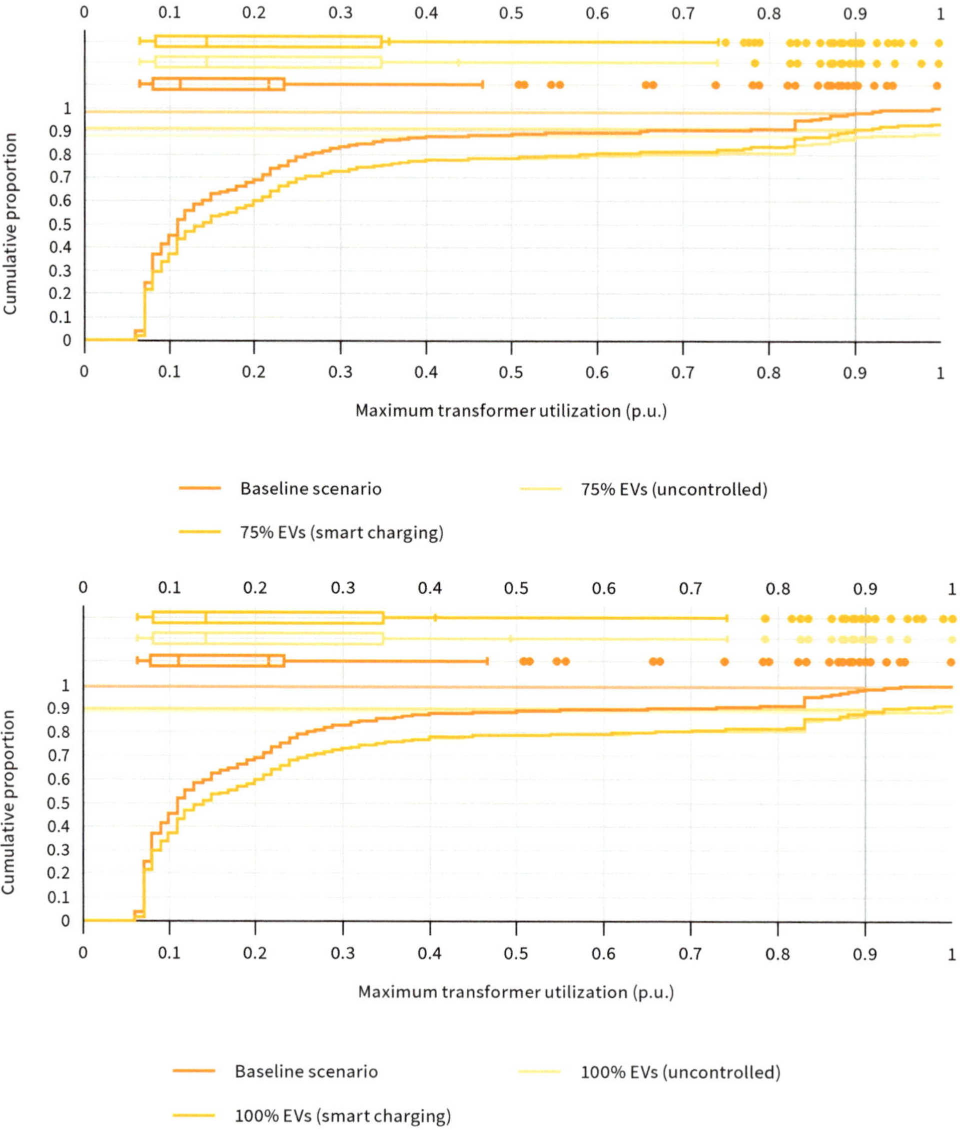

Figure 12. Substation transformer utilization for smart charging. Source: Graphic by authors.

5.4. URBS

For URBS, we set up a model with the parameters defined in Sections 4.4 and 4.5. The model consisted of the grid defined in Section 4.1 plus one additional node representing the "rest of Singapore", with its demand and its fossil generators. Each of the four 66/22-kV substations was connected to this additional node since the grid of the test case was divided into four isolated subgrids. Electricity can be transmitted both ways between Geylang and the rest of Singapore. The transmission capacity of the lines between the four entry points to the Geylang grid and the rest of Singapore was set sufficiently high to allow for unrestricted power exchange, which is realistic given the high robustness of Singapore's grid.

In this case study, PV and electric vehicle chargers could be installed in Geylang only. URBS could decide to increase the transmission capacity of the existing grid lines in the Geylang network if needed. We defined four scenarios with regard to PV and EV chargers to be deployed in Geylang, with Scenario 1 being the reference scenario. See Table 9.

Table 9. Scenarios in URBS. Scenario 1 is the reference scenario.

Scenario	PV	EV Chargers
1	no	no
2	yes	no
3	no	yes
4	yes	yes

Source: Table by authors.

The reference scenario was to test the feasibility of the model, i.e., whether the demand could be satisfied in every time step and whether the existing distribution capacity was sufficient. The costs and emissions of the system were determined as well.

For scenarios 2 and 4, we analyzed how much PV power and additional line capacity URBS decided to install for different costs of PV systems ranging from USD 0.70 per W to USD 0.85 per W. In these scenarios, all PV must be integrated. The results are displayed as subscenarios a, b, c, and d. A typical day regarding the solar insolation profile in Singapore with an average irradiance of 363 W m^{-2} during daytime and a maximum irradiance of 815 W m^{-2} was chosen for the study. URBS determines the cost-optimal solution taking into account costs of fossil generation,

PV installations, and grid upgrades. The lower the price, the higher the installed PV and additional installed distribution capacity.

For scenarios 3 and 4, we chose the fixed EV charging case with and EV penetration of 100%.

In the reference scenario, no additional transmission lines were built. Hence, the test case is feasible. The amount of installed PV and additional transmission capacity for all other scenarios, as well as resulting cost and emission reduction, compared with the reference scenario, are shown in Table 10. For the lowest PV price, the model installs more than 1700 MW of the maximum possible amount of 2023 MW of PV. The new install grid capacity is up to almost 4200 MW for the highest value of installed PV capacity. This number does not depend significantly on the EV penetration of 0 or 100% with fixed charging. Note that in scenario 3, without PV, an additional 150 MW of grid capacity is installed in order to supply the EV charging demand, while the difference of installed grid capacity between cases 2 and 4 is lower than that, which means that some of the PV power can be used directly to charge EVs.

Table 10. URBS scenarios showing installed PV capacity and additional grid capacity for different prices of PV systems and EV penetration levels of 0 or 100%.

Scen.	PV Inst. Cost	EV Penetr.	Inst. PV Cap.	New Grid Cap.	Cost. Red.	Em. Red.
—	(USD/W)	(%)	(MW)	(MW)	(%)	(%)
1	—	0	0	0	0	0
2a	0.85	0	970	780	0.16	2.75
2b	0.80	0	1290	1820	0.30	3.62
2c	0.75	0	1580	2640	0.47	4.14
2d	0.70	0	1710	4170	0.67	4.77
3	—	100	0	150	−0.59	−0.59
4a	0.85	100	1020	870	−0.40	2.31
4b	0.80	100	1340	1900	−0.26	3.17
4c	0.75	100	1510	2690	−0.08	3.63
4d	0.70	100	1720	4060	0.12	4.22

Source: Table by authors.

Cost savings are marginal. When PV is installed, fuel costs decrease since less power from fossil fuels is required, but investment and fixed cost of PV and new gridlines are about as high as the savings. For scenarios 3, 4a, 4b, and 4c, a slight cost increase can be observed. Overall CO_2 emission reduction in Singapore's power generation is up to almost 5%, depending on the amount of PV installed. Due to the additional demand caused by EVs, the emission reduction is lower in scenario 4 and negative in scenario 3, where installation of PV is not allowed.

Figure 13 shows the demand and PV generation in Geylang for one day without and with EV charging. The pattern under the blue demand curve depicts the demand covered by fossil fuels. The white part is covered by PV. The pattern under the red PV curve depicts the amount of PV exported to the rest of Singapore. For both with and without EV charging, the curves appear similar. During the day, when power is generated by PV, the share of demand in Geylang covered by PV is 88% without EV charging and 88% with EV charging. EV chargers are only connected at 43 nodes, while URBS decided to install PV at 302 nodes, such that PV supply and EV charging demand are not necessarily matched, and installing additional grid capacity is costly.

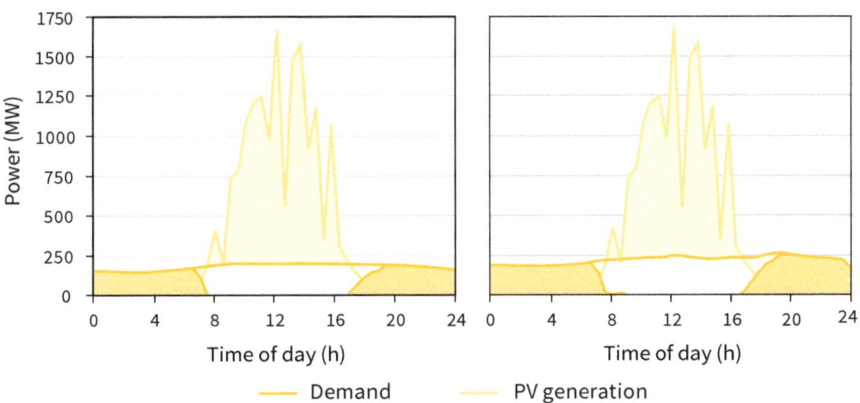

Figure 13. Demand (dark orange) and PV generation (light orange) in Geylang without (**left**) and with (**right**) EV charging. The dark orange area depicts the share of the demand that has to be supplied from Singapore's fossil fuel power plants, while the white area under the dark orange curve represents the amount supplied by PV. The light orange pattern depicts the amount of solar PV that is exported to the rest of Singapore where it replaces fossil generation. Source: Graphic by authors.

6. Discussion

The presented results underline the key capabilities of each of the software tools. For district-scale electric grids, these capabilities can be summarized as follows:

- MATPOWER primarily supports the study of operational problems for the electric grid, with capabilities for both simulation-based and optimization-based analysis. This tool caters to the need for a highly accessible power flow simulation tool with the convenience of scripting directly through MATLAB. While MATPOWER is limited to balanced AC power simulations, this is often sufficient for an initial assessment of grid hosting capabilities upon deployment of renewable generators or additional loads, as demonstrated in Section 5.1. In this regard, the tool is also suitable for the scenario-based study of planning problems in the electric grid. While the MATPOWER's focus was originally on a single-step power flow solution, it has been extended with an ecosystem of optimization-based and multi-period analysis, e.g., through the OPF or MOST interfaces.

- GridLAB-D is a software framework focused on the simulation-based analysis of operational problems for the district-scale electric grid and DERs. The tool enables the utilization of highly detailed models for each subsystem of the district-scale energy system. This is demonstrated in Section 5.2 in terms of the detailed modeling of HVAC loads of the buildings in the synthetic grid. GridLAB-D benefits from the rich ecosystem of tools for model preparation and co-simulation, which were presented in Sections 3.2 and 3.3. Since GridLAB-D caters mainly to simulation-based electric grid analysis, it does not directly enable optimization-based solutions to electric grid operation problems. Yet, an optimization-based control system can indirectly be included through co-simulation, e.g., with TESP. To this end, GridLAB-D can serve as a testbed for novel market frameworks, where the DER dispatch and market clearing are implemented via TESP, and the GridLAB-D simulation acts as a digital twin for the electric grid and DER systems.

- MESMO is a software tool that caters primarily to the optimization-based analysis of operational problems. The tool focuses on supporting the formulation of convex optimization problems for the operation of district-scale electric grids and DERs. With this focus on the convex domain, the tool is well suited for the analysis of market-clearing problems based on decomposition techniques arising from numerical optimization. For example, MESMO directly outputs the DLMP values for operational problems. However, due to the focus on convex modeling, DER models in MESMO are limited to simple state-space expressions. To this end, simulation-based analysis with MESMO is less powerful than in GridLAB-D. Compared with MATPOWER, the scripting interface of MESMO through Python is less mature and less stable. As presented in Section 5.3, MESMO is suitable for the analysis of flexibility potentials in DERs of the electric grid, which is enabled through the highly customizable range of DER models without requiring external model coupling. Note that MESMO also supports the analysis of multi-energy systems, e.g., in terms of thermal grids.

- URBS is a toolbox that is heavily focused on the optimization-based analysis of both planning and operation problems. Hence, it is the only one of the presented tools which directly addresses planning problems. Similar to MESMO, URBS focuses on convex modeling of the electric grid and DERs, although the models are significantly more simplified with a focus on capturing capacity constraints. As presented in Section 5.4, URBS can be employed for determining the optimal deployment of renewable generation, where emission reduction and other objectives can be considered in addition to conventional cost minimization.

Although there is an overlap in the capabilities of the presented tools, there is currently no comprehensive solution that covers the complete feature set for electric grid analysis. Particularly, there is a trade-off between optimization-based tools, which are typically restricted in modeling detail, and simulation-based tools, which favor modeling detail over convex mathematical formulations. This highlights the importance of clarifying the focus of electric grid studies with stakeholders in advance to identify a specific set of features expected for the chosen software platform.

7. Conclusion

This chapter discussed the role of open-source software frameworks in the transition towards affordable electricity. To this end, the chapter introduced the different problem types and requirements for electric grid analysis in the context of studies for the integration of novel DERs, such as renewable generators, EV chargers, flexible loads, and energy storage systems. Along with this, a representative set of open-source tools and their feature sets were introduced and compared. This served to identify and differentiate the key use cases for different software frameworks. The main functionalities of each tool were demonstrated for a synthetic electric grid test case based in Singapore. To conclude, this chapter, and Section 6 in particular, is intended as a guideline to open-source tools for various electric grid simulation and optimization applications.

Author Contributions: Conceptualization, S.T., S.H.; software, S.T., S.H., K.Z., B.A.B.; formal analysis, S.T., S.H., K.Z., T.M.; data curation, S.T., S.H., T.M., B.A.B.; writing–original draft preparation, S.T., S.H.; writing–review and editing, S.T., T.M., K.Z., B.A.B., A.A., M.J.A.;

visualization, S.T., S.H., T.M.; All authors have read and agreed to the published version of the manuscript.

Funding: This work was financially supported by the Singapore National Research Foundation under its Campus for Research Excellence and Technological Enterprise (CREATE) program.

Data Availability Statement: The test case definitions for GridLAB-D, MESMO and urbs (see Section 4) are available open-source, online at: https://github.com/mesmo-dev/testcase_transition_towards_affordable_electricity_2022 (accessed on 4 April 2022).

Acknowledgments: The authors would like to thank Andrej Trpovski, who developed the synthetic distribution grid model for Singapore (Trpovski 2021) that forms the basis for the test case presented in this work.

Conflicts of Interest: The authors declare no conflict of interest.

References

Atabay, Dennis. 2017. An open-source model for optimal design and operation of industrial energy systems. *Energy* 121: 803–21. [CrossRef]

Brown, Simon. 2015. *Software Architecture for Developers—Visualise, Document and Explore Your Software Architecture*. Victoria: Leanpub, vol. 2.

Chassin, David P., Jason C. Fuller, and Ned Djilali. 2014. GridLAB-D: An agent-based simulation framework for smart grids. *Journal of Applied Mathematics* 2014: 1–12. [CrossRef]

Chassin, David P., Kevin Schneider, and Clint Gerkensmeyer. 2008. GridLAB-D: An open-source power systems modeling and simulation environment. Paper presented at the 2008 IEEE/PES Transmission and Distribution Conference and Exposition, Chicago, IL, USA, 21–24 April 2008, pp. 1–5.

Cho, Youngchae, Takayuki Ishizaki, Nacim Ramdani, and Jun-ichi Imura. 2019. Box-based temporal decomposition of multi-period economic dispatch for two-stage robust unit commitment. *IEEE Transactions on Power Systems* 34: 3109–18. [CrossRef]

Coffrin, Carleton, Russell Bent, Kaarthik Sundar, Yeesian Ng, and Miles Lubin. 2018. PowerModels.Jl: An open-source framework for exploring power flow formulations. Paper presented at the IEEE 2018 Power Systems Computation Conference (PSCC), Dublin, Ireland, 11–15 June 2018, pp. 1–8.

Després, Jacques, Nouredine Hadjsaid, Patrick Criqui, and Isabelle Noirot. 2015. Modelling the impacts of variable renewable sources on the power sector: Reconsidering the typology of energy modelling tools. *Energy* 80: 486–95.

Dorfner, Johannes, Konrad Schönleber, Magdalena Dorfner, sonercandas, froehlie, smuellr, dogauzrek, WYAUDI, Leonhard-B, lodersky, and et al. 2019. *Urbs v1.0.1: A Linear Optimisation Model for Distributed Energy Systems*. Munich: Zenodo. [CrossRef]

Dugan, Roger C., and Thomas E. McDermott. 2011. An open source platform for collaborating on smart grid research. Paper presented at the 2011 IEEE Power and Energy Society General Meeting, Detroit, MI, USA, 24–28 July 2011, pp. 1–7.

Energy Market Authority. 2020a. Electricity Generation Capacity by Generator (Sheet 29RSU). Available online: https://www.ema.gov.sg/statistic.aspx?sta_sid=20141211PlH5tAbg9BY8 (accessed on 25 August 2021).

Energy Market Authority. 2020b. Half-Hourly System Demand Data. Available online: https://www.ema.gov.sg/statistic.aspx?sta_sid=20140826Y84sgBebjwKV (accessed on 25 August 2021).

Energy Market Company. 2021. Price Information. Available online: https://www.emcsg.com/marketdata/priceinformation (accessed on 14 June 2021).

EV Database. 2021. Energy Consumption of Full Electric Vehicles. Available online: https://ev-database.org/cheatsheet/energy-consumption-electric-car (accessed on 16 June 2021).

Fleischhacker, Andreas, Georg Lettner, Daniel Schwabeneder, and Hans Auer. 2019. Portfolio optimization of energy communities to meet reductions in costs and emissions. *Energy* 173: 1092–1105. [CrossRef]

Hanif, Sarmad, Kai Zhang, Christoph M. Hackl, Masoud Barati, Hoay Beng Gooi, and Thomas Hamacher. 2019. Decomposition and equilibrium achieving distribution locational marginal prices using trust-region method. *IEEE Transactions on Smart Grid* 10: 3269–81. [CrossRef]

Hilpert, S., C. Kaldemeyer, U. Krien, S. Günther, C. Wingenbach, and G. Plessmann. 2018. The Open Energy Modelling Framework (oemof)—A new approach to facilitate open science in energy system modelling. *Energy Strategy Reviews* 22: 16–25. [CrossRef]

Huang, Qiuhua, Thomas E. McDermott, Yingying Tang, Atefe Makhmalbaf, Donald J. Hammerstrom, Andrew R. Fisher, Laurentiu Dan Marinovici, and Trevor Hardy. 2018. Simulation-based valuation of transactive energy systems. *IEEE Transactions on Power Systems* 34: 4138–47. [CrossRef]

Huber, Matthias, Albert Roger, and Thomas Hamacher. 2015. Optimizing long-term investments for a sustainable development of the ASEAN power system. *Energy* 88: 180–93. [CrossRef]

Hunter, Kevin, Sarat Sreepathi, and Joseph F. DeCarolis. 2013. Modeling for insight using Tools for Energy Model Optimization and Analysis (Temoa). *Energy Economics* 40: 339–349. [CrossRef]

Kleinschmidt, Verena, Sebastian Troitzsch, Thomas Hamacher, and Vedran S. Perić. 2021. Flexibility in distribution systems—Modelling a thermal-electric multi-energy system in MESMO. Paper presented at the IEEE PES Innovative Smart Grid Technologies Conference Europe, Espoo, Finland, 18–21 October 2021.

Klemm, Christian, and Peter Vennemann. 2021. Modeling and optimization of multi-energy systems in mixed-use districts: A review of existing methods and approaches. *Renewable and Sustainable Energy Reviews* 135: 110206. [CrossRef]

Kriechbaum, Lukas, Gerhild Scheiber, and Thomas Kienberger. 2018. Grid-based multi-energy systems—Modelling, assessment, open source modelling frameworks and challenges. *Energy, Sustainability and Society* 8: 35. [CrossRef]

Lacal Arantegui, Roberto, Arnulf Jaeger-Waldau, Marika Vellei, Bergur Sigfusson, Davide Magagna, Mindaugas Jakubcionis, Maria del Mar Perez Fortes, Stavros Lazarou, Jacopo Giuntoli, Eveline Weidner Ronnefeld, and et al. 2014. *ETRI 2014—Energy technology reference indicator projections for 2010–2050*. EUR—Scientific and technical research reports. Luxembourg: Joint Research Center of the European Union.

Land Transport Authority. 2020a. Annual Mileage for Private Motor Vehicles. Available online: https://data.gov.sg/dataset/annual-mileage-for-private-motor-vehicles (accessed on 16 June 2021).

Land Transport Authority. 2020b. Current Vehicle Growth Rate to Continue Until 31 January 2022. Available online: https://www.lta.gov.sg/content/ltagov/en/newsroom/2020/8/news-releases/current-vehicle-growth-rate-to-continue-until-31-january-2022.html (accessed on 16 June 2021).

Land Transport Authority. 2021a. Annual Vehicle Statistics. Available online: https://www.lta.gov.sg/content/dam/ltagov/who_we_are/statistics_and_publications/statistics/pdf/MVP01-1_MVP_by_type.pdf (accessed on 16 June 2021).

Land Transport Authority. 2021b. DataMall. Available online: https://datamall.lta.gov.sg/ (accessed on 16 June 2021).

Lincoln, Richard. 2021. PYPOWER. Github. Available online: https://github.com/rwl/PYPOWER (accessed on 16 July 2021).

Massier, Tobias, Dante Fernando Recalde Melo, Reinhard Sellmair, Marc Gallet, and Thomas Hamacher. 2018. Electrification of road transport in Singapore and its integration into the power system. *Energy Technology* 6: 21–32. [CrossRef]

McDermott, Thomas E., Eric G. Stephan, and Tara D. Gibson. 2018. Alternative database designs for the distribution common information model. Paper presented at the 2018 IEEE/PES Transmission and Distribution Conference and Exposition (T&D), Denver, CO, USA, 16–19 April 2018, pp. 1–9.

Melton, Ronald B., Kevin P. Schneider, Thomas E. McDermott, and Subramanian V. Vadari. 2017. *GridAPPS-D Conceptual Design v1.0*. Technical report. Richland: Pacific Northwest National Laboratory (PNNL).

Melton, Ronald B., Kevin P. Schneider, and Subramanian Vadari. 2019. GridAPPS-D™ a distribution management platform to develop applications for rural electric utilities. Paper presented at the 2019 IEEE Rural Electric Power Conference (REPC), Bloomington, MN, USA, 28 April–1 May 2019, pp. 13–17.

Morstyn, Thomas, Katherine A. Collett, Avinash Vijay, Matthew Deakin, Scot Wheeler, Sivapriya M. Bhagavathy, Filiberto Fele, and Malcolm D. McCulloch. 2020. OPEN: An open-source platform for developing smart local energy system applications. *Applied Energy* 275: 115397. [CrossRef]

Murillo-Sanchez, Carlos E., Ray D. Zimmerman, C. Lindsay Anderson, and Robert J. Thomas. 2013. Secure Planning and Operations of Systems With Stochastic Sources, Energy Storage, and Active Demand. *IEEE Transactions on Smart Grid* 4: 2220–29. [CrossRef]

Pacific Northwest National Laboratory. 2021a. GridAPPS-D. Github. Richland, US-WA. Available online: https://github.com/GRIDAPPSD/ (accessed on 8 April 2022).

Pacific Northwest National Laboratory. 2021b. Transactive Energy Simulation Platform. Github. Richland, US-WA. Available online: https://github.com/pnnl/tesp/tree/master/src/tesp_support/tesp_support (accessed on 8 April 2022).

Pfenninger, Stefan, and Bryn Pickering. 2018. Calliope: A multi-scale energy systems modelling framework. *Journal of Open Source Software* 3: 825. [CrossRef]

Pratt, R. G., and Z. T. Taylor. 1994. *Development and Testing of an Equivalent Thermal Parameter Model of Commercial Buildings from Time-Series End-Use Data*. Technical report. Richland: Pacific Northwest Laboratory (PNNL).

Ramachandran, Srikkanth, Kais Siala, Cristina de la Rúa, Tobias Massier, Arif Ahmed, and Thomas Hamacher. 2021. Carbon footprint of a cost-optimal HVDC connection to import solar energy from Australia to Singapore. *Energies* 14: 7178. [CrossRef]

Ringkjøb, Hans-Kristian, Peter M. Haugan, and Ida Marie Solbrekke. 2018. A review of modelling tools for energy and electricity systems with large shares of variable renewables. *Renewable and Sustainable Energy Reviews* 96: 440–59. [CrossRef]

Schaber, Katrin, Florian Steinke, and Thomas Hamacher. 2012. Transmission grid extensions for the integration of variable renewable energies in Europe: Who benefits where? *Energy Policy* 43: 123–35. [CrossRef]

Schelo, Tom, Antoine Bidel, and Thomas Hamacher. 2021. Biogas Plant Operation under Distribution Locational Marginal Prices. Paper presented at the 2021 IEEE Madrid PowerTech, Madrid, Spain, 28 June–2 July 2021, pp. 1–6.

Siala, Kais. 2021a. *Urbs Model for Mekong Project*. Munich: Zenodo. [CrossRef]

Siala, Kais. 2021b. *Urbs Model for the SunCable Project*. Munich: Zenodo. [CrossRef]

Siala, Kais, A. F. M. Kamal Chowdhury, Thanh Duc Dang, and Stefano Galelli. 2021. *Data for Manuscript "Solar Energy and Regional Coordination as a Feasible Alternative to Large Hydropower in Southeast Asia"*. Munich: Zenodo. [CrossRef]

Solar Energy Research Institute of Singapore (SERIS). 2020. *Update of the Solar Photovoltaic (PV) Roadmap for Singapore*. Technical report. Singapore: SERIS, National University of Singapore.

Stich, Jürgen, Melanie Mannhart, Thomas Zipperle, Tobias Massier, Matthias Huber, and Thomas Hamacher. 2014. Modelling a low-carbon power system for Indonesia, Malaysia and Singapore. Paper presented at the 33rd International Energy Workshop, Beijing, China, 4–6 June 2014, pp. 1–11. ERI-NDRC.

Stich, Jürgen, and Tobias Massier. 2015. Enhancing the integration of renewables by trans-border electricity trade in ASEAN. Paper presented at the 2015 IEEE PES Asia-Pacific Power and Energy Engineering Conference (APPEEC), Brisbane, QLD, Australia, 15–18 November 2015.

Stüber, Magdalena, and Leonhard Odersky. 2020. Uncertainty modeling with the open source framework urbs. *Energy Strategy Reviews* 29: 100486. [CrossRef]

Taylor, Zachary T., Krishnan Gowri, and Srinivas Katipamula. 2008. *GridLAB-D Technical Support Document: Residential End-Use Module Version 1.0*. Technical report. Richland: Pacific Northwest National Lab (PNNL).

Thurner, Leon, Alexander Scheidler, Florian Schafer, Jan-Hendrik Menke, Julian Dollichon, Friederike Meier, Steffen Meinecke, and Martin Braun. 2018. Pandapower—An Open-Source Python Tool for Convenient Modeling, Analysis, and Optimization of Electric Power Systems. *IEEE Transactions on Power Systems* 33: 6510–21. [CrossRef]

Troitzsch, Sebastian, Mischa Grussmann, Kai Zhang, and Thomas Hamacher. 2020. Distribution Locational Marginal Pricing for Combined Thermal and Electric Grid Operation. Paper presented at the IEEE PES Innovative Smart Grid Technologies Conference Europe, The Hague, Netherlands, 26–28 October 2020.

Troitzsch, Sebastian, and Thomas Hamacher. 2020. Control-oriented Thermal Building Modelling. Paper presented at the IEEE PES General Meeting, Montreal, QC, Canada, 2–6 August 2020.

Troitzsch, Sebastian, Sarmad Hanif, Kai Zhang, Andrej Trpovski, and Thomas Hamacher. 2019. Flexible Distribution Grid Demonstrator (FLEDGE): Requirements and Software Architecture. Paper presented at the IEEE Power & Energy Society General Meeting (PESGM), Atlanta, GA, USA, 4–8 August 2019.

Troitzsch, Sebastian, Kai Zhang, Tobias Massier, and Thomas Hamacher. 2021. Coordinated Market Clearing for Combined Thermal and Electric Distribution Grid Operation. Paper presented at the IEEE Power & Energy Society General Meeting (PESGM), Washington, DC, USA, 25–29 July 2021.

Trpovski, Andrej. 2021. Synthetic Grid Generation Using Power System Planning—Case Study of the Power System in Singapore. Ph.D. thesis, Technical University of Munich, Munich, Germany.

van Beuzekom, I., M. Gibescu, and J. G. Slootweg. 2015. A review of multi-energy system planning and optimization tools for sustainable urban development. Paper presented at the 2015 IEEE Eindhoven PowerTech, Eindhoven, The Netherlands, 29 June–2 July 2015, pp. 1–7.

Wang, Hongye, Carlos E. Murillo-Sanchez, Ray D. Zimmerman, and Robert J. Thomas. 2007. On computational issues of market-based optimal power flow. *IEEE Transactions on Power Systems* 22: 1185–93. [CrossRef]

Zimmerman, Ray D., and Carlos E. Murillo-Sánchez. 2019. *MATPOWER User's Manual*. New York: Zenodo. [CrossRef]

Zimmerman, Ray Daniel, Carlos Edmundo Murillo-Sánchez, and Robert John Thomas. 2011. MATPOWER: Steady-state operations, planning, and analysis tools for power systems research and education. *IEEE Transactions on Power Systems* 26: 12–19. [CrossRef]

Zwickl-Bernhard, Sebastian, and Hans Auer. 2021. Open-source modeling of a low-carbon urban neighborhood with high shares of local renewable generation. *Applied Energy* 282: 116166. [CrossRef]

Clean Energy Transition Challenge: The Contributions of Geology

Cristina Rodrigues, Henrique Pinheiro and Manuel Lemos de Sousa

1. Introduction

In order to discuss the global topic of the clean energy transition, it is essential that strategies by world decision makers are clearly understood. International agencies and European Union (EU) institutions have developed and implemented significant programs, as described below. The first European strategy on climate was developed in 1991 through the creation of the European Commission and Climate Program. Later on, in 1997, the Kyoto Protocol was designed, the main goal being to stabilize atmospheric concentrations of GHGs at a level that would prevent dangerous interference with the climate system. Since then, several initiatives have been promoted, such as the reports of the UNFCCC and the first (2000) and second (2005) European Climate Change Programmes, which included the need to identify the most environmentally and cost-effective policies and measures that would allow Europe to cut its GHGs, and to apply carbon capture and storage (CCS) technologies as part of the efforts. The agreements currently in force are as follows:

(i) In 2015, with the Paris Climate Agreement, a global political action plan was developed to put the world on track to avoid climate change, highly supported by the idea of keeping the increase in the global average temperature to well below 2 °C (or ultimately 1.5 °C) above pre-industrial levels. To reach this main goal, it was established that renewable energies should embody a higher share of the global energy matrix, as well as nuclear energy.

(ii) In 2019, the European Green Deal was established, with one of the main targets focusing on climate change, through actions to be developed by the EU. The use of "green hydrogen" and carbon neutrality are seen as priorities for a clean and circular economy.

Given the above context, one should ask how best to respond to the mentioned political agreements whilst taking into account the actual and forecast global and European energy demand, notwithstanding the fact that the world energy supply is, and will continue to be, highly dependent on fossil fuels, for both technological and economic reasons, at least in the short and medium terms. Accelerated, rapid innovation and invention will certainly be the catalysts in developing non-fossil fuel alternatives to energy production, so long as they are sustainable and affordable and can reduce the time frame for achieving the goals of clean energy and almost neutral emissions.

Consequently, in the authors' view, the need to implement the energy transition strategy, meaning to shift the global energy sector from fossil fuel based to zero-carbon energies by the second half of the 21st century, should be a main goal, allowing for growth in the global energy demand to continue whilst addressing climate change concerns and targets. The energy transition has to be implemented in a conscious and coherent manner in order to achieve the clean and circular economy concepts. This means that strong efforts in technology and policy are required, capable of turning non-fossil fuel energy production into competitive and sustainable economically viable sources. Therefore, analysis of the cost of climate change mitigation versus the cost of the energy transition is a must, given the costs of renewable energy solutions, and those of batteries and hydrogen energy (Nordhaus 2018).

If one accepts the general framework described above, and in order to accomplish the main goal of the Paris Climate Agreement, which is to reduce CO_2 emissions in order to keep the global temperature rise well below 2 °C (or even 1.5 °C), we insist that it is indispensable to use the contribution of geology in the application of carbon capture and storage, as well as in underground energy storage technologies, and to do so as soon as possible.

The main goal of this chapter is to discuss the energy and climate sectors, given the need to develop a workable international strategy for a clean energy transition by considering the contribution of geology in the application of carbon capture and storage, as well as in underground energy storage technologies, and to do so as soon as possible.

Energy transition is a well-known subject discussed at all climate and energy meetings. It is also one of the most controversial topics, and therefore an almost impossible mission in seeking to establish successful and acceptable international economic, technical, political and social measures. To better address the energy transition thematic, it is pertinent to highlight and to clarify several related subjects, which are discussed in the present manuscript. This approach begins with a general overview of the international energy and climate strategies implemented in recent decades, and the global and European decisions focusing on the politically binding and non-binding measures established to meet the greenhouse gas (GHG) reduction targets are also highlighted. In the second section, the topic of climate change is presented, with emphasis on the need to contextualize climate change in the geological evolution of planet Earth. The third topic addressed is the energy transition target in order to fulfill the current global energy demand. Finally, the role that geology can play, and will play, in the energy transition strategy is assessed, making it clear that this ambitious target will not be reached without a strong contribution from geology.

2. Overview on International Energy and Climate Strategies

The energy transition was recently established as one of the key solutions to climate change mitigation, which involves shifting from a system based on fossil fuels (oil, natural gas and coal) to one dominated by variable renewable energies. However, the starting point in discussing the energy transition topic goes back, at least, to 1979 with the First World Climate Conference held at Geneva (Conference of Experts on Climate and Mankind), where climate change was recognized as a serious problem. Rightly or wrongly, climate change is considered to be intimately related to the increase in GHGs, of which carbon dioxide (CO_2) is a constituent, despite its relevance, for example, to agricultural production (Dayaratna et al. 2020). CO_2 is acknowledged as playing an important role in the atmospheric temperature of the Earth, and therefore a CO_2 increase can contribute to a gradual warming of the lower atmosphere, especially at high latitudes. It is assumed that anthropogenic activity, including the exploitation and burning of fossil fuels, deforestation and changes in land use, has a fundamental role in the amount of CO_2 increase in the atmosphere and, consequently, in increasing GHG concentrations. The Declaration of the First World Climate Conference established a main goal "to foresee and prevent potential man-made changes in climate that might be adverse to the well-being of humanity" (WMO (World Meteorological Organization) 1979, p. 3). Additionally, it was also advised to implement efforts to reduce fossil fuels in the world energy matrix by including nuclear and renewable energies.

In the late 1980s and early 1990s, several intergovernmental conferences focusing on climate change occurred, in which both scientific and policy subjects were discussed, and the ultimate conclusion reached was the need to establish a global climate action plan.

In 1990, although established in 1988, the Intergovernmental Panel on Climate Change (IPCC) released its First Assessment Report, which scientifically confirmed the climate change issue, allowing governments to adapt their policy decisions. The Second World Climate Conference, also held in 1990, established a framework treaty on climate change, the final declaration of which did not specify any international targets for reducing CO_2 emissions, but a number of principles were defined that were included in the Climate Change Convention. The Intergovernmental Negotiating Committee for a Framework Convention on Climate Change (INC/FCCC) was approved later (December) in 1990.

Only in 1991 was the first European community strategy created on climate change, establishing the need to limit CO_2 emissions and to improve energy efficiency as main objectives. The specific action plan was as follows:

(1) To create a directive to promote electricity from renewable energy;
(2) To promote voluntary commitments by car markets to reduce CO_2 emissions by 25%;
(3) To propose taxation of energy products.

The United Nations Framework Convention on Climate Change (UNFCCC) was signed at Rio de Janeiro in 1992, which was established as the main international treaty on fighting climate change, in order to prevent dangerous human-made interference with the global climate system. A global agreement was required to implement UNFCCC strategies; therefore, in 1995, the first Conference of the Parties (COP1) took place in Berlin, and finally in 1997, in Kyoto (Japan), the Kyoto Protocol, an extension of the UNFCCC, was signed. Nevertheless, the Kyoto Protocol was only implemented in Montreal (Canada) in 2005, with the main goal to stabilize atmospheric concentrations of GHGs at a level that will prevent dangerous anthropogenic interference with the climate system. This was to be achieved by cutting GHGs to 5% below 1990 levels, by the 2008–2012 period. The Kyoto Protocol legally binds industrialized countries and economies in transition and the EU to emission reduction targets. Targets for the first commitment period (2008–2012) of the Kyoto Protocol covered emissions of the six main GHGs, namely CO_2, methane (CH_4), nitrous oxide (N_2O), hydrofluorocarbons (HFCs), perfluorocarbons (PFCs) and sulphur hexafluoride (SF6) (Wigley 1998). It was established that the Party's assigned amount is the maximum amount of emissions, measured as equivalent in CO_2, that a Party could emit during a commitment period.

The COP meetings continue to take place on a regular basis with the main goal to discuss, in detail, the rules for the implementation of the Kyoto Protocol by setting up new funding and planning instruments for adapting and establishing an outline for technology transfer.

As it is well known, the EU has long been committed to international efforts to reduce climate change and felt the responsibility to set an example through the creation of strong policies in Europe. To ensure the development of a comprehensive package of the most environmentally effective policies and measures to reduce GHG emissions, the EU established the European Climate Change Programme (ECCP). Two ECCP plans were developed in the period from 2000 to 2005. The first ECCP, established in 2000, was responsible for examining an extensive range of policy sectors and instruments with potential for reducing GHG emissions, by creating several working groups, namely, energy supply, energy demand, energy efficiency in end-use equipment and industrial processes, transport and research, among others, but the most important and innovative one was emissions trading. Each working group had to identify different options for reducing GHG emissions based on cost effectiveness, seeking the promotion of energy security and air quality as a final target. The second ECCP, launched in 2005, after four years of implementation of the first ECCP, aimed to further identify cost-effective options for reducing GHG emissions that would promote economic growth increase and job creation. Learning from the first period of the ECCP, when priorities allowed identifying five main working groups (energy supply, energy demand, transport, non-CO_2 gases and agriculture), it was possible in the second ECCP period to add new working groups, namely, aviation, CO_2 and cars, adaptation to the effects of climate change, reducing greenhouse gas emissions from ships and carbon capture and storage. In this second program, additional measures were taken, such as promoting the use of renewable energies in heating applications.

As it has become clear, the 21st century has been marked by new energy and climate challenges, in which the EU has played a leading role in identifying potential efficient solutions, although mainly political ones. The European energy/climate strategy has led to the development of several initiatives, namely, "Europe 2020" (EC (European Commission) 2007a), "European Strategic Energy Technology Plan (SET-Plan)" (EC (European Commission) 2007b), "Energy 2020: a strategy for competitive, sustainable and secure energy" (EC (European Commission) 2010), "The EU energy policy: engaging with partners beyond our borders" (EC (European Commission) 2011c), "Energy Roadmap 2050" (EC (European Commission) 2011b), "Directive on energy efficiency" (EC (European Commission) 2011a), "Guidelines for trans-European energy infrastructure" (EC (European Commission) 2011e) and "Smart grids: from innovation to deployment" (EC (European Commission) 2011d) (Rodrigues et al. 2015).

During this period, in order to ensure the security of supply and competitiveness and, at the same time, to promote decarbonization of the energy system, aiming at reducing GHG emissions, mainly CO_2, the EU identified and recommended six European industrial initiatives for implementation: wind, solar, bioenergy, smart grids, nuclear fission and carbon capture and storage (CCS) (EC (European

Commission) 2007b). It was then clear that to meet the goals of the "Limiting Global Climate Change to 2 °C: the way ahead for 2020 and beyond" (EC (European Commission) 2007a) report by reducing CO_2 emissions will not be possible without geological sequestration/storage (Rodrigues et al. 2015; D'Amore and Bezzo 2017). As a result, on 23 April 2009, the European Parliament and the European Council unveiled Directive 2009/31/EC (Directive on Geological Storage of CO_2) (EC (European Commission) 2009), in order to define a regulatory framework for geological sequestration/storage of CO_2 regarding the conditions to deliver "storage permits" proposed by the Kyoto Protocol (2005) whilst promoting the vision of global environmental integration. Nevertheless, it is pertinent to mention that over the years, COP negotiations have been quite controversial. In fact, the COP15 meeting, in 2009 at Copenhagen, was recognized as disappointing for several reasons:

(1) It failed to set the basic targets for reducing global annual emissions of GHGs up to 2050;
(2) It did not secure commitments from countries to meet the emissions targets collectively;
(3) The target agreement was not binding.

It is indisputable that the first period of the Kyoto Protocol (2008–2012) failed, due to deficiencies in the structure of the treaty, such as the time frame of the agreement and the choice to establish a five-year commitment period which would start ten years after being signed, the exemption of developing countries from reduction requirements and the lack of an effective progressive emissions trading system (Rosen 2015). It is well known that the Kyoto Protocol was condemned from the beginning, given that it did not include the world's largest and fastest growing economies, for example, China, which was excluded from binding targets, and the fact that the United States of America (USA) did not sign the agreement.

A second commitment period (2013–2020) to the Kyoto Protocol was agreed in 2012, in the so-called Doha Amendment (UN (United Nations) 2012). This amendment included new commitments for Annex I Parties to the Kyoto Protocol, a revised list of GHGs which included one more GHG (nitrogen trifluoride—NF3) and amendments to several articles of the Kyoto Protocol which needed to be updated. During the first period of the agreement, a 5% reduction in GHGs was established, but the second commitment period was really ambitious by establishing a reduction in GHG emissions by a least 8% below 1990 levels. Flexible market mechanisms were created in the second Kyoto Protocol period, which were based on a trade of emissions permits, namely, international emissions trading, clean development mechanisms and joint implementation by the Parties. As a matter of fact, the commitment period established by the second Kyoto Protocol agreement to meet the GHG emission reduction target was rather ambitious and too short, and when associated with the energy demand increase, it led to a new GHG emissions strategy. In fact, the main goal became the removal of GHGs from the atmosphere, even if GHG emissions are not reduced. In this perspective, these new mechanisms motivated GHG abatement techniques using the most cost-effective processes, such as CCS technologies (UN (United Nations) 2012).

The Paris Agreement, adopted at the Paris Climate Conference (COP21) in 2015, is the first international legally binding global climate change agreement set out to avoid dangerous climate change. This agreement was recognized as a key point between policies and climate carbon neutrality before the end of the 21st century, which implied reaching the following targets (UN (United Nations) 2015):

(1) To limit the increase in the global average temperature to well below 2.0 °C, or even 1.5 °C, above pre-industrial levels;
(2) To reach the global emissions peak as soon as possible, which has already been accomplished in Europe and North America but will, most likely, take longer for developing countries;
(3) To undertake a rapid reduction in emissions using the best available science knowledge to achieve an efficient balance between emissions and removals in the second half of the 21st century;
(4) To strengthen the ability of countries to deal with the impacts of climate change, through appropriate financial programs and new technology frameworks.

In 2019, the European Green Deal (EGD) was proposed to transform the EU into a modern, resource-efficient and competitive economy and therefore the world's first carbon-neutral continent by 2050. The action plan established by the EGD aimed to increase the efficient use of all resources by moving to a clean and circular economy, to restore biodiversity and to cut pollution. In fact, the EGD is not a law that will put all countries on track to achieve climate change goals, but one of its main targets is to propose a European Climate Law to transform this political commitment into a legal obligation through the revision of the EU Regulation (EU) 2018/1999 (EC (European Commission) 2020b). The EGD has been seen as a powerful tool to combine efforts to reduce GHG emissions and, at the same time, to prepare Europe's industry for a climate-neutral economy. In this perspective, hydrogen has been considered as a key priority for addressing both the EGD and Europe's clean energy transition (EC (European Commission) 2020a). The EU Hydrogen Strategy is seen as a prevailing strategy to increase electricity production from renewable energy, and since hydrogen does not emit CO_2, it can play an important role in decarbonizing the industry. However, this energy scenario change will not be an easy task, given that hydrogen energy represents a small fraction of the world energy matrix, and it is still largely produced from fossil fuels, mainly from natural gas and coal. Consequently, the main goal of the hydrogen strategy is to decarbonize hydrogen production and expand its use into sectors where it can replace fossil fuels. The EU proposed an additional program, the Digital Transformation (EU (European Union) 2021), which seems capable of accelerating the energy transition by making power-generating assets more efficient, through grid modernization processes that make the system more secure and resilient, and that assist the industry in providing sustainable and affordable power to final consumers.

Summarizing, the transition energy issue has been subjected to several approaches in recent decades, but it is an international consensus that countries are not doing nearly enough to transition to non-fossil fuel energy sources. Undoubtedly, it is quite a hard task to accomplish a net zero-carbon economy by 2050, and to keep the world temperature increase to 2 °C, or even 1.5 °C. However, it is pertinent to state that this 2 °C target was a political decision, perceived by the public as a realistically achievable and acceptable goal, but it was never clearly advocated and recommended as a safe level of warming through a scientific assessment (Knutti et al. 2015). Additionally, transition energy has negative implications for cohesion and social inclusion of all countries, since several regions (Eastern and Southeastern European countries) will be obliged to make great investments, while other countries (Western Europe and North America) will be encouraged to reduce carbon-intensive industries which will imply infrastructure adaptations. In conclusion, to reach net zero emissions by 2050, the EU needs to agree and ratify a consistent climate change strategy, with strong and rapid investments, which must be accompanied and supported by innovation in science and technology. In fact, several studies are in place that are focused on climate change mitigation, including understanding the influence (if any) of solar variability on surface air temperature (Soon et al. 2015).

3. Climate Change: Is It the Problem?

As previously mentioned, climate change was identified as a serious problem back in 1979, during the First World Climate Conference, which was globally accepted as a result of the increase in GHG emissions. It was recognized that GHGs affected the energy balance of the global atmosphere, ultimately leading to an overall increase in the global average temperature.

First of all, it is pertinent to clearly understand the concept of climate change used worldwide, and to accomplish that goal, the crucial role of two entities must be highlighted, namely, the UNFCCC and the IPCC. However, both entities have different approaches to the climate change definition, and, surely, this inconsistency in the concept is one of the main reasons for the international standoff on a climate policy, leading to a lack of decision making regarding updating policies on climate and energy strategies. The UNFCCC established that "climate change is directly or indirectly attributed to human activity that alters the composition of the global atmosphere and which is in addition to natural climate variability observed over comparable time periods" (UN (United Nations) 1992, p. 3). The IPCC defined climate change as any change in climate over time, whether due to natural variability or induced by human activity (Rahman 2013).

The definition of climate change is a nontrivial and contentious exercise and, therefore, can be understood as the most complex and controversial question in the entire science of meteorology and climatology. In fact, several specialists (Allen 2003; Werndl 2016) argue that there are no strict criteria to justify the use of the expression climate change, and therefore its understanding remains unclear. As a result, the concept of climate change is often roughly employed and, consequently, may lead to considerable confusion regarding the existence and extent of global warming.

Undeniably, climate change means, a priori, a change in the statistical distribution of weather patterns in the long term, which may take place over decades (traditionally 30 years), but such changes and variations are typically studied over significantly longer periods of time, as shown through geological studies covering millions of years. In fact, the Earth's climate has changed in the course of the geological evolution, even before human activity could have played a role in its transformation. Thus, climate change consists of temporary changes in climatic conditions (temperature, precipitation and wind, among others), which can encompass changes in both average conditions and changes in variability. Since planet Earth's climate is naturally variable in the geological time scale, the long-term characteristics—specifically average temperatures—are controlled by the Earth's energy balance over time. In the last million years, the climate has naturally shown fluctuations between warm periods and glacial ages, which are strongly correlated with the natural Milankovitch cycles, established between 1920 and 1942 (Tarling 2010). According to Milankovitch, these oscillations are related to orbital changes, which, in turn, are controlled by three elements: eccentricity, obliquity and precession. These elements also have different periodicities, which affect not only the gravitational field of the Earth's surface but also the intensity and distribution of solar radiation that reaches the upper atmosphere. In this context, what is understood today as climate change, meaning changes produced by anthropogenic activities, is, from a geological time scale perspective, natural climate change (Figure 1). In fact, it appears that the role of human activity in the climate change increase can be addressed as a time scale-dependent subject (Figure 1).

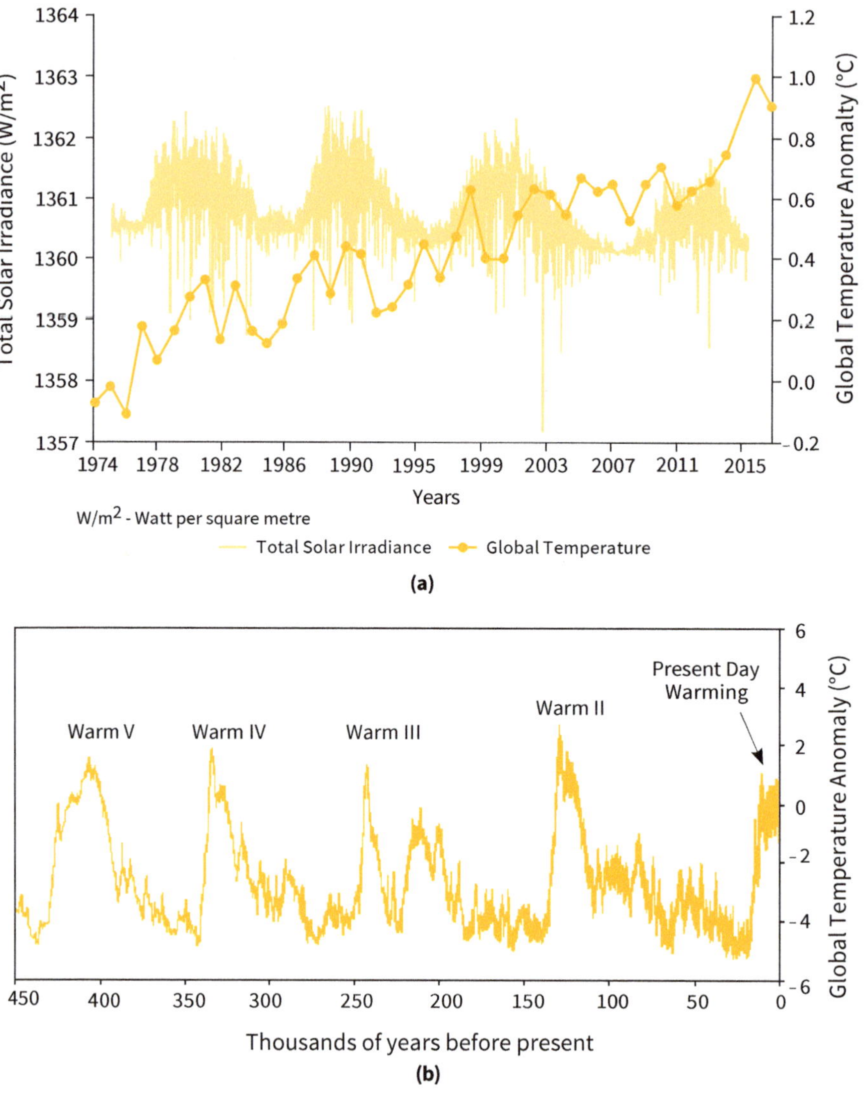

Figure 1. *Cont.*

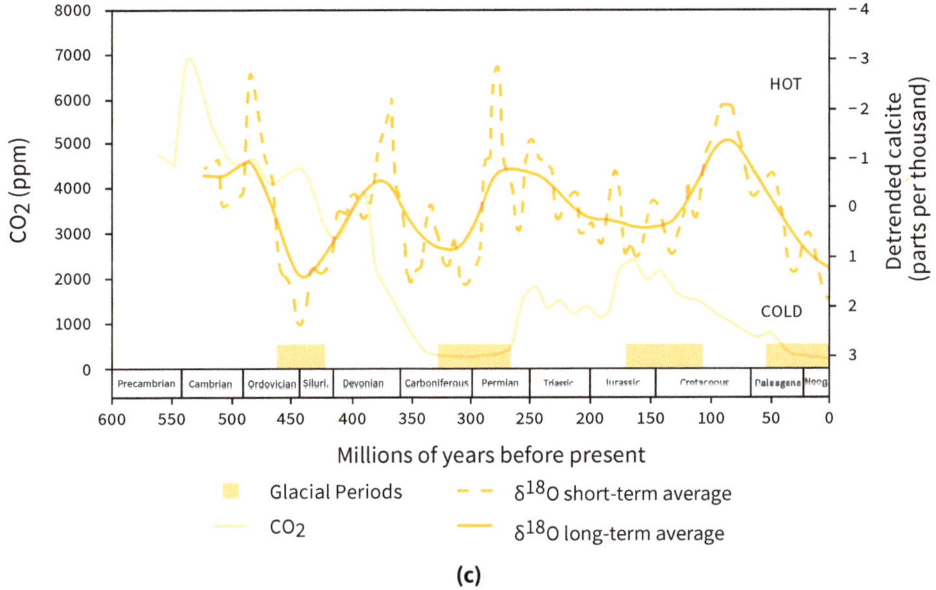

Figure 1. Historical temperature of planet Earth: (**a**) 44-year interval (adapted from Augustin et al. 2004); (**b**) 450,000-year interval (adapted from Augustin et al. 2004); (**c**) 542,000-year interval. Source: Graphics by authors, adapted from Veizer et al. (2000), Berner and Kothavala (2001) and Moore (2019).

Herbert and Fischer (1986) and Park and Herbert (1987), by studying paleoclimatic periodicities in a geologic time series, stated "there is overwhelming evidence for cyclicity at about 2 cycles/m, corresponding to Milankovitch oscillations with periods near from 100 thousand of years" (Park and Herbert 1987, p. 14,037). This type of study will greatly improve our understanding of the role of orbital oscillations in climate change, and, consequently, it could help in defining a time scale which will allow measuring the components and their variability over short periods (Schwarzacher 1993; Giraud et al. 1995; Crowley and Berner 2001; Stenni et al. 2010).

In the present work, the concept of climate change is not addressed with the commonly fatalistic approach, where it is considered to be promoted by GHG emissions, these being responsible for an overall increase in the global average temperature; rather, GHG emissions are seen to correspond to local changes susceptible only to influencing the environment at the scale of humanity's lifetime.

As it stands, one pertinent question arises: why is it that CO_2 is taken as the most dangerous GHG in the atmosphere? According to the Environmental Protection Agency (EPA), and other international entities (IPCC and UNFCCC), the most dangerous GHGs are CO_2, CH_4, nitrous oxide (N_2O) and fluorinated gases (e.g., chlorofluorocarbons, hydrochlorofluorocarbons). It is assumed that an anthropogenic source is responsible for almost all of the increase in GHGs in the atmosphere in recent centuries, mainly attributing CO_2 and CH_4 emission increases to general activities in the fossil fuel sector, whereas those of nitrous oxide and fluorocarbons to other human activities. Table 1 depicts some of the main characteristics of the major GHGs, in which it is quite perceptible that CO_2 is not the most dangerous GHG, since its lifetime can be a few days when it is quickly absorbed by the ocean surface, but some part will remain in the atmosphere for thousands of years; instead, CH_4 and N_2O have a more worrying lifetime average. Therefore, CH_4 plays a more negative role in the atmosphere than CO_2.

Additionally, Figure 2 shows that CH_4 concentrations (between ~300 and 800 ppmv) in the atmosphere have always been higher than CO_2 concentrations (between 160 and 300 ppmv)—actually, more than double. Yet, again, how are these two GHGs' emissions closely related to anthropogenic activities? Several studies using data going back 800,000 years in time, extracted from ice cores located at Dome C in Antarctica (Jouzel et al. 2003; Jouzel et al. 2007; Pol et al. 2010; Kang and Larsson 2013; Persson 2019), have shown what had been proposed by Herbert and Fischer (1986), meaning that consistent fluctuations in CO_2 and CH_4 concentrations exist, and these rising and falling CO_2 and CH_4 concentrations coincide with the onset of ice ages (low CO_2 and CH_4) and interglacial (warm) periods (high CO_2 and CH_4). Indeed, these periodic fluctuations are promoted by changes in the Earth's orbit around the sun, the so-called Milankovitch cycles (Figure 3).

A worrying scenario of an uncontrolled increase in CO_2 emissions due to anthropogenic activities, mainly related to the burning of fossil fuels, has been presented by the IPCC (Intergovernmental Panel on Climate Change) (2013). Actually, Figure 4 displays a common projection of the atmospheric CO_2 concentration measured during the last 800,000 years until the present day, showing that the last few years are clearly marked by a strong increase in the CO_2 concentration. Again, this fatalistic approach is related to the time scale used, 800,000 years, which is negligible from the geological scale point of view. Actually, taking into consideration Figure 1c, the time scale (800,000 years) used in Figure 4 could be understood as a "few minutes" from the geological scale perspective, and consequently, the approximately 4600 million years of planet Earth's evolution, which is intimately connected to the warm and cold periods identified by Milankovitch, is completely neglected.

Table 1. Major GHGs and their main characteristics.

GHG	Major Sources	Average Lifetime in the Atmosphere	Global Warming Potential	
			20 Years	100 Years
Carbon dioxide	Burning fossil fuels (oil, natural gas and coal), solid waste and trees and wood products; changes in land use; deforestation and soil degradation.	A few days to thousands of years	1	1
Methane	Production and transport of fossil fuels; livestock and agricultural practices, mainly rice fields; anaerobic decay of organic waste in municipal solid waste landfills.	12.4 years	84	28–36
Nitrous oxide	Fertilizers, deforestation, burning biomass.	121 years	264	265–298
Fluorinated gases	Industrial processes and commercial (aerosol sprays, refrigerants) and household uses, and they do not occur naturally.	A few weeks to thousands of years	Varies (the highest is sulphur hexafluoride at 15,000)	Varies (the highest is sulphur hexafluoride at 23,500)

Source: Table adapted from IPCC (Intergovernmental Panel on Climate Change) (2013).

It is now pertinent to mention the concept highlighted in 2013 by the IPCC (Intergovernmental Panel on Climate Change) (2013), "Climate Numerical Models", as well as "Earth System Models", which are embodied as the modern environmental and climate science approaches to enable a full understanding of natural systems and their sensitivities (Haywood et al. 2019; Voosen 2021). Different sources of uncertainties in climate change models have been reported, namely, anthropogenic and natural factors. Among the main anthropogenic factors are radiative forcing due to GHGs, and changes in population size and distribution, urbanization, energy system production and consumption and land use. The natural factors are mainly related to major volcanic eruptions, which can be responsible for the injection of small aerosol particles into the stratosphere, and changes in the radiation emitted by the sun (Giorgi 2010; Mitchell et al. 2020; Pielke 2020; Shaviv 2008). The IPCC (Intergovernmental Panel on Climate Change) (2013) stated that the concept of climate models is an attempt to assess the effects, risks and potential impacts associated with anthropogenic GHG emissions, which will allow scientific assessment of mitigation and societal adaptation strategies. Nevertheless, if the idyllic situation of immediately stopping CO_2 emissions were a reality, most of the warming and, consequently, its climate impacts would persist for many centuries. These irreversible changes are often misunderstood and are currently disregarded in most climate models. The irreversibility on time scales is at least hundreds of years, meaning that planet Earth is a multiparameter system in a dynamic equilibrium, and implying that climate change resulting from past emissions, even in the absence of future emissions, constitutes a global commitment for many future generations. Additionally, acknowledging that climate change will simply persist for centuries to millennia due to the long lifetime of CO_2 in the atmosphere is a key point in decision makers' strategies (Knutti et al. 2015).

Another relevant subject, commonly discussed on climate change forums, is the general rise in the sea level around the world (Figure 3). According to the IPCC (Intergovernmental Panel on Climate Change) (2013), the long-term effects of the global average temperature increase are responsible for the general rise in the sea level, resulting in the inundation of low-lying coastal areas and the possible disappearance of some island states, and the melting of glaciers, sea ice and Arctic permafrost. Actually, as already mentioned in this work, climate change corresponds to cyclic natural modifications, such as underwater volcanism, which significantly contributes to the overall increase in temperature, mainly in the oceans. In this regard, on the matter of underwater volcanism, it is worth emphasizing that a recent and remarkable scientific study, with the credibility of MIT (Huppert et al. 2020), although already discussed by others (Johnson et al. 2018), has highlighted what geologists empirically already knew, but that is now supported by rigorous geophysical observations and measurements. As is well known, volcanic islands do not last forever, and their longevity can differ significantly; for example, some islands such as the Canary Islands located in the Atlantic Ocean are more than 20 million years old, while the Galapagos Islands located in the Pacific Ocean have already drowned. Recently, it has been demonstrated that the formation and longevity of volcanic islands are intimately related to the movement of tectonic plates and their relationship with mantle plumes (hotspots). Nevertheless, to determine the actual age of each island, including those that have drowned, the direction and speed at which tectonic plates are moving in relation to the swell uplift underneath need to be measured, as does the length of each swell, which is formed when the mantle plume raises the seafloor. The unavoidable drowning of all volcanic islands over time is a natural phenomenon, which depends on the tectonic plate's speed and the size of the mantle plumes, and therefore it is not, in any way, related to an eventual rise in the sea level.

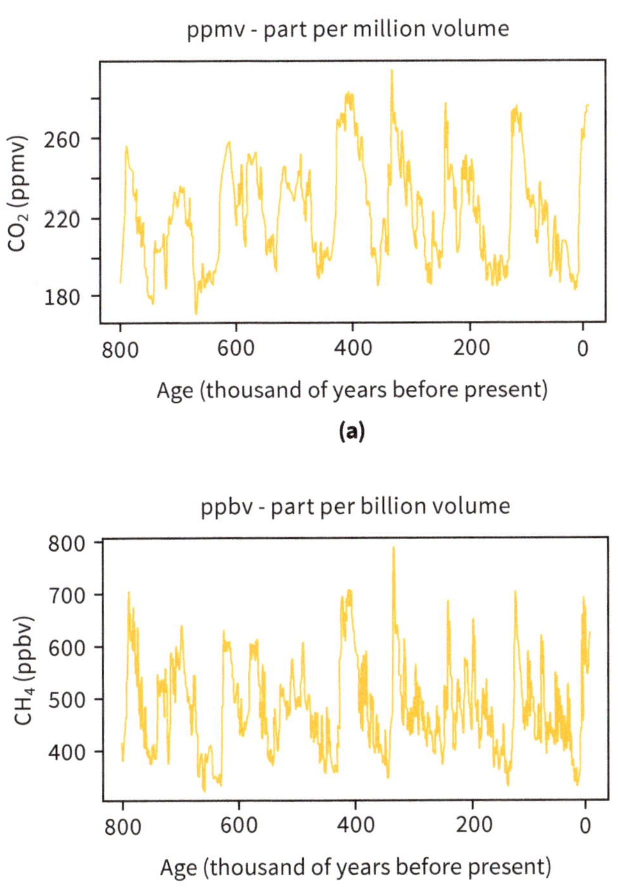

Figure 2. Atmosphere concentrations over the past 800,000 years before the present (1950): (**a**) CO$_2$; (**b**) CH$_4$. Source: Graphics by authors, adapted from Persson 2019.

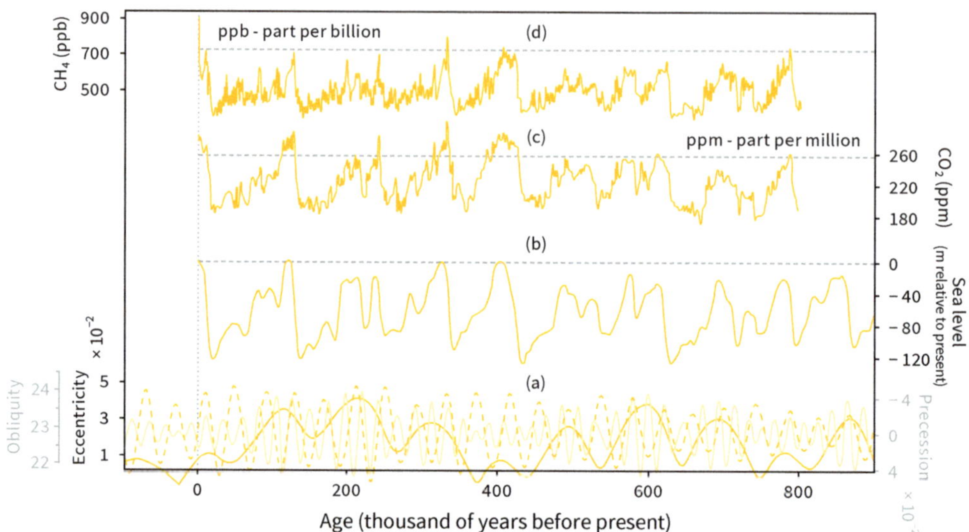

Figure 3. Variability over 800,000 years before the present (1950) of: (**a**) orbital parameters; (**b**) sea level; (**c**) CO_2 concentration; (**d**) CH_4 concentration. Source: Graphic by authors, adapted from Pol et al. (2010).

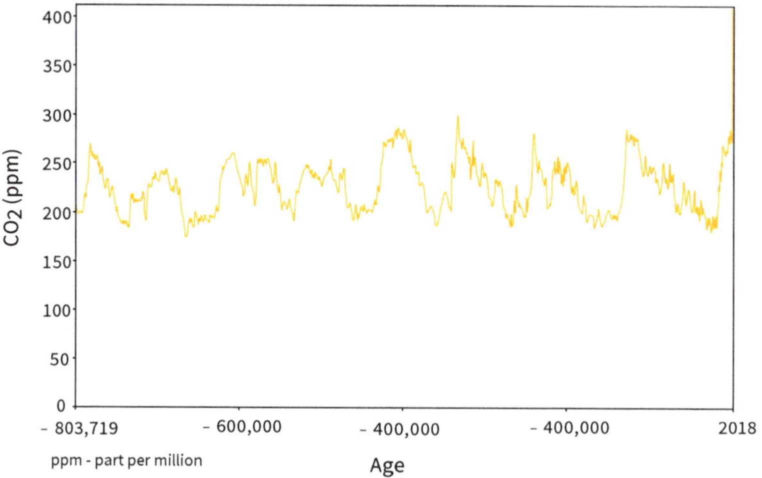

Figure 4. Atmospheric CO_2 concentration during the last 800,000 years. Source: Graphic by authors, adapted from Ritchie and Roser (2017).

One of the key mistakes made when dealing with climate change questions, and consequently the cause of worldwide controversial opinions, is the consideration that the climate change increase is due to anthropogenic sources, resulting from, above all, the use of fossil fuels (Giorgi 2010; Pielke 2020). Even if it is true that fossil fuel activities play a relevant role in the increase in GHG emissions and require mitigation, it is not true that they are the only cause, nor even the most dominant one. Be that as it may, in the second decade of the 21st century, questions related to climate change gained a greater acceptance by private and/or public entities, specifically as a result of the change in the definition or the perspective of the "climate change" concept. Actually, this change consists in a paradigm shift supported by the idea that climate change is no longer a cause of environmental degradation, but a requirement for sustainable development.

The Paris Conference (COP21) represents the culmination of the convergence of efforts of different players, with the main goal of reducing CO_2 emissions, and therefore limiting the global average temperature increase, which will allow meeting the net zero GHG emissions target, in the second part of the 21st century.

4. Energy Transition and Energy Demand

In the Conference of the Parties meetings (COP22—Marrakech, COP23—Bonn, COP24—Katowice and COP25—Madrid) held after the Paris Conference (COP21, 2015), it was emphasized that the future of climate change mitigation involves the processes of decarbonization and energy transition. The urgent need to align the use of fossil fuels and climate goals was also recommended because a radical energy transition requiring a far more predominant use of renewable energies can never be achieved in the short and medium term. With the current state of knowledge, it is not even possible to imagine, in a sustainable (economic and technical) way, a world energy supply exclusively from renewable sources.

The energy transition is not only related to the production of renewable energy, with the main goal to replace fossil fuels in the world energy supply, but it is also seen (IPCC (Intergovernmental Panel on Climate Change) 2013) as a long-term investment opportunity that will transform the entire energy system over the next 30 years and beyond. This means that significant investments across the entire value chain are required, namely, in clean energy generation, transmission and distribution networks, energy storage and electric transport infrastructure.

The EGD (EC (European Commission) 2019) is a set of policy initiatives which combines the twin effort of reducing GHG emissions and preparing Europe's industry for a climate-neutral economy, to be conducted through the implementation of renewable energy sources. The EGD stands for a tech-driven energy revolution that will rapidly replace all hydrocarbons, and Kovac et al. (2021) went further by assuring that such an energy transition process is already being implemented. This unrealistic "new energy economy" scenario is confident that the technologies of wind and solar power and battery storage are undergoing innovative development in computing and communication technologies, to the extent that it will dramatically reduce costs and increase efficiency (Mills 2019). Given the intermittency of renewable energy sources, highly dependent on weather conditions and the number of hours of day light, energy storage has an essential role in the general energy transition framework.

In this context, the European Commission launched a new initiative entitled "A hydrogen strategy for a climate-neutral Europe" (EC (European Commission) 2020a). Hydrogen energy is, today, seen as a key priority to achieve the EGD and, consequently, Europe's clean energy transition. The relevant role of the new carbon-neutral system scenario is, evidently, related to the fact that hydrogen energy production does not emit CO_2, as it uses renewable energy sources, and that almost no air pollution is generated when it is used. Additionally, hydrogen energy can play a crucial role in the renewable energy storage sector, and it can be used as a feedstock, as a fuel, as batteries and in many other industrial applications, such as in the transport, power and building sectors.

However, how is hydrogen energy generated? Hydrogen energy is produced through a chemical process known as electrolysis, which uses an electrical current to separate the hydrogen from the oxygen atoms in water. There are different processes to produce hydrogen energy which are associated with a wide range of emissions, depending on the technology and energy source used, namely: electricity-based hydrogen (any type of electricity source); renewable hydrogen or clean hydrogen (renewable electricity source); fossil-based hydrogen (fossil fuel as an electricity source); fossil-based hydrogen with carbon capture (fossil fuel as an electricity source using carbon capture technologies); low-carbon hydrogen (any type of electricity source associated with fossil-based hydrogen with carbon capture); and hydrogen-derived synthetic fuels (gaseous and liquid fuels on the basis of hydrogen and carbon) (EC (European Commission) 2020a).

Presently, hydrogen represents a small fraction of the global and EU energy matrix, and according to IRENA (International Renewable Energy Agency) (2019), future projections on the breakdown of renewables used in the total final energy consumption in 2050 indicate that hydrogen energy will continue to account for only 3% (Figure 5).

Despite this reality check, for hydrogen energy to contribute to climate neutrality, it will need to achieve a far larger scale, and its production must be a fully decarbonized process, meaning that hydrogen must be generated from renewable hydrogen (clean hydrogen or green hydrogen) using electricity generated from renewable sources. In such a process, hydrogen energy is produced without emitting CO_2 into the atmosphere, with water vapor being the only emission. The renewable hydrogen process is quite complex, requiring the following requisites: (1) the water used must contain salts and minerals to allow electrical conductivity, and (2) two electrodes must be immersed in the water and connected to an electrical power source. Ultimately, the dissociation of hydrogen and oxygen atoms will occur when the electrodes attract ions with opposite charges.

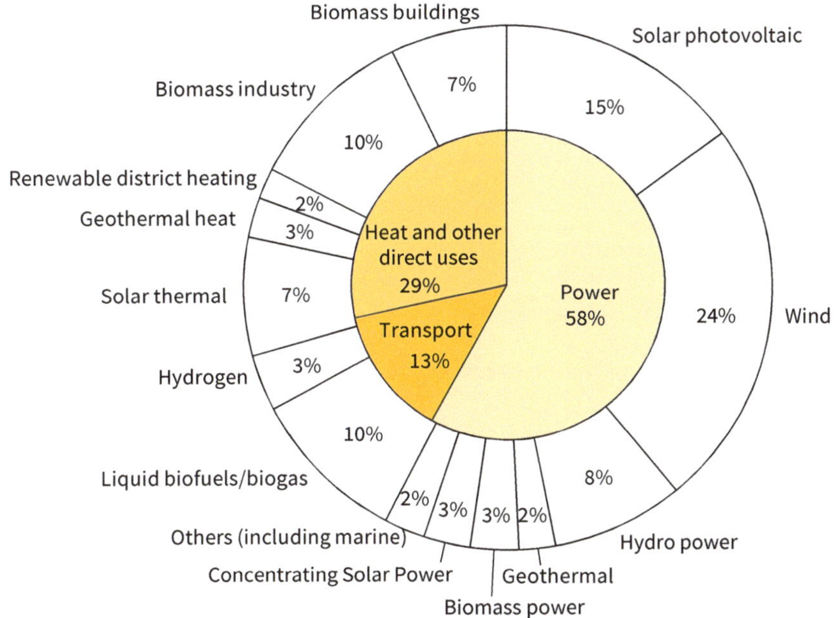

Figure 5. Breakdown of renewables use in total final energy consumption in 2050. Source: Graphic by authors, adapted from Gielen et al. (2019).

However, there are unavoidable questions about the viability of clean hydrogen relating its high production and storage costs (Figure 6), the difficulty of transport over long distances and the energy losses during conversion processes (Bossel et al. 2003), and, today, hydrogen energy is still largely produced from fossil fuels (Figure 7).

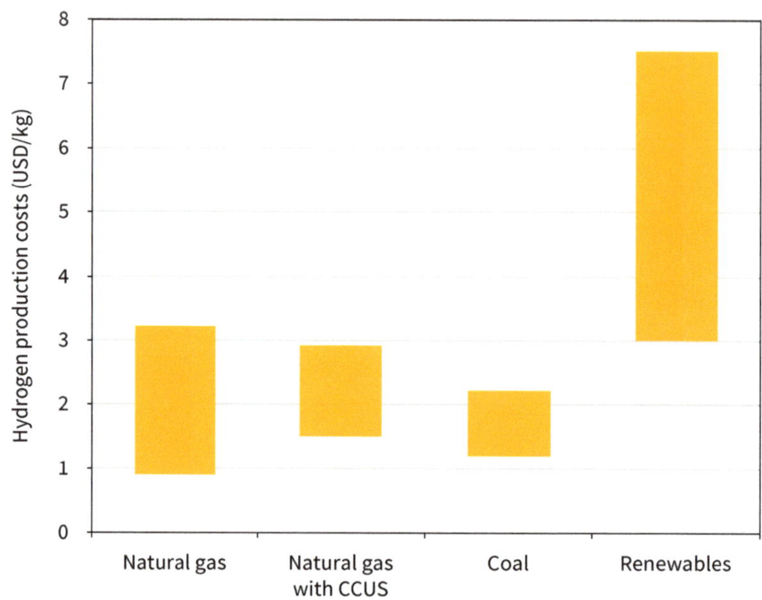

Figure 6. Hydrogen production costs by production source in 2018. Source: Graphic adapted from IEA (International Energy Agency) (2019).

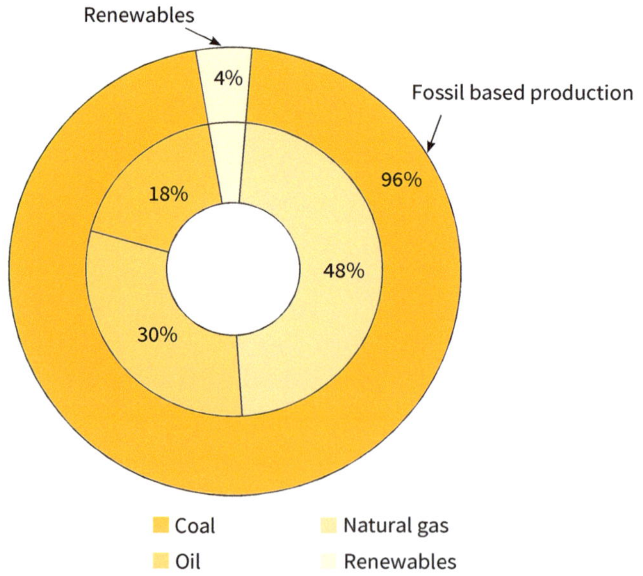

Figure 7. Hydrogen production by major sources. Source: Graphic by authors, adapted from Molloy and Baronett (2019).

Of the four major sources used for commercial production of hydrogen, three require fossil fuels: (1) steam methane reformation (SMR), (2) oil oxidation and (3) coal gasification (Figure 7), and therefore emit CO_2. These three processes are referred to as gray hydrogen production. Nevertheless, if the production of hydrogen is accompanied by CO_2 capture and storage, it is then referred to as blue hydrogen (EC (European Commission) 2020a). However, the effectiveness of CO_2 capture (maximum 90%) needs to be taken into consideration.

The fourth source is renewable electrolysis, generating so-called green hydrogen, the only process that does not emit CO_2 (Figure 8).

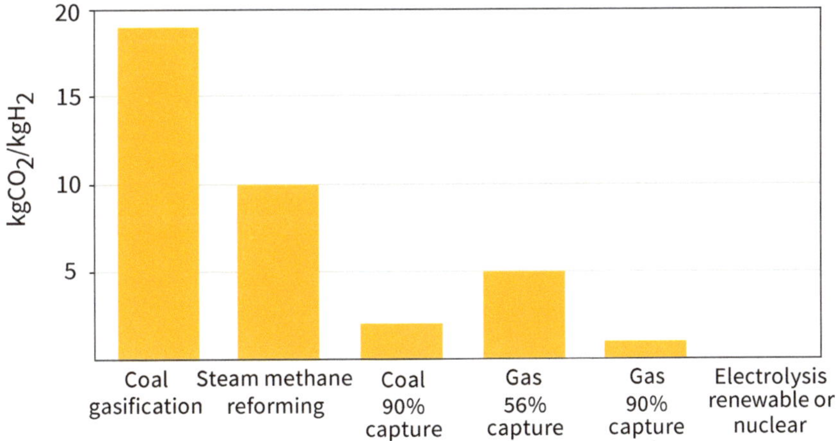

Note: includes only CO2 emissions from combustion and chemical conversion.

Figure 8. CO_2 emissions by hydrogen production process in 2019. Note: Includes only CO_2 emissions from combustion and chemical conversion. Source: Graphic by authors, adapted from Bartlett and Krupnick (2020).

In the extremely complex and high-cost scenario of the hydrogen sector, Germany aims for the European leadership in carbon capture and storage, and to produce hydrogen from natural gas, and has launched a new hydrogen strategy with a clear focus on green hydrogen production (BMWI 2020). To achieve carbon neutrality by 2050, Germany greatly relies on the energy transition strategy, keeping in mind that gaseous and liquid energy sources will continue to be an essential part of Germany's energy supply, with the expectation that hydrogen energy will play a key role in enhancing and completing the energy transition.

There is no doubt that, today, green hydrogen (renewable hydrogen) is not cost competitive against fossil-based hydrogen and, in particular, fossil-based hydrogen with carbon capture and storage, which is highly dependent on natural gas prices. The EC (European Commission) (2020b) stated that costs for renewable hydrogen are declining quickly and will continue to get cheaper, with the cost of the electrolysis process having already reduced by 60% in the last ten years. Therefore, there is an expectation that cost-competitive renewable hydrogen will eventually be achieved if accompanied by cost-competitive renewable energy.

In this optimistic scenario, where the generation of renewable energy will quickly become cheaper, the matter of intermittency comes to mind, as does the imperative need for cost-effective and reliable energy storage.

Schernikau and Smith (2021) highlighted the enormous difficulties and practical issues related to one source of renewable energy—solar photovoltaic panels. They focused on Germany as a case study and showed that supplying Germany's electricity demand entirely from solar photovoltaic panels (located in Spain, the optimal region for the production of solar energy due to the high direct normal solar irradiation), considering several adjustment factors (peak power, backup peak, transmission loss, winter capacity, etc.), would require a total area of approximately 35,000 km^2 (7% of Spain´s surface area) covered with solar photovoltaic panels. Such a scenario would equate to an installed capacity of 2000 GW, which is almost three times more than the worldwide capacity installed in 2020 (715 GW). In addition, one cannot ignore the fact that solar photovoltaic panels last, on average, 15 years and would require replacement every 15 years. Schernikau and Smith (2021) further estimated that the annual silicon and silver requirements for such a scenario would require close to 10% and 30% of the current global production capacity, respectively. This seems to be an unrealistic and unachievable scenario which will worsen once we add estimations of resource requirements for the production and installation of battery backup systems. Advancing this scenario to cover about 40% (200,000 km^2) of Spain with solar photovoltaic panels in order to supply the entire European electricity demand, as suggested by Schernikau and Smith (2021), how would all the energy produced be stored?

In fact, the EC (European Commission) (2020b) stated that energy storage is the key factor in promoting an increase in renewables into the global and EU energy matrix. Battery electricity storage has been established as a vital technology in the world' transition to a sustainable energy system, and its worldwide acceptance is also connected to specific advantages, namely, its fast response capability, sustained power delivery and geographical independence (Yang et al. 2018). Despite the major improvements in the battery sector, it seems that selecting the battery energy storage system sizing methodology, which is clearly dependent on the renewable energy system used, is the biggest challenge of the entire process. Given the difficulty for a single battery energy storage system to produce capable and reliable renewable energy independent of electricity provided through main grids, unless an oversized generator and storage capacities are utilized, new battery systems are being developed. Javed et al. (2020) proposed a hybrid pumped and battery storage (HPBS) system, in which "the battery is only used in order to meet very low energy shortfalls considering the net power deficiency and state of charge, while pumped hydro storage works as the main storage for high energy demand" (Javed et al. 2020, p. 1).

On the subject of batteries, how many would be needed to store the world renewable energy demand? The study by Schernikau and Smith (2021) was a realistic analysis of 14 days of energy storage backup for Germany during the winter period. For this period, Germany will require approximately 45 TWh of battery storage. However, producing the required storage capacity from batteries using current technology would require the full production of 900 Tesla Gigafactories, such as the Nevada Gigafactory (Mills 2019), working at full capacity for an entire year. Additionally, for the annual replacement of batteries, an extra output of 45 Tesla Gigafactories, corresponding to a full production of 2.25 TWh, will be required. To gain a general idea of what these numbers actually mean, the global battery production in 2020 was 0.5 TWh. Another key point on batteries is the future demand, and the rate of demand, for specific raw materials (such as lithium, copper, cobalt, nickel, graphite, rare earths, bauxite, iron and aluminum), leading to a dramatic increase in production, which would significantly affect the global mining sector. The mining chain comprises several processes from prospecting to extraction, transportation and processing, requiring significant amounts of energy. Mills (2019) suggested that the energy equivalent of 100 barrels of oil is required to produce a single battery that can store the equivalent of one barrel of oil. Additionally, today, natural gas accounts for more than 70% of the energy used to produce glass required to build solar photovoltaic planes, and if wind turbines are used to supply half the electricity in the entire world, 2000 Mt of coal would be required to produce the concrete and the steel needed to build the wind equipment (Mills 2020), knowing that the annual production of coal in 2019 was approximately 8000 Mt (IEA (International Energy Agency) 2020). Therefore, how will the clean energy system process work if a dramatic increase in the production of raw materials is required to produce green energy, which in themselves have negative environmental and health impacts, with high amounts of energy being required, which actually originates from fossil fuels?

Despite all the production issues related to battery systems, it is obvious that the raw materials required to produce battery storage systems will soon dominate the global production of minerals. In fact, today's production of lithium batteries already accounts for about 40% of all lithium and 25% of cobalt, meaning that, in the near future, global lithium mining would have to expand by at least 500% (Mills 2019). This analysis leads to another essential question—does planet Earth have enough raw materials to fulfill the production of batteries that the market will demand? The study by Schernikau and Smith (2021) clearly assists in answering this key point, by using the Germany case study. As mentioned, a 14-day battery storage solution for Germany would imply 45 TWh of battery storage production, using Tesla's newest technology, and consequently a 7000–13,000 Mt demand for raw materials. Battery replacement would require a capacity production of 2.25 TWh, implying a 400–700 Mt demand for raw materials. For example, 1,800,000 t of lithium production would be needed, knowing that the 2020 global production was 230,000 t, meaning feeding Germany's battery storage systems would require a lithium supply which is 5.6 times greater than the current global production (Figure 9). In the case of cobalt, another essential mineral used in the production of batteries, 225,000 t of cobalt production would be required, but the current global production is about 120,000 t, meaning Germany's battery storage production would demand 1.9 times more than the 2020 global production (Figure 9). To conclude on this matter, considering the current global production of raw materials, the Germany case study clearly demonstrates that the use of renewables, in the form of solar photovoltaic panels and/or wind energy, is a rather unrealistic solution for Germany or for the world—simply put, the rate of demand for raw materials, coupled with the energy demand that goes with their extraction and processing, current technology and the immediate environmental impact typically associated with mining and processing, is totally impractical and unaffordable, especially given the time frame proposed by international organizations such as the EU and the IPCC (Table 2).

Table 2. Global reserves of most important raw materials by country.

Country	Reserves			
	Cobalt (kt)	Lithium (kt)	Nickel (kt)	Graphite (kt)
Argentina		2000		
Australia	1100	1500	19,000	
Brazil	78	48	10,000	72,000
Canada	240		2900	
Chile		7500		
China	80	3200	3000	55,000
Colombia			1100	
Cuba	500		5500	
Democratic Republic of the Congo	3400			
Guatemala			1800	
India				8000
Indonesia			4500	
Madagascar	130		1600	940
Mexico				3100
New Caledonia	200		8400	
Philippines	250		3100	
Russia	250		7900	
South Africa	30		3700	
Turkey				90,000
United States of America	23	38	160	
Zambia		270		
Zimbabwe		23		
Other countries	610		6500	960
World total	7000	14,000	80,000	230,000

Source: Table adapted from EC (European Commission) (2018).

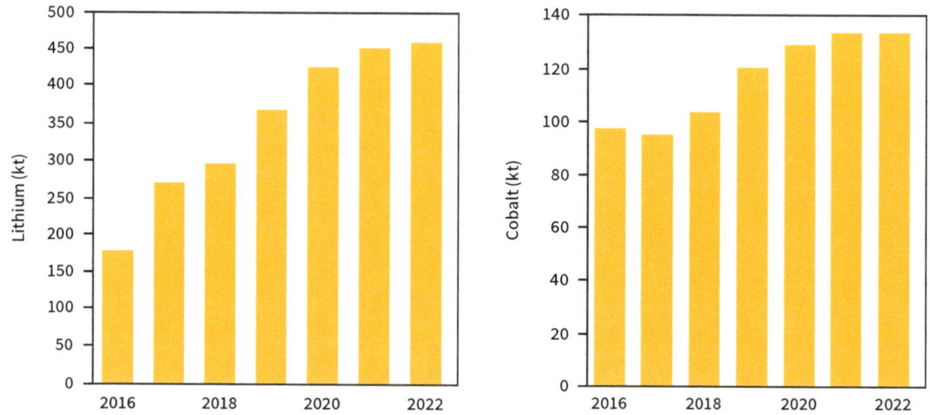

Figure 9. Lithium and cobalt global production from 2016 until 2022. Source: Graphics by authors, adapted from Metso: Outotec (2021).

Ultimately, the world needs to sustain its energy supply and growth while significantly reducing emissions, and the following question remains: how can this be achieved when world energy consumption continues to increase as a result of population increase, industrialization and improved quality of life?

A 30% increase in world energy consumption is expected between 2014 and 2035, mainly related to the rapid growth of emerging economies, amongst which China and India account for half of this increase (Figure 10). Yet, an apparent mistake is consistently made, intentionally or not: that of discussing the global energy matrix based on the EU and/or the USA case studies, relegating the rest of the world, its populations and, above all, its energy needs and development rights to an obscure platform.

International policies, as well as the targets established in international treaties, namely, the Paris Agreement, play a very powerful role in the evolution of the world energy sector and may even cause real deviations in the evolution of the global energy matrix. In the current energy scenario (Figure 11), it is clear that fossil fuels will continue to play an important role in the global energy matrix. GECF (Gas Exporting Countries Forum) (2017) still projects that approximately 75% of global primary energy consumption will be met by fossil fuels (oil, natural gas and coal) in the year 2040, despite a decrease of 6% being projected between 2014 and 2040. There are expectations that a substantial increase in nuclear and renewable energies will occur, corresponding to approximately 25% of the world energy matrix. Renewable energies are the primary energy sources with the highest growth, from 13% in 1990 to 18% in 2040. Despite this promising renewable energy scenario, direct use of fossil fuels is, and is expected to remain, the dominant energy source in the modern world. Mills (2019) suggested that in order to completely replace fossil fuels over the next 20 to 30 years, knowing that half a century was needed for global oil and gas production to expand by 10-fold, global renewable energy production would have to increase by at least 90-fold, and this is an unrealistic proposition even when substantial financial efforts are involved. It is therefore quite obvious that the global energy demand will not allow the Paris Agreement targets to be met, especially with the current energy and climate strategies, which lack binding agreements and carbon markets, as with those proposed in the Kyoto Protocol.

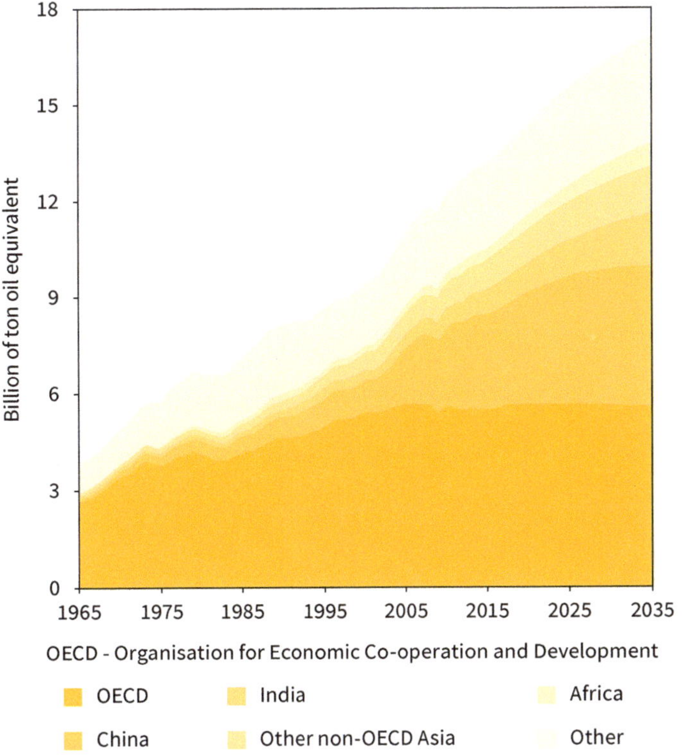

Figure 10. World energy consumption by region. Source: Graphic by authors, adapted from BP (British Petroleum) (2017).

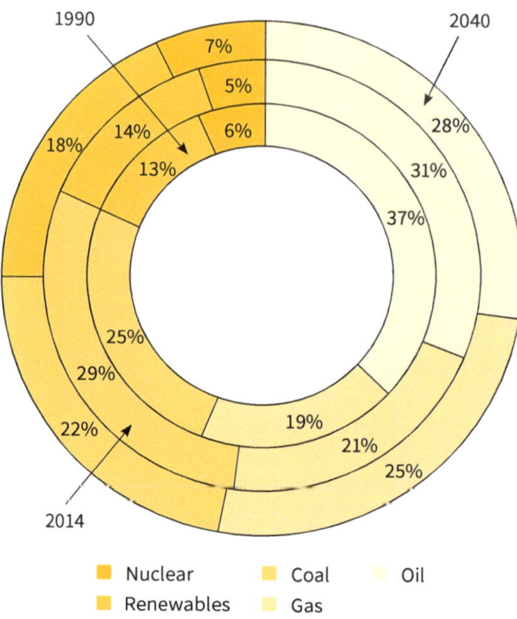

Figure 11. Evolution of shares of global primary energy, between 1990 and 2040. Source: Graphic by authors, adapted from GECF (Gas Exporting Countries Forum) (2017).

Returning to the IEA's assumed goal to achieve net zero CO_2 emissions by 2050, we are of the view that, even with the enormous financial and technological efforts that are continuously made, it is unlikely that total decarbonization of the energy sector in the envisaged timespan will be achieved. In fact, full decarbonization of the energy sector would imply an urgent and rapid deployment of available technologies, but above all, worldwide use of technologies that are not on the market yet. These new technologies are mainly related to battery storage systems, hydrogen electrolyzers and direct air capture and storage systems (IEA (International Energy Agency) 2021). As previously mentioned, there is still much to be conducted in the implementation process of new technologies, from the availability of raw materials to energy efficiency and cost competitiveness. For the implementation of new technologies, three main scenarios on global CO_2 emissions have been proposed and projected, namely, business-as-usual, rapid transition and net zero (BP (British Petroleum) 2020). Additionally, BP (British Petroleum) (2020) and IEA (International Energy Agency) (2021) stated that the global CO_2 emissions peak has already been reached (Figure 12), mainly due to the impact of COVID-19, and they will not return to their pre-pandemic levels. Nevertheless, it is important to keep in mind that the previous projections are significantly different from these new proposals; actually, BP (British Petroleum) (2017) projected an increase in global CO_2 emissions of 13% between 2014 and 2035 (Figure 13), which is far from the targets of achieving a 30% reduction by 2035, as proposed by the Paris Agreement.

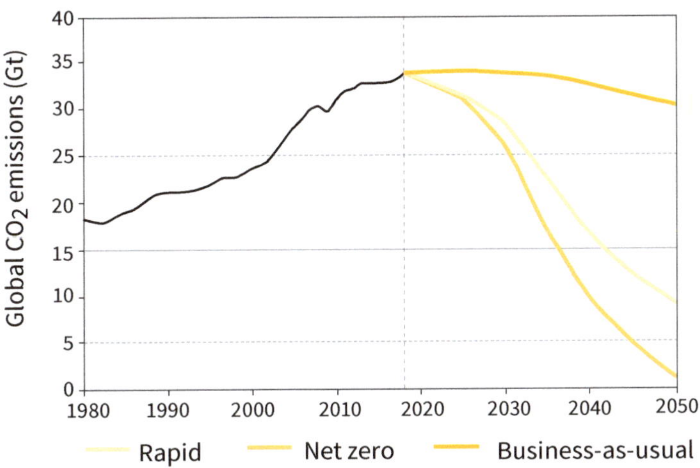

Figure 12. Global CO_2 emission scenarios. Source: Graphic by authors, adapted from BP (British Petroleum) (2020).

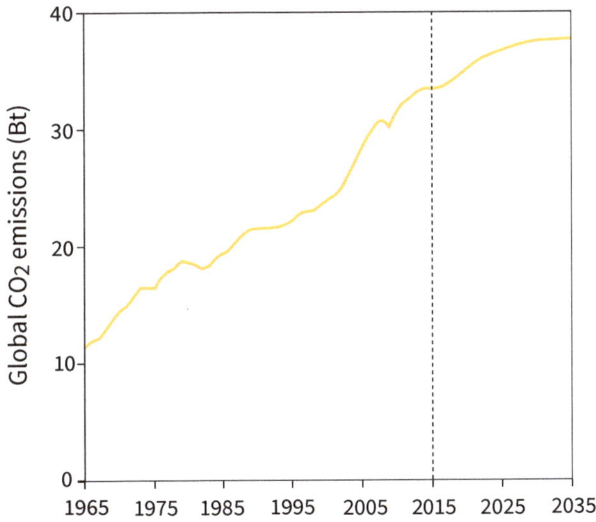

Figure 13. Global CO_2 emission projection. Source: Graphic by authors, adapted from BP (British Petroleum) (2017).

The three CO_2 emission scenarios presented by BP (British Petroleum) (2020) are supported by differences in economies, energy policies and social preferences. One main subject that has been discussed since the Kyoto Protocol is the establishment of carbon prices and, consequently, the carbon market. The business-as-usual scenario assumes carbon price increases of USD 65/t in developed countries and USD 35/t in emerging countries by 2050. Instead, both the rapid transition and net zero scenarios assume a substantial increase of USD 250/t in developed countries and USD 175/t in emerging economies by 2050.

The business-as-usual scenario corresponds to the global CO_2 emission projection if government policies, technologies and social preferences continue working in the usual way, as seen recently. In this scenario, projections address CO_2 emissions above 30 Gt in 2050, showing a slight decline of 10% between 2020 and 2050, which is far from the carbon-neutral target established by the European Commission.

The rapid transition scenario is based on a series of policy measures, led by a significant increase in carbon prices and supported by more targeted sector-specific measures, such as a significant shift away from traditional fossil fuels to non-fossil fuels, led by renewable energy, in order to achieve a more diversified global energy matrix (Figure 14). This rapid scenario projection suggests a 70% decline by 2050, which is consistent with the Paris Agreement targets of limiting the rise in the global average temperature to well below 2 °C above pre-industrial levels (Figure 12).

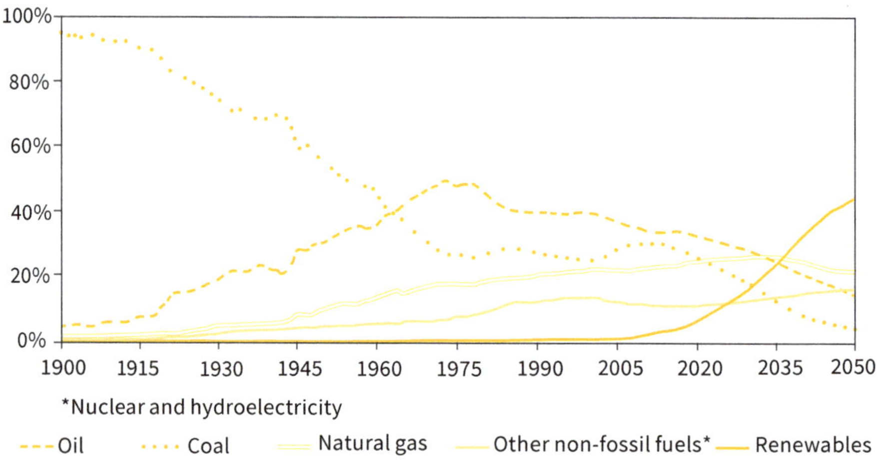

Figure 14. Shares of primary energy in the rapid transition scenario. Source: Graphic by authors, adapted from BP (British Petroleum) (2020).

The net zero scenario assumes that the policies in the rapid transition scenario are reinforced by significant shifts in social behaviors and preferences. Nevertheless, this scenario believes that an accelerated energy transition cannot be achieved based only on government policies, and that binding strategies need to be established. The net zero scenario suggests a deep decline by over 95% in global CO_2 emissions by 2050, which will allow meeting the global average temperature rise of well below 1.5 °C above pre-industrial levels.

At this moment, it is quite obvious that achieving a carbon-neutral energy system by 2050 will be a challenging task, requiring enormous efforts by all stakeholders, spanning social, economic, political and technological points of view. From the technological perspective, besides developments in new energy systems, it is undeniable that carbon capture and geological storage, so-called CCS technologies and specifically carbon capture utilization and storage (CCUS) technologies, must be included, as suggested by the Paris Agreement, as one of the potential solutions to meet the targets of carbon neutrality by 2050.

5. Geology Contribution to an Efficient Energy Transition

The contribution of geology on the path to achieving an efficient energy transition has different approaches. The first and inherent approach deals with the important role that geology plays in understanding climate change in the general context of planet Earth's evolution, which will offer potential tools to calibrate our future climate models. Three other major geological "contributors" are directly linked to reducing GHG emissions in this overall energy transition framework, namely, (1) mineral raw materials to build renewable energy equipment, (2) underground geological structures for hydrogen storage and (3) underground geological storage structures for CO_2 abatement.

The main target of the energy transition is to shift energy production from fossil fuels to non-fossil sources, meaning that the ultimate goal is to base the global energy matrix on CO_2-free energy sources. These CO_2-free energy sources and technologies, such as wind, solar, biomass or geothermal, require the exploration and exploitation of mineral raw materials for their deployment, with the obvious and often seriously detrimental consequences of the environmental impacts. Additionally, raw materials are not only needed for the construction of renewable energy equipment, such as solar photovoltaic panels and aeolian turbines, but also for building battery energy storage systems, considering the variability and intermittency of renewable sources.

Another important contribution from geology is related to the need for energy savings and efficiency, commonly associated with hydrogen energy production, which can be achieved by managing the heat and cold demand using underground storage structures (Dalebrook et al. 2013). Hydrogen energy is considered the key priority of carbon-neutral energy systems, due to its regenerative and environmentally friendly features. Nevertheless, hydrogen energy has two major inherent problems: its production and storage. Hydrogen energy production is complex, and it is not a cost-competitive energy, which must be produced from water or even hydrocarbons (Dalebrook et al. 2013). Hydrogen has a low critical temperature of $-251.15\,°C$, meaning that hydrogen is a gas at ambient temperature and atmospheric pressure; therefore, its storage implies a reduction in an enormous volume of hydrogen gas (Züttel 2003). Despite the significant issues in hydrogen production, hydrogen storage in large quantities is arguably the most challenging part of the entire hydrogen energy chain. Hydrogen storage in itself is not the main problem, a subject already addressed by several authors in recent decades (Yartys and Lototsky 2004; Zhou 2005; Niaz et al. 2015), and six different methods have already been sufficiently implemented, namely, (1) high-pressure gas cylinders (up to 800 bar), (2) liquid hydrogen in cryogenic tanks (at $-252.15\,°C$), (3) adsorbed hydrogen on materials with a large specific surface area (at $-173.15\,°C$), (4) absorbed hydrogen on interstitial sites in a host metal (at ambient pressure and temperature), (5) chemically bonded hydrogen in covalent and ionic compounds (at ambient pressure) or (6) oxidation of reactive metals (e.g., Li, Na, Mg, Al, Zn) with water, and hydrogen is already stored in underground salt cavities in the UK and the USA. Nevertheless, the hydrogen storage issue can be avoided when hydrogen production is supported by fossil-based hydrogen processes, namely, blue and gray hydrogen processes, usually reflecting an efficient alignment between production and consumption, where production occurs at the site of an industrial consumer, and usually where significant hydrogen storage is not required. The main issue is the large-scale storage of hydrogen commonly required for green hydrogen production, which is powered by intermittent renewable energies. For storing temporally large volumes of renewable energy surplus, geological options probably have the lowest cost and represent the best solution (Schoenung 2011; Heilek et al. 2016; Tarkowski 2017; Andersson and Grönkvist 2019; Karakilcik and Karakilcik 2020). Four specific geological options have been evaluated: salt caverns, depleted oil fields, depleted gas fields and deep saline aquifers. In fact, these geological options are commonly used in technological processes of CO_2 abatement, which will be discussed later in this section. Figure 15 shows a scheme for hydrogen production and storage when using 100% renewable energy sources.

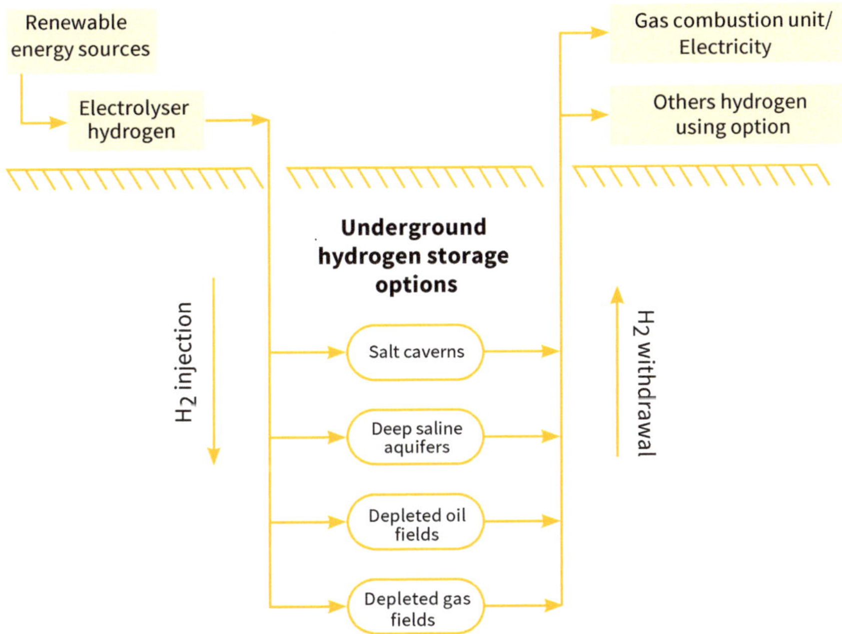

Figure 15. Renewable energy system scheme with an underground geological hydrogen storage facility. Source: Graphic by authors, adapted from Tarkowski (2019).

Salt caverns, which are built in underground salt domes, are the most mature option for geological storage facilities for hydrogen (Ozarslan 2012; Lemieux et al. 2020; Liu et al. 2020). Several advantages are attributed to salt caverns, such as their storage efficiency given that only a small fraction of the hydrogen injected is unable to be extracted from the geological structure, their lack of contaminants and, lastly, one of the most crucial advantages, their high-pressure operating systems, enabling a rapid discharge when hydrogen is needed (IEA (International Energy Agency) 2019). In this context, large-scale underground geological structures will play a crucial role in the hydrogen energy economy as integrated power plants with grids that rely mainly on renewable energy sources.

The last geological contribution, and perhaps the most controversial in recent years to reduce GHG emissions, mainly CO_2, is represented by CCS technologies (Figure 16). CCS technologies (Bui et al. 2018) were well studied at the beginning of the 21st century, resulting in several well-established geologic screening criteria for "pure" sequestration/storage of CO_2. As previously mentioned, a European regulatory framework for geological sequestration/storage of CO_2 (Directive 2009/31/EC, Shogenova et al. 2014) was defined in 2009. Directive 2009/31/EC was prepared for pure sequestration solutions in deep saline aquifers, and it was rapidly understood that they could be one of the largest potential technical storage solutions to reduce CO_2 emissions (Celia et al. 2015; Khan et al. 2021), but economically unviable, mainly due to the lack of investor-motivating measures.

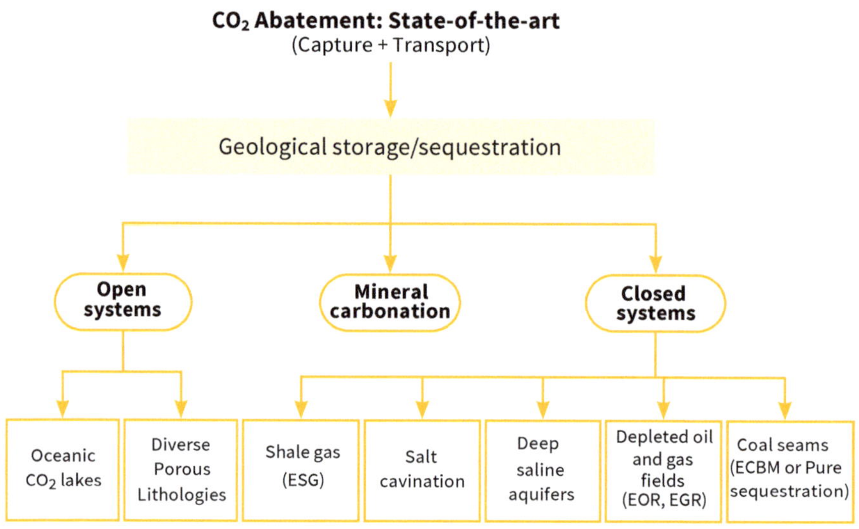

Figure 16. Geological storage solutions for CO_2 abatement. Source: Graphic by authors, adapted from Lemos de Sousa et al. (2008).

The subject became so important that the European Commission requested the European Academies Science Advisory Council (EASAC) to address the subject of carbon capture and storage in Europe (EASAC (European Academies Science Advisory Council) 2013). A similar important concern was developed by Chinese experts (Jiang et al. 2020). It became clear that to address CCS technologies, the key point would be to transform these technologies into an effective economic approach (D'Amore and Bezzo 2017; Kapetaki and Scowcroft 2017; Shogenova et al. 2021). The term "CO_2 utilization" was seen as the obvious solution, which allowed for the definition of three main categories: CCUS—hydrocarbon resource recovery, CCUS—consumptive applications and CCU—reuse (non-consumptive) applications (CSLF (Carbon Sequestration Leadership Forum) 2012).

In CCUS—hydrocarbon resource recovery (EOR), CO_2 is used to enhance hydrocarbon (oil and gas) production, which may partly compensate for the initial cost of CCS and contribute to the implementation of long-term CO_2 storage in other geological facilities, such as deep saline aquifers. The CCUS—consumptive applications involve the formation of minerals, also designated as mineral carbonation, which results in CO_2 storage by "locking up" the carbon component in the structure of the new mineral formed. In CCU—reuse (non-consumptive) applications, the temporarily stored CO_2 is also not directly consumed, and instead, the CO_2 is reused or used only once while generating some additional benefit.

Additionally, in 2015, the Paris Conference (COP21) highlighted that CCS is one of the key promising technologies that can effectively contribute to reducing CO_2 emissions in the power generation sector, even if CO_2 utilization options are used.

Nonetheless, in this section, the main goal is to highlight the role of geological sites in reducing CO_2 emissions, and in such cases, those that allow for CO_2 utilization. Therefore, considering the geological solutions for CO_2 abatement that are presented in Figure 16, four main groups can be identified, namely, (1) depleted oil fields (CO_2 enhanced oil recovery—CO_2-EOR), (2) CO_2 enhanced gas recovery (CO_2-EGR), (3) shale gas reservoirs (CO_2 enhanced shale gas—CO_2-ESG) and (4) coal seams (CO_2 enhanced coalbed methane—CO_2-ECBM). There are several key criteria for CO_2 storage projects that must be reached, which are different for each of the geological solutions but supported by the same general assessment procedures, namely, risk assessment; monitoring, reporting and verification requirements; reservoir simulations; accounting for the amount of CO_2 that can be stored; post-injection monitoring and site closure; economics evaluation; social context analysis; and legal and regulatory issues (Ajayi et al. 2019; MRI (Mitsubishi Research Institute) 2020). Certainly, the amount of CO_2 that can be stored in a geological site is one of the most important criteria in the whole key criteria list; thus, in this context, coal seams could play a major role in the CCUS technology framework.

Coal is a porous medium reservoir characterized by a unique organic microstructure, which allows for a CO_2 storage volume that is much higher than its pore volume capacity (Rodrigues and Sousa 2002), due to its adsorbed inherent features. The dominant adsorption characteristics of coal mean that CO_2 is mainly stored in the internal surface area of pores, in a condensed form, which is very close to the liquid state. Therefore, reservoirs characterized by organic microporous media signify higher internal surface areas and, consequently, higher storage capacities (Rodrigues et al. 2015). The main attribute that justifies coal as the better storage site option and, at the same time, the most permanent and secure solution for CO_2 storage in the medium–long term is its high organic matter content (greater than 50% in weight) (ISO 11760 2005). Besides the CO_2 storage capacity, the CO_2 injectivity rate is also a relevant criterium to select a geological site. The low CO_2 injectivity of a coal seam, due to its low permeability (usually lower than 5 millidarcy—mD), is undeniably an unfavorable key parameter suggesting not to use ECBM as the most economically viable CCUS technology.

CO_2-EOR projects have the largest potential of the various CO_2 utilization options, and they are the most used to date. In fact, they have been used on a commercial scale since the 1970s, totalling more than 100 commercial and pilot/demonstration projects (Ajayi et al. 2019). Yet, due to amazing improvements in the shale gas sector in recent years (Soeder and Borglum 2019), it is possible to consider CO_2-ECBM as an economically viable solution for CCUS technologies. These improvements are intimately related to horizontal drilling technologies, which involve a special form of directional drilling, typically through a formation at a well inclination of 90° from the vertical, using air hammers with rotation, and with directional control by means of bent housing motors. The well is drilled vertically until the reservoir's calculated depth is reached, and then the well is drilled to turn at an angle that is steadily increased until the well becomes parallel with the reservoir (Jiang et al. 2017; Guo et al. 2018). The long laterals of horizontal wells increase the reservoir–well contact, allowing for significant improvement in the reservoir's CO_2 injectivity and therefore in hydrocarbon production, which avoids the commonly but extremely expensive multi-well drilling approach used in the past.

Over the years, CCS technologies have been applied worldwide to both CO_2 pure sequestration and to CO_2 utilization. Nevertheless, due to the costs involved in the entire process, the most implemented type has been the CCUS technologies. According to the Economic Forecasting and Policy Analysis (EPPA) model developed by the Massachusetts Institute of Technology, which takes into account several pre-requisites to perform the assessment of storage capacity (MRI (Mitsubishi Research Institute) 2019), the word's total accessible geological CO_2 storage capacity is estimated to be between 8000 and 55,500 Gt, depending on the estimation scenario. The EPPA model includes the implementation of several CCS technologies, from CO_2 pure sequestration to CO_2 utilization projects. Considering the pre-requisites needed to address an efficient assessment of a site storage capacity, eighteen regions were selected for the model (Table 3). This model presents two distinct approaches: (1) a lower estimation, where a storage capacity factor of 0.037 Gt of CO_2 stored per 1000 km^3 of the sedimentary basin is used, and (2) an upper estimation, where a storage capacity factor of 0.26 Gt of CO_2 per 1000 km^3 of the sedimentary basin is used. Therefore, this EPPA model, taking into consideration the annual global CO_2 emission projections proposed by BP (British Petroleum) (2020), which stated that the global

CO_2 peak was reached in 2020, with annual total CO_2 emissions of approximately 35 Gt (Figure 12), implies that in the lower estimation scenario, the global storage capacity will be able to store the emitted CO_2 for approximately 228 years, and in the upper estimation scenario, CO_2 abatement through CCS technologies would take place over approximately 1585 years.

CCS technologies seem to be a plausible option in the short and medium term to reduce CO_2 emissions in the industrial sector. Actually, these technologies are currently known as the technological solution that will allow fighting for the global climate change targets, meaning to achieve carbon neutrality by 2050, despite several challenges related to costs, infrastructure and incentives that must be overcome. Significant progress has been made in recent decades; in fact, one of the first CCS projects in the USA, initiated in 1972, remains operating, in which CO_2 is captured in a fertilizer facility (Enid Fertilizer) and utilized in an EOR project (CSLF (Carbon Sequestration Leadership Forum) 2019). However, it has been during this last decade that the implementation of CCS projects has increased worldwide. The facility classification system proposed by CSLF identified two major categories based on their annual CO_2 capture capacity, namely, large-scale CCS facilities (capture capacity over 0.4 Mt/year), and pilot and demonstration CCS facilities (capture capacity less than 0.4 Mt/year) (CSLF (Carbon Sequestration Leadership Forum) 2019). Yet, a new CCS facility classification system was proposed by Global Status of Global CCS Institute (2020), which, besides the general pre-requisites list (CSLF (Carbon Sequestration Leadership Forum) 2019), is mainly supported by the commercial return while operating parameter. Therefore, this new classification established the following categories: (1) commercial CCS facilities, and (2) pilot and demonstrative CCS facilities. Today, there are 65 commercial CCS facilities and 34 pilot and demonstration CCS facilities around the world, but mainly located in North America, Europe and Asia. From the 65 commercial CCS facilities, 26 are operating, and they can capture and permanently store around 40 Mt of CO_2 per year. In the present scenario, it is only possible, when using CCS technologies, to provide a CO_2 abatement of approximately 0.11% of the 35 Gt of current annual CO_2 emissions in the entire world.

At this stage, it is quite clear that significant improvements are required throughout the entire energy chain in order to achieve carbon neutrality by 2050.

Gates (2021) was quite emphatic about the different approaches required for meeting reduction targets by 2030 and net zero targets by 2050. If the world is to go for the latter, he proposed both what can be done and what needs to be done. One thing is absolutely essential, and that is innovation at various levels, in order to create and roll out breakthrough technologies that can assist in reaching the ultimate goal by 2050. It is hoped that CCS technologies will play an essential role through this challenge.

Table 3. EPPA global storage capacity by region.

Region	Estimated Storage Capacity (Gt)	
	Lower Estimation (0.037 Gt/1000 km^3)	Upper Estimation (0.26 Gt/1000 km^3)
Africa	1563	10,986
Australia and New Zeland	595	4184
Dynamic Asia	119	834
Brazil	297	2087
Canada	318	2236
China	403	2830
Europe	302	2120
Indonesia	163	1144
India	99	697
Japan	8	59
Korea	3	24
Other Latin America	606	4257
Middle East	492	3454
Mexico	138	967
Other East Asia	272	1911
Other Eurasia	485	3410
Russia	1234	8673
United States of America	812	5708
Global	7910	55,581

Source: Table by authors, adapted from MRI (Mitsubishi Research Institute) (2019).

6. Conclusions

Whilst the world seeks to better understand climate science and significantly improve on climate models, predictions and interpretations of data relevant to so-called climate change, besides addressing the sustainability of resources, and the reduction in emissions, in an effort to protect the environment, factors such as energy economics, security and cost of supply must be properly addressed at all times.

In an ever-changing world, highly dependent on hydrocarbon-based energy sources, seeking to transition to new so-called cleaner sources, energy security concerns are heightened and the risks for disruption increase significantly. Such a transition needs to be as unincumbered as possible and should take into account the energy return on energy investment (EROEI) in order for sensible measures and policies to be put in place and succeed in delivering the end result. Current investments significantly favor renewables, but it seems as if work is being conducted at the expense of a major increase in additional natural resources and space, adding to even more environmental pressures. New renewable technologies will continue to raise the problem of waste disposal and recycling, adding further to the global concern of pollution and waste management, with a direct impact on nature, land and human and animal life. Under-investment in the old economy in favor of the new economy will result in disruption to the energy supply chain, recently referred to as the "revenge of the old economy" (Gillespie 2021), with all the economic impacts and consequences that accompany it.

In conclusion, it is quite obvious that the energy transition system, mainly supported by the idea of replacing fossil fuels, which are still responsible for about 80% of global primary energy in 2021, by renewable sources, is an essential step required by the global energy sector, in order to try to meet the targets of carbon neutrality by 2050, which is considered to be a difficult, if not impossible, task to reach. This subject has been discussed since the 1970s and was strongly raised with the implementation of the first Kyoto Protocol commitment, but the current global climate scenario is far from the main target, that is, net zero emissions.

The general analysis presented in this manuscript seeks to consider the different aspects relating to climate and to global energy demands. The energy transition must be conducted in a gradually and consistent manner to avoid massive disruptions to the energy and human development chains whilst seeking to ameliorate the impacts of climate change. Evidently, innovation and new developments need to come into play to accelerate the pace of change well beyond the impact seen from the current suit of alternative energy sources, namely, renewables. In this approach, fossil fuels will most likely continue to share the global energy matrix, although they will start decreasing, but most likely not in the medium term. To achieve carbon neutrality by 2050, a significant and intense multi-disciplinary effort in all sectors is required, with radical changes or improvements to cause a significant reduction in fossil fuel consumption that can only result from processes that are not available as of yet. This may include the likes of hydrogen, and perhaps an increased nuclear contribution to the energy matrix, but much work is required in a very short period of time (30 years).

The contribution of geology to the energy transition phase, which forms part of the drive to achieve the target of zero carbon by 2050, is multi-faceted. Batteries required to store energy produced from intermittent renewable sources require mineral raw materials. Large-scale facilities required to store large volumes of hydrogen can be provided by geological structures, such as salt caverns, depleted oil and gas fields and deep saline aquifers. Finally, given that a rapid energy transition to a fossil fuel-free energy system is presently impossible, CO_2 will continue to be emitted, and the most efficient solution to reduce CO_2 emissions into the atmosphere is to apply CO_2 abatement technologies, such as CCS technologies, by selecting the best underground geological structures through a set of key criteria.

Author Contributions: Each author contributed with an equal amount of expertise and discussions in the development of the present investigation. All authors have read and agreed to the published version of the manuscript.

Funding: This research received no external funding.

Acknowledgments: The authors are indebted to Fundação Fernando Pessoa for supporting this investigation in the scope of the research program of the FP-ENAS Research Unit. Thanks, are also due to Patrícia Moreira for her competent contributions to the literature research, and to the final organization of the reference list.

Conflicts of Interest: The authors declare no conflict of interest.

References

Ajayi, Temitope, Jorge Salgado Gomes, and Achinta Bera. 2019. A review of CO_2 storage in geological formations emphasizing modeling, monitoring and capacity estimation approaches. *Petroleum Science* 16: 1028–63. [CrossRef]

Allen, Myles. 2003. Liability for climate change. *Nature* 421: 891–92. [CrossRef] [PubMed]

Andersson, Joakim, and Stefan Grönkvist. 2019. Large-scale storage of hydrogen. *International Journal of Hydrogen Energy* 44: 11901–19. [CrossRef]

Augustin, Laurent, Carlo Barbante, Piers R. F. Barnes, Jean Marc Barnola, Matthias Bigler, Emiliano Castellano, Olivier Cattani, Jerome Chappellaz, Dorthe Dahl-Jensen, Barbara Delmonte, and et al. 2004. Eight glacial cycles from an Antarctic ice core. *Nature* 429: 623–28. [PubMed]

Bartlett, Jay, and Alan Krupnick. 2020. *Decarbonized Hydrogen in the US Power and Industrial Sectors: Identifying and Incentivizing Opportunities to Lower Emissions.* Report 20–25. Washington, DC: Resources for the Future.

Berner, Robert A., and Zavareth Kothavala. 2001. GEOCARB III: A revised model of atmospheric CO2 over Phanerozoic time. *American Journal of Science* 301: 182–204. [CrossRef]

BMWI. 2020. *The National Hydrogen Strategy*. Berlin: Federal Ministry for Economic Affairs and Energy, pp. 1–32.

Bossel, Ulf, Baldur Eliasson, and Gordon Taylor. 2003. The Future of the Hydrogen Economy: Bright or Bleak? *Cogeneration and Competitive Power Journal* 18: 29–70. [CrossRef]

BP (British Petroleum). 2017. *BP Energy Outlook: 2017 edition*. London: British Petroleum, p. 103.

BP (British Petroleum). 2020. *BP Energy Outlook: 2020 edition*. London: British Petroleum, p. 157.

Bui, Mai, Claire S. Adjiman, André Bardow, Edward J. Anthony, Andy Boston, Solomon Brown, Paul S. Fennell, Sabine Fuss, Amparo Galindo, Leigh A. Hackett, and et al. 2018. Carbon capture and storage (CCS): The way forward. *Energy & Environmental Science* 11: 1062–176.

Celia, Michael Anthony, Stefan Bachu, Jan Martin Nordbotten, and Karl W. Bandilla. 2015. Status of CO_2 storage in deep saline aquifers with emphasis on modeling approaches and practical simulations. *Water Resources Research* 51: 6846–92. [CrossRef]

Crowley, Thomas J., and Robert A. Berner. 2001. Paleoclimate: Enhanced: CO_2 and Climate Change. *Science* 292: 870–72. [CrossRef]

CSLF (Carbon Sequestration Leadership Forum). 2012. *CSLF Annual Meeting, 24–26 October 2012, Perth, Australia*. Bergen: Carbon Sequestration Leadership Forum.

CSLF (Carbon Sequestration Leadership Forum). 2019. *Carbon Capture, Utilisation and Storage (CCUS) and Energy Intensive Industries (EIIs)*. Chatou: Carbon Sequestration Leadership Forum.

D'Amore, Federico, and Fabrizio Bezzo. 2017. Economic optimisation of European supply chains for CO_2 capture, transport and sequestration. *International Journal of Greenhouse Gas Control* 65: 99–116. [CrossRef]

Dalebrook, Andrew F., Weijia Gan, Martin Grasemann, Séverine Moret, and Gábor Laurenczy. 2013. Hydrogen storage: Beyond conventional methods. *Chemical Communications* 49: 8735–51. [CrossRef] [PubMed]

Dayaratna, Kevin D., Ross McKitrick, and Patrick J. Michaels. 2020. Climate sensitivity, agricultural productivity and the social cost of carbon in FUND. *Environmental Economics and Policy Studies* 22: 433–48. [CrossRef]

EASAC (European Academies Science Advisory Council). 2013. *Carbon capture and storage in Europe*. (EASAC Policy Report 20). Halle (Saale): German National Academy of Sciences Leopoldina, p. 86.

EC (European Commission). 2007a. *Limiting Global Climate Change to 2 °C: The Way Ahead for 2020 and Beyond*. [COM (2007) 2 Final]. Brussels: European Commission, p. 13.

EC (European Commission). 2007b. *Towards a European Strategic Energy Technology Plan.* [COM (2006) 847 Final]. Brussels: European Commission, p. 12.

EC (European Commission). 2009. *Directive 2009/31/EC of the European Parliament and of the Council of 23 April 2009 on the Geological Storage of Carbon Dioxide.* Brussels: European Commission.

EC (European Commission). 2010. *Energy 2020: A Strategy for Competitive, Sustainable and Secure Energy.* [COM (2010) 639 Final]. Brussels: European Commission, p. 20.

EC (European Commission). 2011a. *Energy Efficiency Plan 2011.* [COM (2011) 109 Final]. Brussels: European Commission, p. 16.

EC (European Commission). 2011b. *Energy Roadmap 2050.* [COM (2011) 885 Final]. Brussels: European Commission, p. 20.

EC (European Commission). 2011c. *On Security of Energy Supply And international Cooperation—"The EU Energy Policy: Engaging with Partners beyond Our Borders".* [COM (2011) 539 Final]. Brussels: European Commission, p. 19.

EC (European Commission). 2011d. *Proposal for a Regulation of the European Parliament and of The Council on Guidelines for Trans-European Energy Infrastructure and Repealing Decision No 1364/2006/EC.* [COM (2011) 658 Final]. Brussels: European Commission, p. 48.

EC (European Commission). 2011e. *Smart Grids: From Innovation to Deployment.* [COM (2011) 202 Final]. Brussels: European Commission, p. 12.

EC (European Commission). 2018. *Report on Raw Materials for Battery Applications.* [SWD (2018) 245/2 Final]. Brussels: European Commission, p. 47.

EC (European Commission). 2019. *The European Green Deal.* [COM (2019) 640 Final]. Brussels: European Commission, p. 24.

EC (European Commission). 2020a. *A Hydrogen Strategy for a Climate-Neutral Europe.* [COM (2020) 301 Final]. Brussels: European Commission, p. 23.

EC (European Commission). 2020b. *Proposal for a Regulation of the European Parliament and of The Council Establishing the Framework for Achieving Climate Neutrality and Amending Regulation (EU) 2018/1999 (European Climate Law).* [COM (2020) 80 Final-2020/0036 (COD)]. Brussels: European Commission, p. 25.

EU (European Union). 2021. Regulation (EU) 2021/694 of the European Parliament and of the Council of 29 April 2021: Establishing the Digital Europe Programme and repealing Decision (EU) 2015/2240. *Official Journal of European Union* 2021: L166/1–L166/34.

Gates, Bill. 2021. *How to Avoid a Climate Disaster—The Solutions We Have and the Breakthroughs We Need.* New York: Penguin Random House, p. 257. ISBN 978-0-241-44830-4.

GECF (Gas Exporting Countries Forum). 2017. *Global Gas Outlook 2017.* Doha: Gas Exporting Countries Forum, p. 96.

Gielen, Dolf, Francisco Boshell, Deger Saygin, Morgan D. Bazilian, Nicholas Wagner, and Ricardo Gorini. 2019. The role of renewable energy in the global energy transformation. *Energy Strategy Reviews* 24: 38–50. [CrossRef]

Gillespie, Todd. 2021. Energy Crisis Is Old Economy's Revenge, Says Goldman's Currie. *Bloomberg News*. Available online: https://www.bnnbloomberg.ca/energy-crisis-is-old-economy-s-revenge-says-goldman-s-currie-1.1658513 (accessed on 25 October 2021).

Giorgi, Filippo. 2010. Uncertainties in climate change projections, from the global to the regional scale. *EPJ Web of Conferences* 9: 115–29. [CrossRef]

Giraud, Fabienne, Luc Beaufort, and Pierre Cotillon. 1995. Periodicities of carbonate cycles in the Valanginian of the Vocontian Trough: A strong obliquity control. *Geological Society Special Publications* 85: 143–64. [CrossRef]

Global CCS Institute. 2020. *The Global Status of CCS 2020 - CCS Vital to achieve Net-Zero*. Melbourne: Global CCS Institute, p. 44.

Guo, Chaohua, Mingzhen Wei, and Hong Liu. 2018. Study of gas production from shale reservoirs with multi-stage hydraulic fracturing horizontal well considering multiple transport mechanisms. *PLoS ONE* 13: e0188480. [CrossRef]

Haywood, Alan M., Paul J. Valdes, Tracy Aze, Natasha Barlow, Ariane Burke, Aisling M. Dolan, Anna S. von der Heyd, Daniel J. Hill, Stewart S. R. Jamieson, Bette Lou Otto-Bliesner, and et al. 2019. What can Palaeoclimate Modelling do for you? *Earth Systems and Environment* 3: 1–18. [CrossRef]

Heilek, Christian, Philipp Kuhn, and Maximilian Kühne. 2016. The Role of Large-Scale Hydrogen Storage in the Power System. In *Hydrogen and Fuel Cell*. Edited by Johannes Töpler and Jochen Lehmann. Berlin: Springer, pp. 21–37.

Herbert, Timothy D., and Alfred G. Fischer. 1986. Milankovitch climatic origin of mid-Cretaceous black shale rhythms in central Italy. *Nature* 321: 739–43. [CrossRef]

Huppert, Kimberly L., J. Taylor Perron, and Leigh H. Royde. 2020. Hotspot swells and the lifespan of volcanic ocean islands. *Science Advances* 6: 1–8. [CrossRef]

IEA (International Energy Agency). 2019. *The Future of Hydrogen Seizing Today's Opportunities*. Report Prepared by the IEA for the G20 Japan. Paris: International Energy Agency, p. 203.

IEA (International Energy Agency). 2020. *Renewables 2020: Analysis and Forecast to 2025*. Paris: International Energy Agency, p. 171.

IEA (International Energy Agency). 2021. *Net Zero by 2050: A Roadmap for the Global Energy Sector*. Special Report. Paris: International Energy Agency, p. 223.

IPCC (Intergovernmental Panel on Climate Change). 2013. *Climate Change 2013: The Physical Science Basis*. Working Group I Contribution to the IPCC Fifth Assessment Report. Cambridge: Cambridge University Press, p. 203. Available online: www.ipcc.ch/report/ar5/wg1 (accessed on 28 June 2021).

IRENA (International Renewable Energy Agency). 2019. *Global Energy Transformation: A Roadmap to 2050*. Abu Dhabi: International Renewable Energy Agency, p. 52.

ISO 11760. 2005. *Classification of Coals*. Geneva: International Organization for Standardization, p. 9.

Javed, Muhammad Shahzad, Dan Zhong, Tao Ma, Aotian Song, and Salman Ahmed. 2020. Hybrid pumped hydro and battery storage for renewable energy based power supply system. *Applied Energy* 257: 114026. [CrossRef]

Jiang, Tingxue, Xiaobing Bian, Haitao Wang, Shuangming Li, Changgui Jia, Honglei Liu, and Haicheng Sun. 2017. Volume fracturing of deep shale gas horizontal wells. *Natural Gas Industry B* 4: 127–33. [CrossRef]

Jiang, Kai, Peta Ashworth, Shiyi Zhang, Xi Liang, Yan Sun, and Daniel Angus. 2020. China's carbon capture, utilization and storage (CCUS) policy: A critical review. *Renewable and Sustainable Energy Reviews* 119: 1–15. [CrossRef]

Johnson, Markes E., B. Gudveig Baarli, Mário Cachão, Eduardo Mayoral, Ricardo S. Ramalho, Ana Santos, and Carlos M. da Silva. 2018. On the rise and fall of oceanic islands: Towards a global theory following the pioneering studies of Charles Darwin and James Dwight Dana. *Earth-Science Reviews* 180: 17–36. [CrossRef]

Jouzel, Jean, Françoise Vimeux, Nicolas Caillon, Gilles Delaygue, Georg Hoffmann, Valerie Masson-Delmotte, and Frederic Parrenin. 2003. Magnitude of isotope/temperature scaling for interpretation of central Antarctic ice cores. *Journal of Geophysical Research* 108: 4361. [CrossRef]

Jouzel, Jean, Valerie Masson-Delmotte, Olivier Cattani, Gabrielle Dreyfus, Sonia Falourd, Georg Hoffmann, Bénédicte Minster, Julius Nouet, Jean-Marc Barnola, Jérôme Chappellaz, and et al. 2007. Orbital and Millennial Antarctic Climate Variability over the Past 800,000 Years. *Science* 317: 793–96. [CrossRef] [PubMed]

Kang, Jian, and Rolf Larsson. 2013. What is the link between temperature and carbon dioxide levels? A Granger causality analysis based on ice core data. *Theoretical and Applied Climatology* 116: 537–48. [CrossRef]

Kapetaki, Zoe, and John Scowcroft. 2017. Overview of Carbon and Storage (CCS) demonstration project business models: Risks and Enablers on the sides of the Atlantic. *Energy Procedia* 114: 6623–30. [CrossRef]

Karakilcik, Hatice, and Mehmet Karakilcik. 2020. Underground Large-Scale Hydrogen Storage. In *Accelerating the Transition to a 100% Renewable Energy Era*. Edited by Tanay Sidki Uyar. Istanbul: Springer, Volume 74, pp. 375–92.

Khan, Chawarwan, Julie K. Pearce, Suzanne D. Golding, Victor Rudolph, and Jim R. Underschultz. 2021. Carbon Storage Potential of North American Oil & Gas Produced Water Injection with Surface Dissolution. *Geosciences* 11: 123–32.

Knutti, Reto, Joeri Rogelj, Jan Sedláček, and Erich M. Fischer. 2015. A scientific critique of the two-degree climate change target. *Nature Geoscience* 9: 13–18. [CrossRef]

Kovac, Ankica, Matej Paranos, and Doria Marcius. 2021. Hydrogen in energy transition: A review. *International Journal of Hydrogen Energy* 46: 10016–35. [CrossRef]

Lemieux, Alexander, Alexi Shkarupin, and Karen Sharp. 2020. Geologic feasibility of underground hydrogen storage in Canada. *International Journal of Hydrogen Energy* 45: 32243–59. [CrossRef]

Lemos de Sousa, Manuel, Cristina Fernanda Rodrigues, Maria Alzira Dinis, and Gisela Marta Oliveira. 2008. *Overview on CO_2 Geological Sequestration*. Porto: Universidade Fernando Pessoa.

Liu, Wei, Zhixin Zhang, Jie Chen, Deyi Jiang, Fei Wu, Jinyang Fan, and Yinping Li. 2020. Feasibility evaluation of large-scale underground hydrogen storage in bedded salt rocks of China: A case study in Jiangsu province. *Energy* 198: 117348. [CrossRef]

Metso: Outotec. 2021. Battery Metals. Available online: https://www.mogroup.com/commodities/battery-metals/ (accessed on 28 June 2021).

Mills, Marker P. 2019. *The "New Energy Economy": An Exercise in Magical Thinking*. New York: Manhattan Institute, p. 23.

Mills, Marker P. 2020. *Mines, Minerals, And "Green" Energy: A Reality Check*. New York: Manhattan Institute, p. 19.

Mitchell, Dann M., Yuen Tung Eunice Lo, William J. M. Seviour, Leopold Haimberger, and Lorenzo M. Polvani. 2020. The vertical profile of recent tropical temperature trends: Persistent model biases in the context of internal variability. *Environmental Research Letters* 15: 1–11. [CrossRef]

Molloy, Patrick, and LeeAnn Baronett. 2019. The Truth about Hydrogen, the Latest, Trendiest Low-Carbon Solution. Available online: https://www.greenbiz.com/article/truth-about-hydrogen-latest-trendiest-low-carbon-solution (accessed on 28 June 2021).

Moore, Patrick. 2019. *The Positive Impact of Human CO2 Emissions on the Survival of Life on Earth*. Winnipeg: Frontier Centre for Public Policy, p. 30.

MRI (Mitsubishi Research Institute). 2019. *Think & Act. Corporate Communications Division*. Tokyo: Mitsubishi Research Institute, Available online: https://www.mri.co.jp/en/about-us/info/group_report/index.html (accessed on 5 July 2021).

MRI (Mitsubishi Research Institute). 2020. *FY2019 Study on the Infrastructure Development Project for Acquisition of JCM Credits (International Cooperation in CCUS) Report*. Environment and Energy Division. Tokyo: Mitsubishi Research Institute, p. 310.

Niaz, Saba, Taniya Manzoor, and Altaf Hussain Pandith. 2015. Hydrogen storage: Materials, methods and perspectives. *Renewable and Sustainable Energy Reviews* 50: 457–69. [CrossRef]

Nordhaus, William. 2018. Projections and Uncertainties about Climate Change in an Era of Minimal Climate Policies. *American Economic Journal: Economic Policy* 10: 333–60. [CrossRef]

Ozarslan, Ahmet. 2012. Large-scale hydrogen energy storage in salt caverns. *International Journal of Hydrogen Energy* 37: 14265–77. [CrossRef]

Park, Jeffrey, and Timothy D. Herbert. 1987. Hunting for Paleoclimatic Periodicities in a Geologic Time Series with an Uncertain Time Scale. *Journal of Geophysical Research: Solid Earth* 92: 14027–40. [CrossRef]

Persson, Erik K. 2019. What Is the Link Between Temperature, Carbon Dioxide and Methane? A Multivariate Granger Causality Analysis Based on Ice Core Data from Dome C in Antarctica. Master's thesis, Uppsala University, Uppsala, Sweden.

Pielke, Roger. 2020. *Understanding the Great Climate Science Scenario Debate.* New York: Forbes, Available online: www.forbes.com/sites/rogerpielke/2020/02/03/understanding-the-great-climate-science-scenario-debate/?sh=5f3251bb26a7 (accessed on 11 October 2021).

Pol, Katy, Valérie Masson-Delmotte, Sigfús Johnsen, Matthias Bigler, Olivier Cattani, Gael Durand, Sonia Falourd, Jean Jouzel, Bénédicte Minster, Frédéric Parrenin, and et al. 2010. New MIS 19 EPICA Dome C high resolution deuterium data: Hints for a problematic preservation of climate variability at sub-millennial scale in the "oldest ice. " *Earth and Planetary Science Letters* 298: 95–103. [CrossRef]

Rahman, Muhammad Ishaq-ur. 2013. Climate Change: A Theoretical Review. *Interdisciplinary Description of Complex Systems* 11: 1–13. [CrossRef]

Ritchie, Hannah, and Max Roser. 2017. CO_2 and Greenhouse Emissions. Our World in Data CO_2 and Greenhouse Gas Emissions Database. Available online: https://ourworldindata.org/atmospheric-concentrations (accessed on 28 June 2021).

Rodrigues, Cristina F., and Manuel J. Lemos de Sousa. 2002. The measurement of coal porosity with different gases. *International Journal of Coal Geology* 48: 245–51. [CrossRef]

Rodrigues, Cristina F., Maria A. P. Dinis, and Manuel J. Lemos de Sousa. 2015. Review of European energy policies regarding the recent "carbon capture, utilization and storage" technologies scenario and the role of coal seams. *Environmental Earth Sciences* 74: 2553–61. [CrossRef]

Rosen, Amanda M. 2015. The Wrong Solution at the Right Time: The Failure of the Kyoto Protocol on Climate Change. *Politics & Policy* 43: 30–58.

Schernikau, Lars, and William Hayden Smith. 2021. How Many km^2 of Solar Panels in Spain and How Much Battery Backup Would It Take to Power Germany. Available online: https://ssrn.com/abstract=3730155 (accessed on 2 May 2021).

Schoenung, Susan M. 2011. *Economic Analysis of Large-Scale Hydrogen Storage for Renewable Utility Applications.* Alburquerque: Sandia National Laboratories, p. 41.

Schwarzacher, Walther. 1993. Milankovitch cycles in the pre-Pleistocene stratigraphic record: A review. *Geological Society Special Publications* 70: 187–94. [CrossRef]

Shaviv, Nir J. 2008. Using the oceans as a calorimeter to quantify the solar radiative forcing. *Journal of Geophysical Research* 113: 1–13. [CrossRef]

Shogenova, Alla, Kris Piessens, Sam Holloway, Michelle Bentham, Roberto Martínez, Kristin M. Flornes, Niels E. Poulsen, Adam Wójcicki, Saulius Sliaupa, Ludovít Kucharič, and et al. 2014. Implementation of the EU CCS Directive in Europe: Results and development in 2013. *Energy Procedia* 63: 6662–70. [CrossRef]

Shogenova, Alla, Nicklas Nordback, Daniel Sopher, Kazbulat Shogenov, Auli Niemi, Christopher Juhlin, Saulius Sliaupa, Monika Ivandic, Adam Wojcicki, Jüri Ivask, and et al. 2021. Carbon Neutral Baltic Sea Region by 2050: Myth or Reality? Paper present at the 15th Greenhouse Gas Control Technologies Conference, Abu Dhabi, United Arab Emirates, March 15–18; p. 12.

Soeder, Daniel J., and Scyller J. Borglum. 2019. The evolutionary U.S. shale plays. In *The Fossil Fuel Revolution: Shale Gas and Tight Oil*. Amsterdam: Elsevier, pp. 109–35.

Soon, Willie, Ronan Connolly, and Michael Connolly. 2015. Re-evaluating the role of solar variability on Northern Hemisphere temperature trends since the 19th century. *Earth-Science Reviews* 150: 409–52. [CrossRef]

Stenni, Barbara, Valérie Masson-Delmotte, Enricomaria Selmo, Hans Oerter, Hanno Meyer, Regine Röthlisberger, Jean Jouzel, Olivier Cattani, Sonia Falourd, Hubertus Fischer, and et al. 2010. The deuterium excess records of EPICA Dome C and Dronning Maud Land ice cores (East Antarctica). *Quaternary Science Reviews* 29: 146–59. [CrossRef]

Tarkowski, Radoslaw. 2017. Perspectives of using the geological subsurface for hydrogen storage in Poland. *International Journal of Hydrogen Energy* 42: 347–55. [CrossRef]

Tarkowski, Radoslaw. 2019. Underground hydrogen storage: Characteristics and prospects. *Renewable and Sustainable Energy Reviews* 105: 86–94. [CrossRef]

Tarling, Donald Harvey. 2010. Milankovitch Cycles in Climate Change, Geology and Geophysics. Paper present at the 6th International Symposium on Geophysics, November 11–12, Tanta, Egypt, pp. 1–8.

UN (United Nations). 1992. *United Nations Framework Convention on Climate Change*. (Document FCCC/Informal/84: GE.5-62220 (E) 200705). New York: United Nations, p. 24.

UN (United Nations). 2012. *Kyoto Protocol to the United Nations Framework Convention on Climate Change, Kyoto, 11 December 1997. Doha Amendment to the Kyoto Protocol, Doha, 8 December 2012. Adoption of Amendment to the Protocol.* New York: United Nations, p. 41, [Reference: C.N.718.2012.TREATIES-XXVII.7.c (Depositary Notification)].

UN (United Nations). 2015. *Paris Agreement.* New York: United Nations, p. 25.

Veizer, Ján, Yves Godderis, and Louis M. François. 2000. Evidence for decoupling of atmospheric CO_2 and global climate during the Phanerozoic eon. *Nature* 408: 698–701. [CrossRef] [PubMed]

Voosen, Paul. 2021. Climate panel confronts implausibly hot models. *Science* 373: 474–75. [CrossRef] [PubMed]

Werndl, Charlotte. 2016. On defining climate and climate change. *The British Journal for the Philosophy of Science* 67: 337–64. [CrossRef]

Wigley, Tom M. L. 1998. The Kyoto Protocol: CO_2, CH_4 and climate implications. *Geophysical Research Letters* 25: 2285–88. [CrossRef]

WMO (World Meteorological Organization). 1979. *World Climate Conference—A Conference of Experts on Climate and Mankind.* Geneva: Declaration and Supporting Documents, p. 50.

Yang, Yuqing, Setphen Bremner, Chris Menictas, and Merlinde Kay. 2018. Battery energy storage system size determination in renewable energy systems: A review. *Renewable and Sustainable Energy Reviews* 91: 109–25. [CrossRef]

Yartys, Volodymyr A., and Mykhaylo V. Lototsky. 2004. An Overview of Hydrogen Storage Methods. In *Hydrogen Materials Science and Chemistry of Carbon Nanomaterials.* NATO Science Series II: Mathematics, Physics and Chemistry. Edited by Turhan Nejat Veziroglu, Svetlana Yu. Zaginaichenko, Dmitry V. Schur, Bogdan Baranowski, Anatoliy Petrovich Shpak and Skorokhod Valery Volodimirovich. Dordrecht: Springer, Volume 172, pp. 75–104.

Zhou, Li. 2005. Progress and problems in hydrogen storage methods. *Renewable and Sustainable Energy Reviews* 9: 395–408. [CrossRef]

Züttel, Andreas. 2003. Materials for hydrogen storage. *Materials Today* 6: 24–33. [CrossRef]

Use of Storage and Renewable Electricity Generation to Reduce Domestic and Transport Carbon Emissions—Whole Life Energy, Carbon and Cost Analysis of Single Dwelling Case Study (UK)

Vicki Stevenson

1. Introduction

The IPCC investigated the relationship between likely atmospheric CO_2 by the year 2100 and the resulting climate change. The report found that the likelihood of staying below a 1.5 °C temperature increase in the 21st century (relative to 1850–1900) was more unlikely than likely even if atmospheric CO_2 did not exceed 450 ppm CO_2 by 2100 (Intergovernmental Panel on Climate Change 2014). The case for reducing our atmospheric carbon emissions is clear.

Globally, transport and building energy use are both major causes of CO_2 emissions. Focusing on the UK, direct carbon emissions from homes were 64 MT CO_2 in 2017 (Climate Change Committee 2019a), while UK road transport emissions were 118 MT CO_2, representing a 6% increase since 1990 (Office for National Statistics 2019).

Net-zero energy buildings have been established as an aim globally, at the European level and in the UK (World Green Building Council n.d.; European Parliament 2010; Government Property Agency 2020). In the UK, battery storage (including electric vehicle charging) has potential to reduce the need to upgrade local electricity grids and enable greater deployment of renewable electricity generation.

2. Literature Review

The UK (Climate Change Committee 2019b) has acknowledged that large-scale onshore wind, offshore wind and solar PV are now the cheapest forms of electricity generation in the UK and expects their contribution to UK electricity to rise to 50–65% by 2030 and potentially even higher afterwards. This will be needed to reduce the UK grid electricity carbon intensity to 100 g CO_2/kWh by 2030, then even further to 10 g CO_2/kWh by 2050. However, this level of penetration requires additional energy storage infrastructure to (1) make best use of renewable generation, (2) ensure that peak demand continues to be met and (3) bring electricity system costs down. Currently, the UK has the equivalent of 3 GW of pumped hydro storage, but only 0.4 GW of battery devices.

In the UK, the important role of storage is presented in the recent Heat and Buildings Strategy (HM Government 2021). Previously, it was not specifically considered, but there have been references to "enabling smart homes" and commitments to introducing time of use tariffs and creating an opportunity for domestic demand load shifting (HM Government 2017) which support the implementation of domestic storage. Although battery storage is eligible for an interest-free loan in Scotland (Home and Energy Scotland n.d.), there are currently no nationwide domestic energy storage incentives (Zakeri et al. 2021).

2.1. Benefits and Financial Viability of Domestic Energy Storage

Recent academic analysis of domestic energy storage has focused on its benefits to the electricity grid and its financial viability. The benefits of domestic energy storage paired with on-site renewable energy generation include the following:

- Enables load management/load shifting (Sheha and Powell 2018);
- Peak load management (Jankowiak et al. 2020; Koskela et al. 2019; Gardiner et al. 2020);
- Balancing grid frequency (Gardiner et al. 2020);
- Reduced grid import/export (Zakeri et al. 2021; Jankowiak et al. 2020; Dong et al. 2020);
- Reduced consumption from the electricity grid (Jankowiak et al. 2020);
- Improves electricity self-consumption and self-sufficiency (Jankowiak et al. 2020; Koskela et al. 2019; Gardiner et al. 2020; Dong et al. 2020).

On a community scale, storage can improve the voltage quality of the local distribution grid and is a cheaper alternative to distribution and transmission network expansion (Dong et al. 2020). However, domestic battery storage is commercially unviable—this has been tested across a wide range of markets (Zakeri et al. 2021; Sheha and Powell 2018; Gardiner et al. 2020; Dong et al. 2020). This explains why the rate of storage with PV is still very low (Zakeri et al. 2021). Although there are many benefits to increased domestic energy storage (Zakeri et al. 2021; Jankowiak et al. 2020; Koskela et al. 2019; Gardiner et al. 2020; Dong et al. 2020), there is a gap in the linkages and markets which would enable the prosumer to gain financially from providing these benefits (Gardiner et al. 2020).

Fewer studies are available on the environmental impact of energy storage systems. (Üçtuğ and Azapagic 2018) analysed a system in Turkey and found that it generates 4.7–8 times more energy than it consumes while having a 1.6–82.6-fold lower impact than grid electricity. It has been found that community installations require less battery capacity than individual installations (Dong et al. 2020; Mair et al. 2021); this shows that embodied carbon of battery storage could be reduced significantly if implemented on a larger scale than single dwellings.

2.2. Policy

California and Germany are making headway in encouraging domestic energy storage. California offers a USD 400/kWh subsidy and a 30% tax credit—despite this, adoption has been low until recently (Gardiner et al. 2020). Germany's success is attributed to subsidies and low-cost loans (Gardiner et al. 2020; Uddin et al. 2017). Australia has little policy support, but increasing electricity prices are driving a surge in domestic battery storage (Gardiner et al. 2020).

A range of policy options have been analysed for the UK (Zakeri et al. 2021; Jankowiak et al. 2020; Gardiner et al. 2020). These include the following:

- Arbitrage (import electricity at a low price and export at a higher price)—This relies on dynamic purchase and export tariffs. Although time of use tariffs exist for purchase, domestic export tariffs are fixed. Even though this improves the financial case for storage, it does not make it profitable unless a policy is implemented to widen the price gap of off-peak and peak hours (Zakeri et al. 2021). Although storage capacity needs to be retained for PV-generated electricity (reducing arbitrage availability), this does not have an operational cost and thus is only a disadvantage if it displaces grid electricity at a negative purchase price.
- Peak shaving tariff—Remuneration of GBP 0.24/kWh of peak shaved would improve profitability of battery storage and incentivise prosumers to purchase a larger battery and operate it in a way that benefits the electricity grid (Jankowiak et al. 2020). Enabling monetisation of a peak shaving service is also supported by (Gardiner et al. 2020).
- Capital subsidy—A 30% capital subsidy would make storage break even assuming a time of use tariff is used; otherwise, a 50–60% subsidy would be required (Zakeri et al. 2021). Another analysis of subsidy found that GBP 2660 was required to support the purchase of a 4 kWh battery along with a 4 kW PV system—this is significantly more than most international programmes (Gardiner et al. 2020).

- Storage tariff—This could be used to reward owners for electricity discharged at peak times. The current electric vehicle subsidy could be repositioned to only support vehicles with vehicle-to-grid capability (Zakeri et al. 2021).
- VAT reduction to 5% for retrofit installation only has a modest impact on the investment case (Gardiner et al. 2020).

There is a consensus that policies which monetise multiple services outperform subsidies, and that battery owners need to be rewarded for the benefits they provide to the electricity grid (Gardiner et al. 2020).

2.3. Research Gap

There is a lack of information and supporting data which can be used to inform prosumers on the energy and carbon payback of their potential investment in PV renewable energy generation and energy storage. This study addresses this by analysing the financial, energy and carbon payback based on a single dwelling case study.

3. Methodology

The key components to the methodology are as follows:

- Description of the case study including capital cost (Section 3.1);
- Calculation of embodied energy and carbon of the case study (Section 3.2);
- Calculation of expected energy, carbon and financial payback of the case study before installation (Section 3.3);
- Process to analyse embodied energy, embodied carbon and financial payback (Section 3.4).

3.1. Case Study Description

The dwelling is a split-level detached house in South Wales (UK) which was built in 2006. In November 2013, it received an EPC rating of C (78 points), with a recommendation that it could achieve an EPC rating of B (84 points) if 2.5 kWp of PV was installed.

On 30 July 2014, a 3.6 kWp vertical, monocrystalline, 2-string PV array with an SMA 3600 TL single-phase inverter and a Sunny web unit was commissioned. The south gable façade was chosen for PV installation as plans for an attic conversion were being considered. As the PV system was at street level, it incorporated matt glass to reduce potential reflection nuisance to neighbours, and it was shaped to maximise the space in the peak of the gable. These factors resulted in a higher cost

(GBP 10,610 including 5% VAT) than that of a roof-mounted system. The system has a PV orientation within 5° of south, vertical inclination and a shading factor of 95% (due to potential shading from nearby trees in winter) and was based on the MCS irradiance dataset of 668 kWh/kWp.

In 2019, the decision was made to purchase an electric vehicle, and thus increased renewable generation was planned. As the vehicle was likely to be off site during the day, energy storage was also planned.

On 28 April 2020, a Tesla Powerwall 2 AC with Gateway 2 was installed in an attached garage. This provides 13.5 kWh of storage with a 10-year warranty for 80% of the capacity at a cost of GBP 8040.

On 16 May 2020, a 6.12 kWp monocrystalline PV array with a Solar Edge HD 4 kW inverter and optimisers was commissioned. The west-facing roof was chosen for PV installation as the east roof suffered from shading in the morning. Initially, a 6.8 kWp system had been specified; however, Western Power set an export limit of 5.18 kW (including the existing façade PV system) and reduced the roof PV array to 6.12 kWp despite the inclusion of the Powerwall and optimisers which could be used to control the export. The system cost GBP 7803 including 5% VAT and warranty for the inverter. The system has a PV orientation within 5° of west, 35° inclination and a shading factor of 1 and was based on an irradiance dataset of 778 kWh/kWp.

The electric vehicle (EV) arrived on 11 September 2020 having been significantly delayed by restrictions related to the COVID-19 pandemic. The vehicle cost GBP 29,710 including 20% VAT and after deducting the GBP 2500 UK government plug-in grant. In addition, a Zappi charger was purchased to enable home charging and make use of potential excess from the PV electricity generation. The Zappi charger cost GBP 1219 including 20% VAT and after deducting the GBP 350 UK government Electric Vehicle Homecharge Scheme grant. The EV consumption is indicated as 270 Wh/mile (Electric Vehicle Database n.d.) which is equivalent to 167.8 Wh/km, 5.95 km/kWh or 3.7 mile/kWh. Although the UK government considers electric vehicles as having zero CO_2 emissions for vehicle tax purposes, there will be CO_2 emissions from the electricity unless the car is entirely charged from renewable sources. Assuming the 2018 annual average electricity carbon intensity of 260.25 g CO_2/kWh (Electricity Info n.d.), this results in EV emissions of 43.9 g CO_2/km. This is expected to decrease as electricity carbon intensity reduces in line with future targets (Energy UK 2016). The electric vehicle replaced a diesel car with published emissions of 109 g CO_2/km (Vehicle Certification Agency n.d.). However, the diesel car performance over a 10-year period proved to be 125.8 g CO_2/km based on actual fuel efficiency.

3.2. Embodied Energy and Carbon

Embodied energy and carbon data were sought for PV and storage and then applied to the case study.

3.2.1. Photovoltaics

There is a range of embodied energy and carbon data available for renewable energy generation and storage technologies (Reddaway 2016; Circular Ecology n.d.; Finnegan et al. 2018; Ito et al. 2011; Kristjansdottir et al. 2016; Feng and van den Bergh 2020; Wong et al. 2016; Kurland 2019; Hausfather 2019), but the data must be treated with care as they often relate to different materials, locations, assumed service life, balance of system, year of analysis, functional units, analysis approach and analysis boundaries.

Over the period 1998–2014, the embodied energy of monocrystalline (m-Si) PV modules was found to range between 2397 and 11,673 MJ/m^2, while polycrystalline (p-Si) PV modules ranged between 2699 and 5150 MJ/m^2 (Ito et al. 2011; Wong et al. 2016). The data are more stable over the period 2009–2014 and averaging these data returns values of 3597 MJ/m^2 for m-Si and 2848 MJ/m^2 for p-Si. More recent data for a fully installed PV system in Australia show values of 4185 kWh/kWp for m-Si and 3708 kWh/kWp for p-Si (Reddaway 2016). Although polycrystalline PV (as a byproduct) has a lower embodied energy than monocrystalline PV, it also has a lower efficiency (13.9% vs. 14.3%) Ito et al. 2011, meaning monocrystalline PV is often chosen for dwellings (as in this case study) because it is more space efficient.

Over the period 2011 to 2016, embodied carbon of monocrystalline (m-Si) PV was found to range between 150 and 300 kg CO_2/m^2 (Finnegan et al. 2018; Ito et al. 2011; Kristjansdottir et al. 2016). Additional 2016 data show values of 2560 kg CO_2/kWp for m-Si (Circular Ecology n.d.). Feng analysed the CO_2 emissions of a range of PV module types manufactured in China, the EU and the USA and found that the carbon intensity of the electricity in the country was the dominating factor, resulting in EU production having the lowest embodied carbon, followed by that of the USA and then China (Feng and van den Bergh 2020).

Analysis of embodied energy of PV systems in Norway (Kristjansdottir et al. 2016) and Australia (Reddaway 2016) indicated that panel manufacturing accounts for over 80% of emissions, with the next highest impact being from the inverter.

3.2.2. Lithium Batteries

In the UK, 2020 saw 108,205 battery electric vehicles sold, representing a 180% rise on the previous year (Attwood 2021). This represents a significant increase in lithium-ion battery production. Already 10% of Tesla's batteries get reused, while 60% get recycled (Forfar 2018), and stationary power storage applications are seen as a valuable use of second-life batteries (Engel et al. 2019). Stationary power storage requires a lower current density than EVs, meaning it can use batteries with 80–85% of their original capacity (Pagliaro and Meneguzzo 2019). Nissan and Renault have partnered with organisations to reuse battery packs (Engel et al. 2019), while Tesla already has a stationary power storage product—the Powerwall.

Kurland's analysis of energy use in lithium-ion battery production indicated significant improvements from 2014 (163 kWh/kWhc) to 2019 (50–65 kWh/kWhc). Key factors include the following:

- Increased scale, although this did not seem to have an impact after increasing the size of the facilities with an annual capacity above 2 GWh;
- All electricity production in comparison to utilising steam from fossil fuels;
- For Tesla and Northvolt Ett, embodied carbon has reduced due to the use of renewable electricity (Kurland 2019).

The importance of the grid carbon intensity was also reflected in Hausfather's analysis of lifecycle carbon emissions of EV batteries, which found that US- or Europe-sourced batteries had approximately 20% lower lifecycle emissions than Asia-sourced batteries. This was attributed to the widespread use of coal for electricity generation in Asia (Hausfather 2019). Lifecycle carbon emissions of batteries analysed between 2017 and 2019 ranged between 61 and 106 kg CO_2/kWh, with an average of 100 kg CO_2/kWh (Hausfather 2019).

3.2.3. Embodied Energy and Carbon Data Applied to Case Study

Due to embodied energy and carbon data from the literature being based on different functional units, these are now considered in relation to the case study elements to allow comparison. Embodied energy data are shown in Table 1, with the calculated results for both PV systems and storage presented in kWh for comparison. Similarly, embodied carbon data are shown in Table 2, with the calculated results presented in kg CO_2 for comparison.

Table 1. Embodied energy data calculated for façade PV system, roof PV system and battery.

Element/Functional Unit		PV System—Austral 2016	PV Modules + Inverter—Japan 2011	PV Modules Ave 2009–2014	Tesla Battery (Large Scale 2019)
		4185 kWh/kWp	3986 MJ/m^2	3597 MJ/m^2	57.5 kWh/kWc
Roof PV	6.12 kWp	25,612 kWh [1]	-	-	-
	32.5 m^2	-	36,664 kWh	32,514 kWh	-
Roof inverter	4 kWp	-		-	-
façade PV	3.6 kWp	15,066 kWh [1]	-	-	-
Tesla Powerwall	13.5 kWh	-	-	-	776 kWh

[1] Data used in calculations. Source: Table by author, adapted from Reddaway (2016); Ito et al. (2011); Wong et al. (2016); Kurland (2019).

Table 2. Embodied carbon data calculated for façade PV system, roof PV system and battery.

Element/Functional Unit		UK Panels	PV Modules Ave 2011–2016	Battery—Large Scale 2019
		2560 kg CO$_2$/kWp	225 kg CO$_2$/m^2	100 kg CO$_2$/kWh
Roof PV	6.12 kWp	15,667 kg CO$_2$ [1]	-	-
	32.5 m^2	-	7322 kg CO$_2$	-
Roof inverter	4 kWp	-	-	-
façade PV	3.6 kWp	9216 kg CO$_2$ [1]	-	-
Tesla Powerwall	13.5 kWh	-	-	1350 kg CO$_2$ [1]

[1] Data used in calculations. Source: Table by author, adapted from Circular Ecology (n.d.); Finnegan et al. (2018); Ito et al. (2011); Kristjansdottir et al. (2016); Hausfather (2019).

As an accurate area measurement is not available for the façade PV system, the only embodied energy data which can be used for both case study PV arrays are from the Australian PV system. It is acknowledged that this is possibly an underestimate of the total embodied energy for the PV system.

The embodied carbon data, which are available for both PV arrays, are based on UK panels and are significantly higher than the alternative data. Given the potential assumption differences between embodied energy and carbon data, this variance was accepted as inevitable, and the higher embodied carbon value was used for further calculations in this paper.

The embodied energy and carbon data for the Powerwall were considered to be the most appropriate estimates from the data available and were used for further calculations in this paper.

3.3. Expected Energy, Carbon and Financial Payback before Installation

Data provided before installation were used to calculate a simplified energy carbon and financial payback.

3.3.1. Façade PV

Based on the installer information, the system was expected to produce 2295 kWh annually. To pay back the 15,066 kWh embodied energy (Table 1) would take 6.56 years.

The carbon intensity of the UK grid electricity in 2013 (European Environment Agency 2020) was 467 g CO_2e/kWh. To pay back the 9216 kg CO_2 embodied carbon (Table 2) based on the predicted energy generation would take 8.6 years.

The PV system was registered for the UK Government Feed in Tariff (FiT) scheme. This paid 14.9 p/kWh with an assumed 50% export qualifying for an additional 5.5 p/kWh. Based on installer information, the PV system was anticipated to make an annual electricity cost saving of GBP 359.09, with an annual FiT contribution of GBP 627.39 (total GBP 986 per year), leading to a payback of the GBP 10,610 cost over 10.75 years.

3.3.2. Roof PV

The system was expected to produce 4761 kWh annually. To pay back the 25,612kWh embodied energy (Table 1) would take 5.37 years.

The carbon intensity of the UK grid electricity in 2019 (European Environment Agency 2020) was 228 g CO_2e/kWh. To pay back the 15,667 kg CO_2 embodied carbon (Table 2) based on the predicted energy generation would take 14.43 years.

The PV system was assumed to save GBP 675 in electricity costs with an additional GBP 145 in export sales, leading to a financial payback time of 9.5 years. However, it was not cost effective to take an export tariff for the roof PV system instead of the FiT scheme for the façade PV system, meaning the financial payback time increased to 11.56 years.

3.3.3. Tesla Powerwall

It is more complex to calculate the energy, carbon and financial payback for the Tesla Powerwall as more factors are involved.

The first factor in the financial payback for the Tesla Powerwall is the electricity tariff used. The tariffs considered were as follows:

- Bulb standard tariff (from Bulb energy statement of 30 June 2020)—14.35 p/kWh
- Bulb EV tariff (Bulb n.d.):

  - 00:00–07:59   9.03 p/kWh (off peak)
  - 08:00–16:59 and 20:00–23:50   13.53 p/kWh (standard)
  - 17:00–18:59   30.22 p/kWh (peak)

- Octopus Go (from Octopus energy statement of 6 July 2020):

  - 00:30–04:29   5.00 p/kWh (off peak)
  - 04:30–00:29   14.02 p/kWh (standard)

- Octopus Agile (Energy-Stats n.d.):

  - Tracking tariff which varies at half-hourly intervals depending on the predicted combination of supply and demand, published at 17:00 the day before use;
  - Initial analysis of diurnal electricity cost trends (Figure 1) indicated that the period 16:00–18:59 had a significantly higher tariff than the rest of the day, with a slightly cheaper period of 01:00–04:49. Averaging costs for these periods over the year resulted in a mean Agile tariff of:

    - 16:00–18:59   23.00 p/kWh (peak)
    - 19:00–15:59   7.00 p/kWh (standard)

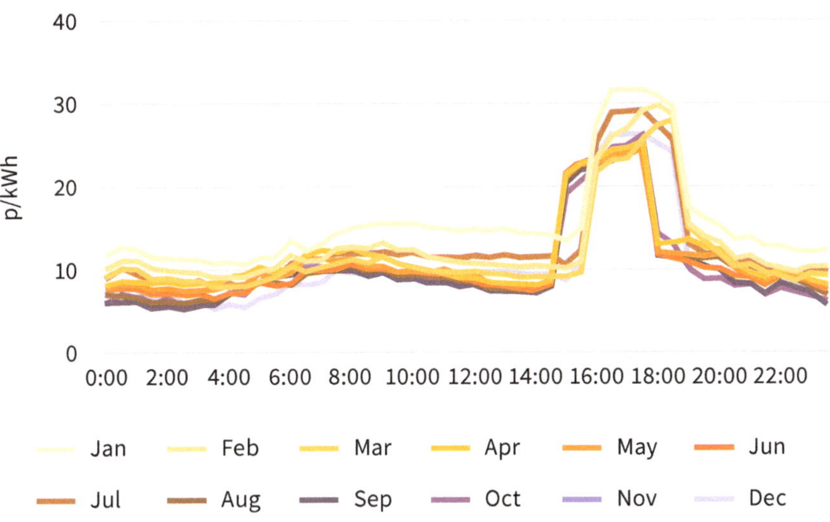

Figure 1. Half-hourly costs for Octopus Agile tariff 2019. Source: Graphic by author.

It was assumed that no grid electricity would be required in the period 16:00–18:59 to avoid the most expensive periods, but that the Powerwall could charge at any other time of day. Electricity consumption was based on actual consumption data from 1 January to 31 December 2019. Energy consumption for the planned EV was assumed to be 42.6 kWh per week for 48 weeks (allowing 4 weeks in the year without significant commuting). It was assumed that the EV would charge during the off-peak period. These assumptions resulted in an assumed consumption for 2019 of:

- Peak electricity   1834.53 kWh
- Standard electricity   10,225.76 kWh
- Off-peak electricity   4708.76 kWh (2044.8 kWh for EV)

Table 3 indicates the 2019 electricity cost for the following:

- Each tariff;
- The potential saving over the Bulb standard tariff;
- The cost saving of using a Powerwall via load shifting from grid (and financial payback in years).

Table 3. Electricity tariff options and financial benefit for load shifting.

Tariff	Octopus Go	Octopus Agile	Bulb EV	Bulb Standard
2019 electricity cost (GBP)	2017.54	1544.01	2437.75	2477.42
Saving vs. Bulb standard (GBP)	459.88	933.42	39.68	-
Powerwall load shifting saving (GBP)	165.47	293.52	388.74	-
Powerwall load shifting payback (y)	48.6	27.4	20.7	-

Source: Table from author, adapted from Bulb energy statement of 30 June 2020; Bulb (n.d.); Octopus energy statement of 6th July 2020; Energy-Stats (n.d.).

On its own, load shifting to avoid peak costs does not justify the financial cost of a Powerwall. However, in 2019, 172.571 kWh of PV energy was exported to the grid from the gable PV system, and with the planned installation of an additional PV system, the likely export of PV (or throttling of generation) could be mitigated by using the Powerwall.

The embodied energy of the Tesla Powerwall is 776 kWh (Table 1), while the embodied carbon is 1350 kg CO_2 (Table 2). Insufficient data were available to enable calculation of energy and carbon payback at this stage.

3.4. Process to Analyse Embodied Energy, Embodied Carbon and Financial Payback of the Case Study

Payback is calculated for each of the elements in relation to energy, carbon and capital cost:

Energy payback (years) = Embodied Energy/Annual Energy Generation[1]

Carbon payback (years) = Embodied Carbon/Annual CO_2 savings

Financial payback (years) = Capital Cost/Annual financial benefit

The embodied energy, embodied carbon and capital cost have already been presented in Sections 3.1 and 3.2.

Annual energy generation was obtained through monitoring (Section 3.4.1) up until December 2020. Beyond this, the performance was based on the period July–December 2020 (justified in Section 4.1.2).

Annual CO_2 savings were derived from monitored energy generation and analysis of existing UK electricity grid carbon intensity data (Section 3.4.2) and predicted grid carbon intensity (Section 4.2.2). This includes the capability of the Powerwall to store off-peak low-carbon electricity to use at higher carbon periods. For the EV, the CO_2 savings are based on the reduction in carbon emissions in comparison to the diesel car it replaced (Section 4.4).

Annual financial benefit was derived from monitored energy generation and electricity tariff, FiT and export payments (Section 4.3.1).

3.4.1. Monitoring

The monitoring options for each element of the renewable generation and storage system are discussed in the following paragraphs, summarised in Table 4 and illustrated in Figure 2.

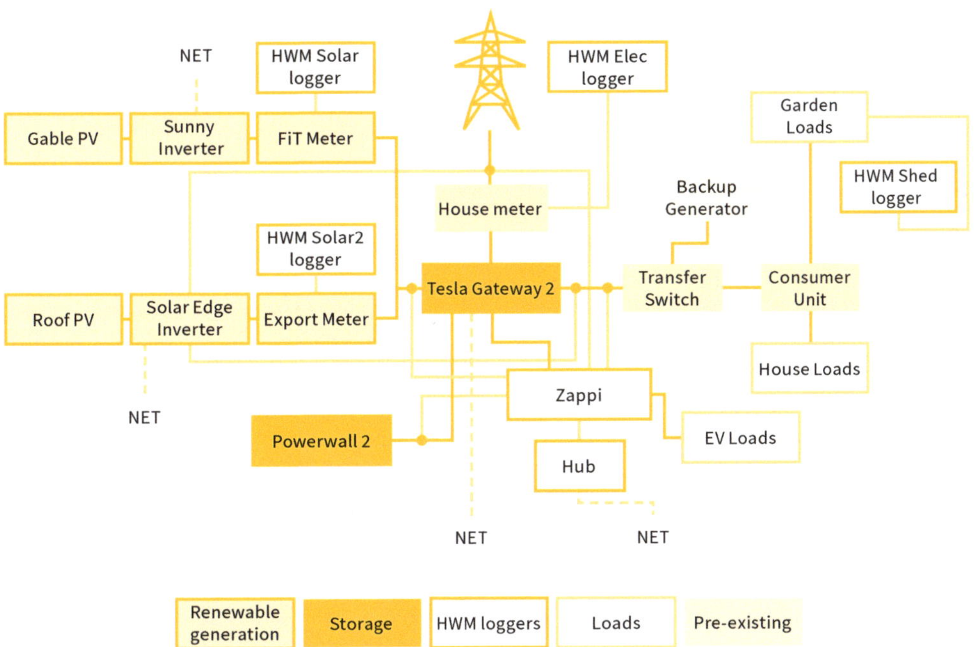

Figure 2. Schematic of equipment for monitoring renewable generation and storage. Source: Graphic by author.

Table 4. Summary of monitoring equipment.

Monitor	Data	Source	Start Date/Time Zone	Interval	Accuracy
Car dashboard	EV efficiency and distance travelled [2]	EV	11 September 2020 n/a	Manual	-
Export meter	Roof PV generation	Wired between inverter and incoming grid	16 May 2020 n/a	Manual	Schedule 7: Electricity Act 1989
FiT meter	Façade PV generation	Wired between inverter and incoming grid	30 July 2014 n/a	Manual	Schedule 7: Electricity Act 1989
HWM Sarn Solar [1]	Façade PV generation	Pulse counter on FiT meter	19 December 2014 BST/GMT	15 min	1 pulse/Wh
HWM Sarn Solar 2 [1]	Roof PV generation	Pulse counter on export meter	17 May 2020 BST/GMT	15 min	1 pulse/Wh
Solaredge.com	Roof PV generation [2]	Solar Edge inverter	16 May 2020 BST/GMT	15 min	1% var vs. Utility bill
	Grid import Grid export	Current clamp below Gateway 2			
Sunnyportal.com	Façade PV generation [2]	Sunny inverter	30 July 2014 BST/GMT	60 min	-
Tesla app	Total PV generation Powerwall import [2] Powerwall export [2] Grid import Grid export	Tesla Gateway 2 and current clamp below Gateway 2	29 April 2020 BST/GMT	5 min	1% var vs. Utility bill
Utility billing data	Grid import [2]	House electricity meter	24 May–19 October 2020	30 min	Schedule 7: Electricity Act 1989

[1] Equipment loaned by HWM-Water Ltd., Wales, UK; [2] Data used for monitoring. Source: Table by author.

Façade PV—Energy Generation

The Feed in Tariff export meter meets the requirements of Schedule 7 of the Electricity Act 1989; however, only manual readings can be taken from it. The Sunny web unit transmits data directly from the inverter to Sunnyportal.com, which are recorded at hourly intervals. Unfortunately, a data service change resulted in lost data in 2018. The HWM "Sarn Solar" logger was installed in December 2014 and records data at 15 min intervals. Unfortunately, this suffered some data loss in 2016 and 2017. In years with no apparent data loss, the HWM "Sarn Solar" logged electricity generation figures 95.3–97.7% of those logged by Sunnyportal.com. Sunnyportal.com was used as the primary source of data for the façade PV system. HWM "Sarn Solar" was used for the periods of data loss by Sunnyportal.com.

Roof PV—Energy Generation

The export meter meets the requirements of Schedule 7 of the Electricity Act 1989; however, only manual readings can be taken from it. Data are transmitted directly from the inverter to monitoring.solaredge.com and recorded at 15 min intervals. The HWM Sarn Solar 2 logger is battery operated and suffered from some data loss in September 2020. Prior to this (17 May–31 August 2020), the HWM Sarn Solar 2 and solaredge data were within 1% of each other. As electricity generation reduced in October–December 2020, the discrepancy increased to 2–3%. Solaredge was used as the primary source of data for the roof PV system.

Total PV—Energy Generation

A current clamp on cables between the utility meter and Gateway 2 transmits data to the Tesla app, which are recorded at 5 min intervals.

Electricity Import to Site for Electricity Stored by Powerwall and EV

The utility meter meets the requirements of Schedule 7 of the Electricity Act 1989, with half-hourly data available from 24 May 2020 until 19 October 2020. However, no data after that date have been published.

Data from a current clamp on cables between the utility meter and Gateway 2 are available at 15 min intervals via monitoring.solaredge.com. Over the period 24 May to 19 October, solaredge and the average daily utility data were within 1% of each other.

Data from a separate current clamp on cables between the utility meter and Gateway 2 are available at 5 min intervals via the Tesla app. Over the period 24

May–19 October, the Tesla app and the average daily utility data were within 1% of each other.

Utility billing data were used as the primary source of data for electricity import to the site. When utility billing data were not available, the Tesla data were averaged to half-hourly intervals and used.

To Powerwall/From Powerwall—Electricity Stored by Powerwall

The electricity directed to and from the Powerwall is available at 5 min intervals via the Tesla app.

EV Efficiency and Distance Travelled

Electric vehicle efficiency and distance travelled data were read manually from the car dashboard at monthly intervals.

The monitoring options are summarised in Table 4, and their relationship to the case study is illustrated in Figure 2.

Figure 2 illustrates the relationship of the monitoring equipment to the house, renewable generation and storage infrastructure.

3.4.2. UK Electricity Grid Carbon Intensity Data

The UK electricity grid carbon intensity was required to calculate the carbon payback. High-resolution (half-hourly) data were required for the Powerwall analysis. There are different sources of UK grid carbon intensity data depending on the time period and resolution required. These are explained below and presented in Tables 5 and 6 and Figure 3.

Average annual data for 1990–2019 are available from the (European Environment Agency 2020). Data for 2014–2019 are presented in Table 5. The progress in decarbonising the UK grid electricity is clear from these data.

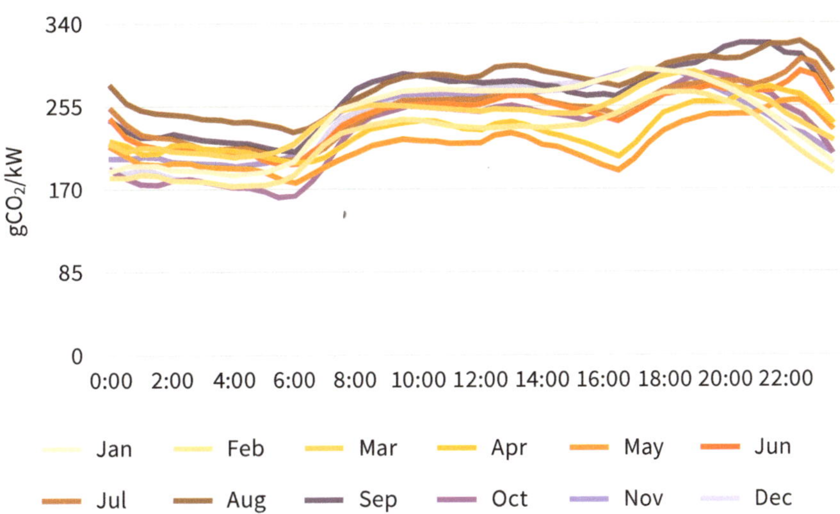

Figure 3. Monthly averaged UK grid electricity carbon intensity 2020 (half-hourly data). Source: Graphic by author.

Table 5. Average annual carbon intensity data for UK grid: 2014–2018.

Year	UK Electricity Generation g CO₂e/kWh
2013	467
2014	425
2015	380
2016	294
2017	264
2018	250
2019	228

Source: Table by author, adapted from European Environment Agency (2020).

Daily carbon intensity data from 12 September 2017 (Electricity Info n.d.) can be used to produce monthly average figures. Average monthly data since January 2018 are shown in Table 6. The average annual data for 2018 and 2019 are observed to approximate to the annual data in Table 5. The year 2020 saw a significant drop in the industrial and service sectors' electricity consumption compared to previous years. This reduced the requirement for carbon-intensive electricity generation (Department for Business, Energy & Industrial Strategy 2021).

Table 6. Average monthly carbon intensity data for UK grid: January 2018–March 2021.

Period	Carbon Intensity of UK Electricity Supply g CO$_2$/kWh			
	2018	2019	2020	2021
January	254	281	209	191
February	293	217	176	179
March	319	197	189	185
April	228	213	168	-
May	216	210	142	-
June	238	215	172	-
July	248	223	185	-
August	219	188	204	-
September	223	178	194	-
Octorber	234	202	162	-
November	261	240	177	-
December	390	202	185	-
Annual Average	**260**	**214**	**180**	-

Source: Table by author, adapted from Electricity Info (n.d.).

Half-hourly UK electric grid carbon intensity was calculated. Half-hourly power data detailing the various sources feeding the electricity grid (BMRS n.d.a.) were multiplied by the carbon intensity for each fuel type to calculate the half-hourly carbon emissions for the UK grid electricity. The mean carbon intensity for each fuel type (World Nuclear Association n.d.) is illustrated in Table 7 (Gridwatch n.d.). Finally, the half-hourly carbon emissions for the UK grid electricity were divided by the UK energy demand (BMRS n.d.b.) to obtain the carbon intensity per MW electricity.

Table 7. Mean carbon intensity of fuels used in UK electricity production.

Source	Tonnes CO_2e/GWh
Combined cycle gas turbine	499
Nuclear	29
Biomass	45
Coal	888
Wind	26
Solar	85
Oil	733
Open cycle gas turbine	499
Hydroelectric	26
Pumped hydro	415
Interconnector (import)	273
Other	273

Source: Table by author, adapted from World Nuclear Association (n.d.).

Half-hourly data averaged for each month of 2020 are shown in Figure 3. These data confirm that electricity consumed between approximately 00:30 and 06:00 has a lower carbon intensity than that consumed in the rest of the day.

4. Results

The energy, carbon and financial paybacks of the façade PV system, roof PV system, Tesla Powerwall and EV battery are considered in relation to actual performance since purchase.

4.1. Energy Payback

4.1.1. Façade PV

Up to 31 March 2021, the façade PV system generated 12,209.1 kWh. The average annual generation from 2015 to 2020 was 1881.9 kWh, approximately 18% lower than predicted. The monthly yield is illustrated in Figure 4.

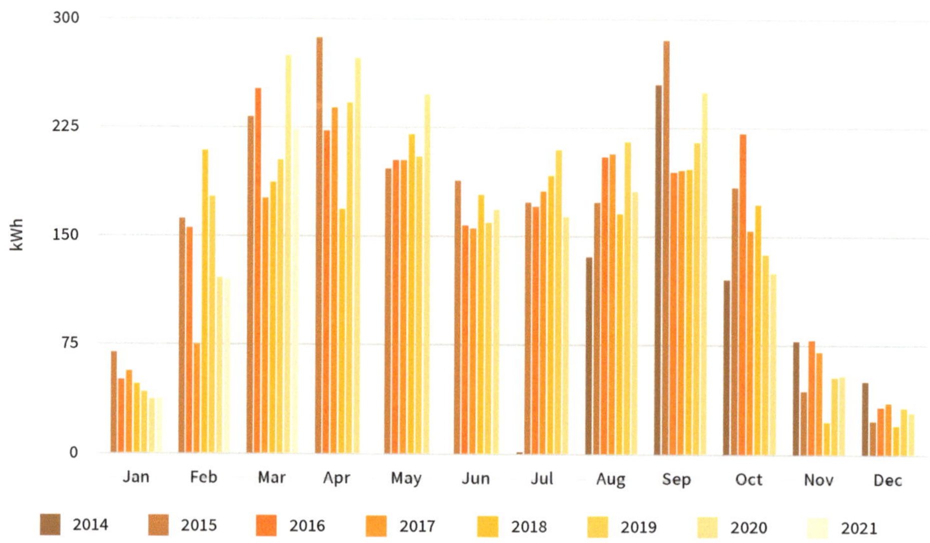

Figure 4. Monthly renewable energy generation of the façade PV system since installation in 2014. Source: Graphic by author.

Based on past energy generation trends, the façade PV system will achieve energy payback between August 2022 and March 2023 depending on weather conditions—at least 8 years to achieve energy payback. Even though this is slower than anticipated based on installer data, it is well within the expected service life of the PV system.

4.1.2. Roof PV

As of 31 March 2021, the roof PV system generated 4156.6 kWh of renewable electricity (illustrated in Figure 5). As the roof PV system was installed in May 2020, there was not a full year of renewable generation data to analyse. Instead, the period 22 June–21 December 2020, which represents a half year between the highest sun availability in summer to the lowest sun availability in winter, was analysed.

Referring to the renewable energy generation of the façade PV system for the period July–December (illustrated in Figure 4), the ranking (best to worst) was 2016, 2015, 2019, 2017, 2020 and then 2018. Table 8 presents the façade PV system's renewable energy generation annually and for the period July–December. For the full year, 2020 ranked third; therefore, the poor performance between July and December can be attributed to weather conditions rather than purely to the expected performance degradation of the façade PV system. As the July–December 2020 period ranks fifth since 2015, the data can be used to predict payback, without being unduly optimistic.

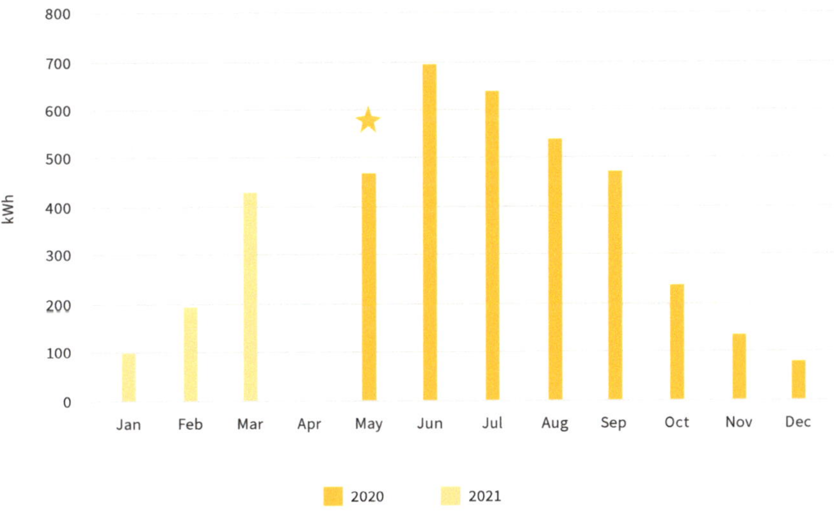

Figure 5. Monthly renewable energy generation of the roof PV system since installation in May 2020. NB: The star highlights that May 2020 data only represents 15 days. Source: Graphic by author.

Table 8. Renewable energy generation of façade PV system (annual and $\frac{1}{2}$ year from July to December).

Year	kWh/year	% Normalised to 2015	Ranking	kWh (½ year)	% Normalised to 2015	Ranking
2015	2021.57	100.0	1	883.58	100.0	2
2016	1945.47	96.2	2	904.05	102.3	1
2017	1751.35	86.6	6	845.44	95.7	4
2018	1752.78	86.7	5	739.57	83.7	6
2019	1893.87	93.7	4	864.57	97.8	3
2020	1926.39	95.3	3	803.49	90.9	5

Source: Table by author.

The roof PV system's generation from 22 June to 21 December 2020 was 2231.2 kWh. To pay back the 25,612 kWh embodied energy of the roof PV system would take 5.6 years. This is close to the expected figure of 5.37 years and well within its service life.

4.1.3. Tesla Powerwall

The energy payback of the Tesla Powerwall is dependent on its ability to capture renewable energy generated by the façade and roof PV systems which would otherwise have been exported or throttled.

There was not a full year of operational data to analyse, and thus the same half-year period as for the roof PV system was analysed. During this period, 21.44 kWh of renewable energy was captured which would otherwise have been exported/throttled (23.26 kWh for the full installation period until 31 March 2021). Based on this, we can assume 42.88 kWh would be captured over a year, resulting in an energy payback in 18.1 years. Although this is beyond the warranty period, the system may still be operating for that period of time. Energy attributed to the Powerwall was not attributed to the roof PV system in order to avoid double counting.

4.2. Carbon Payback

4.2.1. Façade PV

The annual UK electricity grid carbon intensity was taken from Table 5 (European Environment Agency 2020) for 2014–2017, as more detailed data could not be found for this period. For 2018–2021, the monthly data shown in Table 6 (Electricity Info n.d.) were used.

A total of 3.24 tonnes of CO_2 emissions has been avoided by the façade PV system since installation in 2014. The annual CO_2 saving has varied from 768.2 kg CO_2e in 2015 (which combined the highest UK electricity grid carbon intensity with the highest electricity generation) down to 342.2 kg CO_2e in 2020, which had the second highest electricity generation since installation but the lowest electricity grid carbon intensity (as illustrated in Figure 6).

A projection of the future carbon payback based on average energy generation thus far, in combination with an assumed linear progression of the UK grid electricity to 100 g CO_2/kWh by 2030, and then to 10 g CO_2/kWh by 2050, is illustrated in Figure 7. Following this trend, it will take until 2126 to reach a carbon payback of 9216 kg CO_2. This significantly exceeds the expected service life.

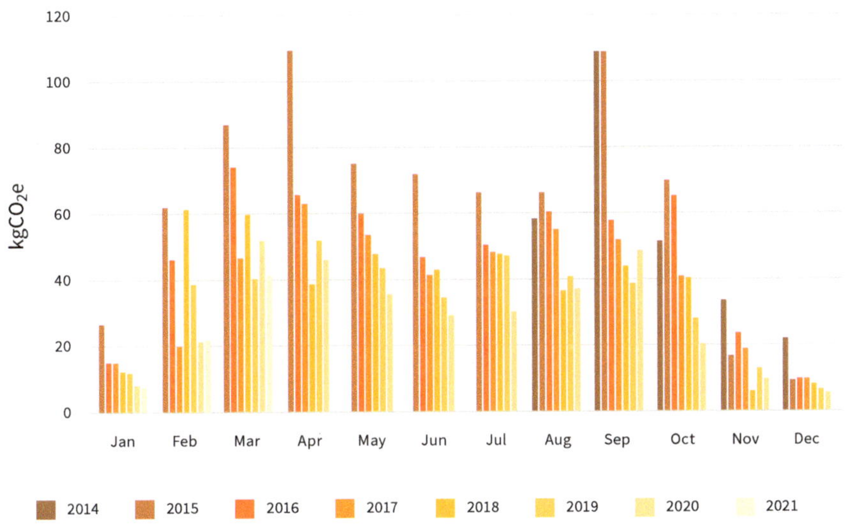

Figure 6. Monthly CO_2 emissions avoided (kg CO_2e) by the façade PV system since installation in 2014. Source: Graphic by author.

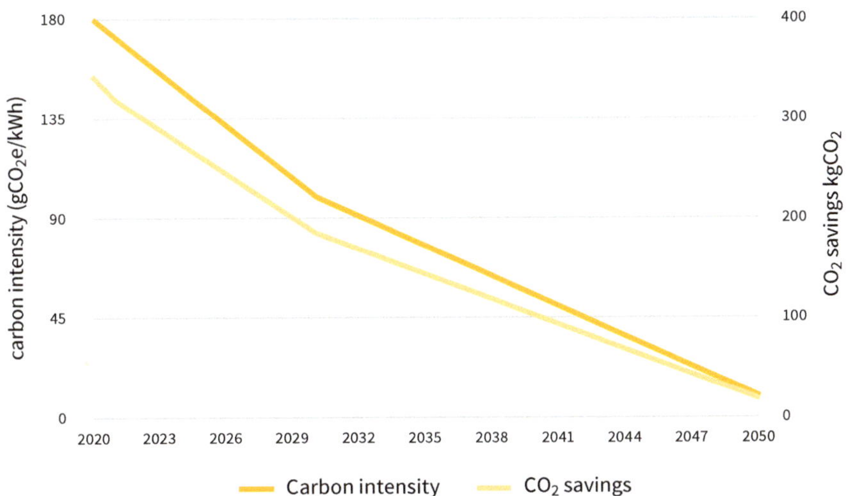

Figure 7. Projected UK electricity grid carbon intensity and annual façade PV system carbon savings until 2050. Source: Graphic by author.

4.2.2. Roof PV

The monthly UK electricity grid carbon intensity was taken from Table 6 (Electricity Info n.d.). The CO_2 emissions avoided by the roof PV system since installation equal 0.71 tonnes CO_2e.

Using the projection of the UK grid carbon intensity illustrated in Figure 7, and assuming annual renewable energy generation by the roof PV system of 4562.4 kWh, it will take until 2036 to reach a carbon payback of 9216 kg CO_2. The carbon payback period has been extended significantly due to sharing the CO_2 saving with the Tesla Powerwall in the first few years when the electricity carbon intensity is higher. Despite this, the roof PV system achieves carbon payback in its expected service lifetime.

4.2.3. Tesla Powerwall

The carbon payback of the Tesla Powerwall is dependent on two elements:

1. Its ability to capture renewable energy which exceeds the current demand of the site for use at another time;
2. Its ability to capture grid electricity with a low carbon intensity for use when grid electricity has a higher carbon intensity (load shifting).

By the end of March 2021, 332.5 kg of CO_2 emissions had been avoided by capturing renewable energy for use at another time. However, half of this is attributed to the roof PV system until the Powerwall's payback is achieved (to avoid double counting); thus, based on the half-year data, 211.85 kg of CO_2 emissions will be avoided annually. This indicates that a significant proportion of the renewable energy generated by the roof PV system is not used immediately. After the Powerwall's payback is achieved, the full CO_2 saving is applied to the roof PV system.

On days when significant renewable generation is expected from the façade and roof PV systems, storage of excess renewable energy should take priority over load shifting. However, from November to February, there is a significant drop in excess renewable energy generation, and load shifting can be utilised instead. The Powerwall has settings which allow it to selectively charge at off-peak times and use the stored energy to minimise grid import at peak times. Although these settings are based around electricity costs, the Octopus tariff's off-peak cost period (00:30–04:30) fits within the lower-carbon intensity period illustrated in Figure 3, while the peak period (16:00–19:00) is associated with a higher carbon intensity (Figure 3).

The carbon saving from load shifting is based on the difference between the average carbon intensity between 16:00 and 19:00 and the carbon intensity of the stored energy while allowing for energy loss between storage and return (referred to as round-trip energy efficiency). Table 9 indicates the monthly energy storage in the Powerwall, the resulting round-trip efficiency and the electricity source. Using the half-year data, the average round-trip efficiency was 90.2%. The round-trip efficiency would be lower for a Powerwall located in a colder place (e.g., outdoors).

Table 9. Energy stored in the Powerwall, round-trip efficiency and electricity source.

Date	From Powerwall (kWh)	Round-Trip Energy Efficiency (%)	Electricity Source
29 April–15 May 2020	11.6	30.1	Façade PV
16–25 May 2020	122.7	86.2	Façade PV + Roof PV
26 May–30 June 2020	492.6	91.1	
July 2020	422.4	90.0	
August 2020	405.5	88.4	Façade PV + Roof PV + Off-peak grid
September 2020	520.7	91.7	
October 2020	497.1	90.3	
November 2020	389.7	89.3	
December 2020	418.1	89.3	Façade PV + Roof PV + Off-peak grid (minimising peak rate import)
January 2021	415.9	89.1	
$\frac{1}{2}$ year (22 June–21 December 2020)	2567 [1]	90.2	-

[1] Data used for calculation. Source: Table by author.

By the end of March 2021, 157.2 kg of CO_2 emissions had been avoided by load shifting off-peak energy for use at another time. Based on the half-year data, 166.4 kg of CO_2 emissions will be avoided annually. However, care should be taken to avoid using load shifting at periods when there is less than a 10% difference in the carbon intensity of off-peak and peak electricity, as the carbon reduction benefit will be outweighed by the energy loss during storage and release.

Combining both elements of carbon saving from the Powerwall results in 378.3 kg CO_2e savings per year. Even allowing for the decreasing carbon intensity of grid electricity in the future, the carbon payback of the Powerwall will be achieved by 2025.

4.3. Financial Payback

4.3.1. Façade PV

There are three elements contributing to the financial payback of the façade PV system:

1. The FiT was 14.9 p/kWh at installation. Retained FiT statements show that the rate was 15.78 p/kWh in 2018, 16.2 p/kWh in 2019 and 16.56 p/kWh in 2020. The UK Retail Prices Index (RPI) (Office for National Statistics n.d.) was used to determine the approximate tariff for 2015–2017.
2. The export payment—5.5 p/kWh for an assumed 50% export.
3. Saving on electricity payments. Retained electricity bills show that the rate (excluding service charge) was as follows:

a.	September 2017–April 2018	12.45 p/kWh + 5% VAT
b.	May–June 2018	12.48 p/kWh + 5% VAT
c.	July–October 2018	13.62 p/kWh + 5% VAT
d.	October 2018–April 2020	13.63 p/kWh + 5% VAT
e.	May–October 2020	13.35 p/kWh + 5% VAT

Prior to 2017, electricity rates were derived based on the Consumer Price Index for electricity (Statista n.d.) relative to the 2017 tariff. At the time of calculation, no evidence could be found of an anticipated increase in electricity costs, meaning the average of 2015–2020 was used for future years.

Financial savings since installation total GBP 3719.38 (approximately GBP 598/year), and this equates to a payback period of 18 years. The expected payback was affected by the lower than expected energy generation. The proportional contributions are as follows: FiT 49.5%, export 8.7% and electricity cost savings 41.8%.

4.3.2. Roof PV

The only contribution to the financial payback of the roof PV system is the saving on electricity payments. Retained electricity bills show that the rate (excluding service charge) was 13.35 p/kWh + 5% VAT. Financial savings since installation total GBP 557.73. Based on the half-year data, the assumed annual renewable energy generation will be 4476 kWh. Relating this to the current electricity tariff, annual savings will be GBP 639.53; however, GBP 105.86 annually is attributed to the Powerwall, resulting in a financial payback of GBP 7803 within 15.4 years.

4.3.3. Tesla Powerwall

As with the carbon payback, the financial payback of the Tesla Powerwall is dependent on two elements:

1. Its ability to capture renewable energy which exceeds the current demand of the site for use at another time;
2. Its ability to capture grid electricity with a low cost (and carbon intensity) for use when grid electricity has a higher cost (and carbon intensity) known as "load shifting".

The renewable energy element of the cost saving is shared equally between the Powerwall and the roof PV system until the roof PV system's payback is achieved, whereon the full cost saving applies to the Powerwall. By the end of March 2021, a GBP 174.00 saving had been made based on storing the roof PV system's generation for later use. Based on the half-year period, the annual cost saving will be GBP 211.71, which will be equally shared between the Powerwall and the roof PV system.

The cost saving from load shifting is based on the Octopus Go tariff which has an off-peak period (00:30–04:30). Accounting for the average 9.8% energy loss during storage and return, the cost saving up to 31 March 2021 was GBP 192.27. Based on the half-year period, the annual cost saving will be GBP 296.29.

Combining both elements of the cost saving from the Powerwall results in GBP 402.15 savings per year, achieving a financial payback of GBP 8040 by 2039.

4.4. Electric Vehicle

The electric vehicle has a 35.5 kWh battery which has an embodied carbon value of 3550 kg CO_2 assuming 100 kg CO_2/kWh (Hausfather 2019).

The monthly distance travelled by the EV, along with average energy efficiency, electricity consumption and carbon emissions, for the period is presented in Table 10, along with a comparison of the carbon emissions which would have resulted from the replaced diesel vehicle (Figure 8).

Table 10. Electric vehicle monthly distance travelled, energy efficiency, electricity consumption and carbon emissions.

Month	Distance (km)	Energy Efficiency (km/kWh)	Electricity Consumption (kWh)	EV Carbon Emissions (kg CO$_2$) [1]	Diesel Carbon Emissions for Same Distance (kg CO$_2$)
October 2020	850	4.82	176.2	28.5	92.6
November 2020	1379	4.67	295.5	52.3	150.3
December 2020	1447	4.50	321.2	59.4	157.7
January 2021	780	4.34	179.4	34.3	85.0
February 2021	943	4.50	209.4	37.5	102.8
March 2021	1337	4.83	276.9	51.2	145.7

[1] Carbon emissions are based on the monthly average for the UK electricity grid. Source: Table by author, adapted from Electricity Info (n.d.).

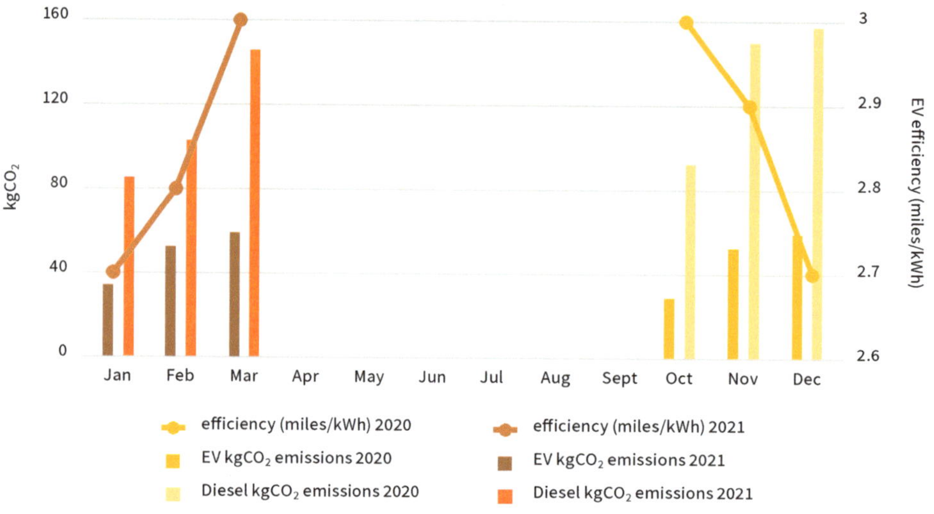

Figure 8. Carbon emission comparison between electric vehicle and diesel vehicle, plus electric vehicle efficiency. Source: Graphic by author.

Based on monthly UK electricity carbon emissions (Table 6), the EV saved 471 kg of CO$_2$ in six months of operation in comparison to the diesel car it replaced. On this basis, the electric vehicle batteries will achieve carbon payback within 7.5 years.

Cost payback has not been considered as it is difficult to assess how much more expensive the EV is than a comparable diesel car, although it is acknowledged that electric variants are generally more expensive than their diesel equivalents. Energy payback is not considered as the car consumes energy.

5. Discussion

The calculated energy, carbon and financial paybacks of each part of the renewable generation and storage system are summarized in Table 11.

Table 11. Calculated energy, carbon and financial paybacks of façade PV system, roof PV system, Tesla Powerwall and EV battery.

Element	Energy Payback (y)	Carbon Payback (y)	Financial Payback (y)
Façade PV	8.0	132.0	18.0
Roof PV	5.6	16.0	15.4
Tesla Powerwall	18.1	5.0	19.0
EV Battery	N/A	7.5	N/A

Source: Table by author.

5.1. Energy Payback

The energy payback of both PV systems is well within the expected service lifetime. The energy payback of the Tesla Powerwall is more complex as its ability to facilitate energy payback relies on the specification of the renewable generation it is partnered with. In this case, the energy payback is 18.1 years, which is significantly beyond the warranty period, but there is potential for the system to still be operating in that time period. It should be considered that the captured energy would have been higher (and energy payback shorter) if the full 6.8 kWp roof PV system had been permitted.

5.2. Carbon Payback

The carbon payback is less certain than the energy payback because the calculations vary depending on the future potential decarbonisation of the UK electricity grid. The decarbonisation of the grid in recent years has had a noticeable impact on the carbon payback of the façade PV system, which was initially expected to achieve carbon payback within 8.6 years.

Assuming that targets to reach 100 g CO_2/kWh by 2030 and then 10 g CO_2/kWh by 2050 are achieved, the roof PV system will achieve carbon payback within its service life. However, the carbon payback of the façade PV system is much longer.

Some people may argue that the reduced carbon intensity of the electricity grid negates the requirement for renewable energy installation. For this reason, it needs to be acknowledged that the decarbonisation of the electricity grid is only possible if renewable energy generation is increased. Approximately 162 TWh of low-carbon generation is required to meet the 100 g CO_2/kWh target by 2030 while covering the planned retiral of nuclear plants and increased electrification of transport and heating (Evans 2020). It should also be considered that the electricity decarbonisation contributes to reduce embodied carbon of new renewable energy generation systems. This trend is already being observed in the decreasing embodied carbon data for batteries (Hausfather 2019).

The carbon payback of the Tesla Powerwall benefitted from its year-round capability to save carbon. In sunny weather, the maximum carbon saving comes from "saving" solar energy for a higher-carbon intensity period. However, even in low-sun conditions during the winter, it is able to powershift lower-carbon intensity electricity generated overnight to a higher-carbon intensity period. This enables a carbon payback of 5 years which is well within its service life.

Under the 2020 electricity grid carbon intensity, the electric vehicle battery is likely to reach carbon payback in less than 8 years (which is the warranty period for the EV battery). There is potential for this to be even earlier as the data analysed do not consider the following factors:

- EV grid charging was scheduled from 00:30 to 04:30. Analysis of 2020 data (as illustrated in Figure 3) has shown that the carbon intensity during this period is nearly 20% lower than that during the rest of the day (202.239 g CO_2 compared to 249.975 g CO_2).
- Not all of the electricity came from the electricity grid. Periods of solar PV generation being stored directly in the car were observed in October 2020, and February, March and April 2021. Unfortunately, the amount could not be logged with the current equipment.
- The period analysed was during cold months which are known to adversely affect vehicle energy efficiency.
- As decarbonisation of the electricity grid improves, the carbon emissions of the electric vehicle will reduce further.

5.3. Financial Payback

The financial payback of the façade PV system will be reached within its service life due to the FiT contribution. The roof PV system was a simpler installation and will reach financial payback within its service life without any subsidy. However, the roof PV system can only reach financial payback in tandem with the Tesla Powerwall; otherwise, much of the energy generated would be exported to the grid with no benefit to the owner. This is why its financial benefit has been shared with the Tesla Powerwall. Even with the payback contribution from the PV system, the Tesla Powerwall will not achieve financial payback within its expected service life. This agrees with the findings in the literature (Zakeri et al. 2021; Sheha and Powell 2018; Gardiner et al. 2020; Dong et al. 2020).

There is an element of uncertainty in the financial payback calculations—particularly over an extended period. There is potential for fluctuation in electricity prices which could significantly extend or reduce the payback period.

5.4. Behaviour

Visual displays were used to ensure that the status of solar generation and the battery can be considered when scheduling use of appliances such as the washing machine in order to minimise grid electricity imports.

6. Conclusions

Although the façade PV system's financial payback was only made viable by the FiT contribution, this subsidy had wider benefits as it transformed the UK market and established PV as a viable renewable energy generation technology in the UK.

The energy, carbon and financial payback for simple roof-installed PV generation covering the base load is clear.

The installation of PV generation beyond the base load can be justified as long as suitable storage capability is also installed to reduce stress on the grid and maximise benefit to the owner. However, this can impact on financial payback, as previously indicated in the literature (Zakeri et al. 2021; Sheha and Powell 2018; Gardiner et al. 2020; Dong et al. 2020). It should be noted that the main benefits of storage installed with PV are to the electricity grid (Zakeri et al. 2021; Sheha and Powell 2018; Jankowiak et al. 2020; Koskela et al. 2019; Gardiner et al. 2020; Dong et al. 2020); however, owners of the storage are not currently rewarded for providing these benefits. In fact, the payback of this system has been adversely affected by the reduction in the intended PV installation from 6.8 to 6.12 kWp (as required by Western Power). The literature has indicated that policy changes which monetise the benefits offered by energy storage to the grid will outperform subsidies (Gardiner et al. 2020).

The ability of a PV system to achieve carbon payback through its own generation will vary depending on the solar efficiency of its installation and future electricity grid decarbonisation. As renewable generation is contributing to decarbonisation of the electricity grid—which, in turn, reduces the embodied carbon of the equipment produced in future—it is important that solar-efficient installations continue.

The carbon payback of the electric vehicle battery is clear and will benefit from future electric grid decarbonisation.

The combination of renewable generation, energy storage and swapping to an electric vehicle is likely to avoid 1655.6 kg CO_2 per year operational emissions at the 2020 UK electric grid carbon intensity. Obviously, this is a small decrease on the national scale, but there is potential to replicate this approach in other buildings.

This study has been limited by a relatively short monitoring period. Future work to review the data after an extended period would be beneficial.

Funding: This research received no external funding.

Acknowledgments: I would like to express my eternal gratitude to my husband, Andrew Earp, for co-funding the PV systems, the Tesla Powerwall and the electric vehicle, as well as for managing the monitoring equipment and prototyping the visual displays. I would like to acknowledge the loan of sensors by HWM.

Conflicts of Interest: The author declares no conflict of interest.

References

Attwood, James. 2021. Analysis: 2020 UK Car Sales Hit 28-Year Low, EV Market Grows Rapidly. Available online: https://www.autocar.co.uk/car-news/industry-news/analysis-2020-uk-car-sales-hit-28-year-low-ev-market-grows-rapidly (accessed on 3 February 2021).

BMRS. n.d.a. Generation by Fuel Type. Available online: https://www.bmreports.com/bmrs/?q=generation/fueltype/current (accessed on 1 February 2021).

BMRS. n.d.b. Rolling System Demand. Available online: https://www.bmreports.com/bmrs/?q=demand/rollingsystemdemand/historic (accessed on 1 February 2021).

Bulb. n.d. About Bulb's Smart Tariff. Available online: https://help.bulb.co.uk/hc/en-us/articles/360017795731-About-Bulb-s-smart-tariff (accessed on 7 April 2021).

Circular Ecology. n.d. Embodied Carbon of Solar PV: Here's Why It Must Be Included in Net Zero Carbon Buildings. Available online: https://circularecology.com/solar-pv-embodied-carbon.html (accessed on 3 February 2021).

Climate Change Committee. 2019a. UK Housing: Fit for the Future? Available online: https://www.theccc.org.uk/publication/uk-housing-fit-for-the-future/ (accessed on 29 April 2021).

Climate Change Committee. 2019b. Net Zero—Technical Annex: Integrating Variable Renewables into the UK Electricity System. Available online: https://www.theccc.org.uk/wp-content/uploads/2019/05/Net-Zero-Technical-Annex-Integrating-variable-renewables.pdf (accessed on 29 April 2021).

Department for Business, Energy & Industrial Strategy. 2021. Energy Trends UK, October to December 2020 and 2020. Available online: https://assets.publishing.service.gov.uk/government/uploads/system/uploads/attachment_data/file/972790/Energy_Trends_March_2021.pdf (accessed on 1 February 2021).

Dong, Siyuan, Enrique Kremers, Maria Brucoli, Rachael Rothman, and Solomon Brown. 2020. Techno-enviro-economic assessment of household and community energy storage in the UK. *Energy Conversion and Management* 205: 112330. [CrossRef]

Electric Vehicle Database. n.d. Honda-E: Battery Electric Vehicle. Available online: https://ev-database.uk/car/1171/Honda-e#:~{}:text=is%2090%20mph.-,Battery%20and%20Charging,on%20a%20fully%20charged%20battery (accessed on 3 February 2021).

Electricity Info. n.d. National Grid Carbon Intensity Archive. Available online: https://electricityinfo.org/carbon-intensity-archive/#data (accessed on 1 February 2021).

Energy-Stats. n.d. Download Historical Pricing Data. Available online: https://www.energy-stats.uk/download-historical-pricing-data/ (accessed on 4 February 2021).

Energy UK. 2016. Pathways for the GB Electricity Sector to 2030. Available online: https://www.energy-uk.org.uk/publication.html?task=file.download&id=5722 (accessed on 8 April 2021).

Engel, Hauke, Patrick Hertzke, and Giulia Siccardo. 2019. Second-Life EV Batteries: The Newest Value Pool in Energy Storage. Available online: https://www.mckinsey.com/industries/automotive-and-assembly/our-insights/second-life-ev-batteries-the-newest-value-pool-in-energy-storage (accessed on 3 February 2021).

European Environment Agency. 2020. Greenhouse Gas Emission Intensity of Electricity Generation. Available online: https://www.eea.europa.eu/data-and-maps/daviz/co2-emission-intensity-6#tab-googlechartid_googlechartid_chart_111_filters=%7B%22rowFilters%22%3A%7B%7D%3B%22columnFilters%22%3A%7B%22pre_config_date%22%3A%5B2018%5D%7D%3B%22sortFilter%22%3A%5B%22ugeo%22%5D%7D (accessed on 7 April 2021).

European Parliament. 2010. Directive 2010/31/EU of the European Parliament and of the Council of 19 May 2010 on the Energy Performance of Buildings (Recast). Available online: https://eur-lex.europa.eu/LexUriServ/LexUriServ.do?uri=OJ:L:2010:153:0013:0035:EN:PDF (accessed on 29 April 2021).

Evans, Simon. 2020. Analysis: UK Low-Carbon Electricity Generation Stalls in 2019. Available online: https://www.carbonbrief.org/analysis-uk-low-carbon-electricity-generation-stalls-in-2019 (accessed on 12 April 2021).

Feng, Liu, and Jeroen C. J. M. van den Bergh. 2020. Differences in CO_2 emissions of solar PV production among technologies and regions: Application to China, EU and USA. *Energy Policy* 138: 111234. [CrossRef]

Finnegan, Stephen, Craig Jones, and Steve Sharples. 2018. The embodied CO_2e of sustainable energy technologies used in buildings: A review article. *Energy and Buildings* 181: 50–61. [CrossRef]

Forfar, John. 2018. Tesla's Approach to Recycling Is the Way of the Future for Sustainable Production. Available online: https://medium.com/tradr/teslas-approach-to-recycling-is-the-way-of-the-future-for-sustainable-production-5af99b62aa0e (accessed on 3 February 2021).

Gardiner, Dan, Oliver Schmidt, Phil Heptonstall, Rob Gross, and Iain Staffell. 2020. Quantifying the impact of policy on the investment case for residential electricity storage in the UK. *Journal of Energy Storage* 27: 101140. [CrossRef]

Government Property Agency. 2020. Net Zero and Sustainability Design Guide—Net Zero Annex. Available online: https://assets.publishing.service.gov.uk/government/uploads/system/uploads/attachment_data/file/925231/Net_Zero_and_Sustainability_Annex_ _August_2020_pdf (accessed on 29 April 2021).

Gridwatch. n.d. GB Electricity National Grid CO_2e Output per Production Type. Available online: https://gridwatch.co.uk/co2-emissions?src=lk01 (accessed on 1 February 2021).

Hausfather, Zeke. 2019. How Electric Vehicles Help to Tackle Climate Change. Available online: https://www.carbonbrief.org/factcheck-how-electric-vehicles-help-to-tackle-climate-change#:~{}:text=The%20IVL%20researchers%20now%20estimate,upper%20bound%20of%20146%20kg.&text=Literature%20review%20of%20lifecycle%20greenhouse, per%20kWh%20of%20battery%20capacity (accessed on 3 February 2021).

HM Government. 2017. Upgrading Our Energy System: Smart Systems and Flexibility Plan. Available online: https://assets.publishing.service.gov.uk/government/uploads/system/uploads/attachment_data/file/633442/upgrading-our-energy-system-july-2017.pdf (accessed on 20 November 2021).

HM Government. 2021. Heat and Buildings Strategy. Available online: https://assets.publishing.service.gov.uk/government/uploads/system/uploads/attachment_data/file/1032119/heat-buildings-strategy.pdf (accessed on 19 November 2021).

Home and Energy Scotland. n.d. Home and Energy Scotland Loan Overview. Available online: https://www.homeenergyscotland.org/find-funding-grants-and-loans/interest-free-loans/overview/ (accessed on 19 November 2021).

Intergovernmental Panel on Climate Change. 2014. *Climate Change 2014 Synthesis Report Summary for Policymakers*. Paris: IPCC.

Ito, Masakazu, Mitsuru Kudo, Masashi Nagura, and Kosuke Kurokawa. 2011. A comparative study on life cycle analysis of 20 different PV modules installed at the Hokuto mega-solar plant. *Progress in Photovoltaics* 19: 878–86. [CrossRef]

Jankowiak, Corentin, Aggelos Zacharopoulos, Caterina Brandoni, Patrick Keatley, Paul MacArtain, and Neil Hewitt. 2020. Assessing the benefits of decentralised residential batteries for load peak shaving. *Journal of Energy Storage* 32: 101779. [CrossRef]

Koskela, Juha, Antti Rautiainen, and Pertti Järventausta. 2019. Using electrical energy storage in residential buildings—Sizing of battery and photovoltaic panels based on electricity cost optimization. *Applied Energy* 239: 1175–89. [CrossRef]

Kristjansdottir, Torhildur F., Clara S. Good, Marianne R. Inman, Reidun D. Schlanbusch, and Inger Andresen. 2016. Embodied greenhouse gas emissions from PV systems in Norwegian residential Zero Emission Pilot Buildings. *Solar Energy* 133: 155–71. [CrossRef]

Kurland, Simon D. 2019. Energy use for GWh-scale lithium-ion battery production. *Environmental Research Communications* 2: 012001. [CrossRef]

Mair, Jason, Kiti Suomalaienen, David M. Eyers, and Michael W. Jack. 2021. Sizing domestic batteries for load smoothing and peak shaving based on real-world demand data. *Energy and Buildings* 247: 111109. [CrossRef]

Office for National Statistics. 2019. Road Transport and Air Emissions. Available online: https://www.ons.gov.uk/economy/environmentalaccounts/articles/roadtransportandairemissions/2019-09-16 (accessed on 29 April 2021).

Office for National Statistics. n.d. Retail Prices Index: Long Run Series: 1947–2019. Available online: https://www.ons.gov.uk/economy/inflationandpriceindices/timeseries/cdko/mm23 (accessed on 5 February 2021).

Pagliaro, Mario, and Francesco Meneguzzo. 2019. Lithium battery reusing and recycling: A circular economy insight. *Heliyon* 5: E01866. [CrossRef] [PubMed]

Reddaway, Andrew. 2016. Energy Flows: How Green Is My Solar? Available online: https://renew.org.au/renew-magazine/solar-batteries/energy-flows-how-green-is-my-solar/#:~{}:text=Embodied%20energy,-Manufacturing%20a%20solar&text=Estimating%20the%20energy%20use%20is,including%20an%20allowance%20for%20transport (accessed on 29 April 2021).

Sheha, Moataz N., and Kody M. Powell. 2018. An economic and policy case for proactive home energy management systems with photovoltaics and batteries. *The Electricity Journal* 32: 6–12. [CrossRef]

Statista. n.d. Consumer Price Index (CPI) of Electricity Annually in the United Kingdom (UK) from 2008 to 2020. Available online: https://www.statista.com/statistics/286548/electricity-consumer-price-index-cpi-annual-average-uk/#:~{}:text=This%20statistic%20shows%20the%20Consumer,electricity%20was%20measured%20at%20124.3 (accessed on 7 April 2021).

Üçtuğ, Fehmi G., and Adisa Azapagic. 2018. Environmental impacts of small-scale hybrid energy systems: Coupling solar photovoltaics and lithium-ion batteries. *Science of The Total Environment* 643: 1579–89. [CrossRef] [PubMed]

Uddin, Kotub, Rebecca Gough, Jonathan Radcliffe, James Marco, and Paul Jennings. 2017. Techno-economic analysis of the viability of residential photovoltaic systems using lithium-ion batteries for energy storage in the United Kingdom. *Applied Energy* 206: 12–21. [CrossRef]

Vehicle Certification Agency. n.d. Car Fuel Data, CO_2 and Vehicle Tax Tools. Available online: https://carfueldata.vehicle-certification-agency.gov.uk/ (accessed on 31 January 2021).

Wong, J. H., M. Royapoor, and C. W. Chan. 2016. Review of life cycle analyses and embodied energy requirements of single-crystalline and multi-crystalline silicon photovoltaics systems. *Renewable and Sustainable Energy Reviews* 58: 608–18. [CrossRef]

World Green Building Council. n.d. The Net Zero Carbon Buildings Commitment. Available online: https://www.worldgbc.org/thecommitment#:~{}:text=The%20Net%20Zero%20Carbon%20Buildings%20Commitment%20(the%20Commitment)%20challenges%20business,carbon%20in%20operation%20by%202050 (accessed on 29 April 2021).

World Nuclear Association. n.d. Comparison of Lifecycle Greenhouse Gas Emissions of Various Electricity Generation Sources. Available online: http://www.world-nuclear.org/uploadedFiles/org/WNA/Publications/Working_Group_Reports/comparison_of_lifecycle.pdf (accessed on 1 February 2021).

Zakeri, Behnam, Samuel Cross, Paul E. Dodds, and Giorgio C. Gissey. 2021. Policy options for enhancing economic profitability of residential solar photovoltaic with battery energy storage. *Applied Energy* 290: 116697. [CrossRef]

Part 2: Technology and Carbon Economy

Sustainable Energy Future with Materials for Solar Energy Collection, Conversion, and Storage

Juvet Nche Fru, Pannan I. Kyesmen and Mmantsae Diale

1. Introduction

During the last 12 years, halide perovskites (HaP) have emerged as the fastest-emerging third-generation solar cell material, competing well with silicon and other thin-film technologies. The power conversion efficiency (PCE) of these easy-to-process and low-cost solar cells has risen from 3.58% to 25.6% from 2009 to 2021 (Zheng and Pullerits 2019; Jeong et al. 2021), already exceeding that of commercially available thin-film photovoltaics (PV) and rivaling that of the best single-junction silicon solar cells. Moreover, HaP have a wide range of applications including solar cells, water splitting and carbon dioxide (CO_2) reduction, light-emitting diodes, photodiodes in photodetectors, gas sensing, lasers and solar batteries. However, instability, toxicity of lead and solvents, poor laboratory-to-laboratory reproducibility, and scalability remain bottlenecks blocking the commercialization of this technology. Among all these difficulties, instability and short lifespan are the major impediments to the commercialization of HaP solar cells. It is crucial to understand the causes of instabilities and develop strategies that will stabilize this low-cost technology and facilitate its transfer to the market.

The production of solar hydrogen by water splitting through the PEC process was initially demonstrated by Fujishima and Honda in 1972 (Fujishima and Honda 1972). Fujishima and Honda used titanium dioxide (TiO_2) as a semiconductor photoanode and achieved a low quantum efficiency of 0.1%. A contributing factor to the low quantum efficiency was the inability of TiO_2 to absorb photons in the visible spectrum due to its large bandgap of 3.0 eV. The use of nitrides, chalcogenides, metal sulfides, and metal oxides as photoelectrodes for PEC water splitting has been explored for decades (Wang et al. 2017; Tee et al. 2017). Despite several decades spent in search of suitable materials, no single semiconducting material has been found to fulfill all the required performance benchmarks for efficiency, durability, and cost (Shen et al. 2014). Metal oxides are among the most promising candidates for use as photoanodes in PEC devices for hydrogen production. Their low cost, ease of preparation, lattice manipulation flexibility, and stability in the PEC environment make them attractive (Eftekhari et al. 2017).

Popular metal oxide photoelectrodes for water splitting are TiO_2 (Eidsvåg et al. 2021), $\alpha\text{-}Fe_2O_3$ (Kyesmen et al. 2021), bismuth vanadate ($BiVO_4$), and tungsten trioxide (WO_3) (Kafizas et al. 2017). Among these, $\alpha\text{-}Fe_2O_3$ is a promising material for use as a photoelectrode in PEC water splitting due to its low bandgap, availability, low cost, non-toxicity, and stability in aqueous environments. It can absorb light in the visible region due to its bandgap of ~2.0 eV and promises a maximum theoretical photocurrent and solar-to-hydrogen (STH) efficiency of ~14 mA/cm^2 and ~17%, respectively (Dias et al. 2014; Murphy et al. 2006). In addition, $\alpha\text{-}Fe_2O_3$ is the most common crystal structure of the oxides of iron, and it is easy to process (Yilmaz and Unal 2016). However, the efficiency of $\alpha\text{-}Fe_2O_3$ is yet to attain the theoretically predicted value due to its poor conductivity, high electron–hole recombination, inefficient charge separation (Lee et al. 2014; Xi and Lange 2018), high overpotential, and low absorption coefficient, requiring films with a thickness of over 400 nm for sufficient photon utilization (Sivula et al. 2011). Numerous approaches have been employed in dealing with the challenges associated with the use of $\alpha\text{-}Fe_2O_3$ for PEC water splitting. These strategies include nanostructuring (Ito et al. 2017), doping (Feng et al. 2020), formation of heterostructures (Natarajan et al. 2017), the use of co-catalysts (Eftekhari et al. 2017), plasmonic enhancement effects (Archana et al. 2015), and the use of light-harvesting bio-molecules (Ihssen et al. 2014).

In this chapter, we present the challenges of using HaP and $\alpha\text{-}Fe_2O_3$ for the direct conversion of solar energy into electricity and hydrogen fuels, respectively, with a special focus on the up-to-date strategies that have been engaged towards overcoming them. The instability of perovskite solar cells is influenced by the Goldschmidt tolerance, chemical composition, and defects in halide perovskites. The use of additives to achieve large grain sizes with few grain boundaries and to passivate the surface and boundaries of HaP is effective in improving the stability of HaP solar cells. In addition, protecting back-metal contacts from reacting with the halide perovskites, passivation of 2D perovskites to form 2D/3D mixed-dimensional perovskites, encapsulation of the devices/modules, and use of MA-free perovskites as strategies for the long-term stability and lifetime of perovskite solar cells are presented. For $\alpha\text{-}Fe_2O_3$ films, the simultaneous engagement of strategies such as nanostructuring, doping, formation of heterostructures, use of co-catalysts, and plasmonic enhancement effects has shown great promise in enhancing their photocatalytic hydrogen production. The concurrent use of multiple strategies for the enhancement of the solar-to-hydrogen conversion (STH) efficiency of α-Fe$_2$O$_3$-based photoanodes is mostly implemented through the systematic use of interface engineering. More research is still needed to realize the anticipated commercialization of solar hydrogen production and photovoltaic technologies using α-Fe$_2$O$_3$ and halide perovskites, respectively.

2. Developments Towards Sustainability of Perovskite Solar Cells

2.1. Stability of Perovskite Solar Cells

Perovskite solar cells/panels in operation must withstand eternal environmental conditions including heat, moisture, oxygen, hail and external stresses such as heat/cold cycles, light/dark cycles. Stability can be regarded as the ability to maintain constant performance while operating under these conditions.

2.1.1. Goldschmidt's Tolerance Factor and Intrinsic Structural Stability of 3D HaP

Three-dimensional (3D) HaP solar cells are highly efficient but very unstable. In the AMX$_3$ form for 3D halide perovskites, A stands for a monovalent cation such as cesium (Cs$^+$), methylammonium (CH$_3$NH$_3$$^+$, MA), and formamidinium (H$_2$NCHNH$_2$$^+$, FA), M represents a divalent cation such as lead (Pb^{2+}) and tin (Sn^{2+}), and X is a halide anion such as bromide (Br$^-$), iodide (I$^-$), and chloride (Cl$^-$). Some 3D HaP have mixtures of the different A-cations, M-cations, and/or X-anions. The cubic perovskite crystal structure has an M-cation in the 6-fold coordination position, enclosed by a corner-linked octahedron of X-anions, called the MX$_6$ octahedral framework, and A-cations in the 12-fold coordination positions, as shown in Figure 1a. The size of the A-cation is larger than that of the M-cation, large enough to fit into the 12-fold coordinated voids of the MX$_6$ octahedral inorganic framework, to maintain the cubic symmetry, as shown in Figure 1b. HaP can reversibly transition between cubic, tetragonal, and orthorhombic crystal structures at different temperatures (Thomson 2018).

The ideal cubic symmetry of the 3D HaP structure is normally distorted in practice. Possible distortions include M-cation displacement from its central position and tilting of the MX$_6$ octahedron, depending on the sizes of the A-cation and M-cation. The degree of distortion is determined by Goldschmidt's tolerance factor (t), given by Equation (1),

$$t = \frac{\sqrt{2}\,(R_A + R_X)}{2\,(R_M + R_X)} \tag{1}$$

where R_A is the ionic radius of the A-cation, R_M is the ionic radius of the M-cation, and R_X is the ionic radius of the X-anion. It can also predict whether a combination of anions and cations will form a stable HaP structure. For the ideal cubic symmetry, the ionic size requirement for stability is quite stringent, and the A-cation and M-cation adjust their equilibrium bond distances to the X-anions, without distortion of the unit cell, such that $t = 1$. The tolerance factor ranges from 0.8 to 1.0 for practical HaP because the cubic symmetry is distorted slightly to accommodate a wide range of cations and anions. For instance, the MX_6 octahedron may distort by tilting to reduce the coordination number from 12, so that a smaller sized A-cation can be accommodated, thereby decreasing t.

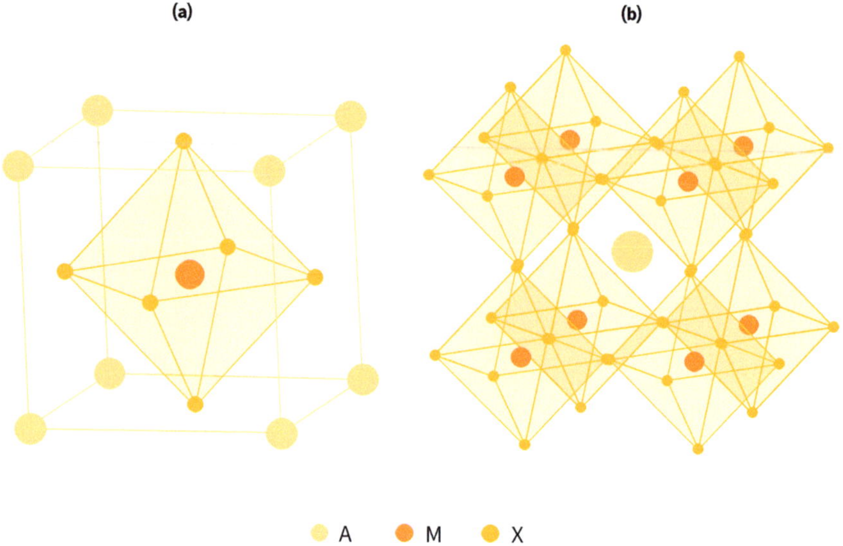

Figure 1. Cubic crystal structure (**a**) and A-Cation in 12-fold coordinated voids of the MX_6 octahedral inorganic framework (**b**) of HaP with general formula AMX_3. Source: Liu et al. (2015b).

2.1.2. Environmental/External Factors Responsible for the Degradation of HaP Solar Cells

Heat, light, moisture, oxygen and electrical bias, are among the environmental factors responsible for degradation of HaP. The action of light and heat on methylammonium lead tri-iodide may cause the evaporation of volatile components such as ammonia (NH_3) and iodine gas (I_2) (Juarez-Perez et al. 2018). As a result, the halide perovskite is irreversibly degraded/decomposed. Light soaking of HaP can cause negative effects such as ion migration (Zhao et al. 2017), halide segregation (Hoke et al. 2015), and photodecomposition (Kim et al. 2018).

2.1.3. Impact of Defects on the Stability of Perovskite Materials

The presence of defects in HaP reduces the charge carrier lifetime and impacts stability. Defects can be located at the interface between the active layers or in the bulk of the halide perovskites. These defects can be point (zero dimensional), line (one dimensional), surface (two dimensional), and volume (three dimensional) defects. Point defects in the widely studied $MAPbI_3$ include native defects such as positive iodine vacancies ($I_V{}^+$), negative iodine vacancies ($I_V{}^-$), neutral iodine vacancies (I_V), negative lead vacancies ($Pb_V{}^{-2}$), positive lead interstitials ($Pb_i{}^{2+}$), iodide interstitials (I_i), positive methylammonium interstitials ($MA_i{}^+$), negative methylammonium vacancy ($MA_V{}^-$), and impurities such as Au interstitials (Yang et al. 2016; Sherkar et al. 2017; Motti et al. 2019). Yang and co-workers (Yang et al. 2016) showed that the bulk $I_V{}^+$ have low formation energies, low diffusion barriers, and fast hopping rates, making them primarily responsible for ionic conductivity in $MAPbI_3$. They also showed that the diffusion barrier and formation energy of gold (Au) interstitial impurities in $MAPbI_3$ are low, leading to possible diffusion of Au into $MAPbI_3$ devices with biased $Au/MAPbI_3$ interfaces. Meanwhile, defects such as $Pb_V{}^{-2}$, $Pb_i{}^{2+}$, and $MA_V{}^+$ have very high activation energies, implying that their formation may require very high temperatures or strong irradiation conditions to form, thus not likely participating in the defects (Motti et al. 2019). Cation substitutions such as MA_{Pb} and Pb_{MA} and substitution anti-sites including MA_I, Pb_I, I_{MA}, and I_{Pb} are also present in $MAPbI_3$ (Jin et al. 2020; Yang et al. 2017b). The nature of Schottky defects and Frenkel-type defects has also been studied in HaPs. Dewinggih and co-workers (Dewinggih et al. 2017) showed that iodine vacancy/interstitial ($I_V{}^+/I_V{}^-$) Frenkel pair trapping centers are abundant in $MAPbI_3$ and are annihilated under illumination conditions which increases photoluminescence quantum efficiency. Kim and co-workers (Kim et al. 2014) showed that the formation energies of Schottky defects (neutral vacancy pairs) such as PbI_2 and MAI in $MAPbI_3$ are relatively low. Fortunately, these defects are not trap states that can reduce the carrier lifetime. A Schottky couple in HaP has very low formation energies and originates from halide vacancy coupling with the metal vacancies (Motti et al. 2019).

Planar defects include grain boundaries (GBs), surfaces or perovskite/transport layer interfaces, stacking faults, and twin boundaries. GBs are interfaces between two grains in polycrystalline materials, as shown in Figure 2a. It has been shown that degradation in HaP starts at the surface and grain boundaries (Shao et al. 2016). This is because GBs are sources of high defect densities, trap accumulation sites, infiltration sites for water vapor, and fast pathways for ion migration due to reduced steric hindrance (Shao et al. 2016). Grain boundaries absorb moisture and oxygen from the environment and cause HaP degradation (Wu et al. 2021). The transformation of the perovskite phase to a non-perovskite phase is initiated at boundaries which are active sites accumulating chemical species (Yun et al. 2018). GBs serve as trapping centers for charge carriers, leading to non-radiative recombination that reduces the carrier lifetime, and also causing hysteresis in the current–voltage characteristic (Uratani and Yamashita 2017). DeQuilettes and co-workers (DeQuilettes et al. 2015) measured the photoluminescence intensities and carrier lifetimes from different grains and grain boundaries of the same $MAPbI_3$ thin film. They concluded that grain boundaries are dimmers and show the fastest non-radiative recombination. The surfaces of HaP are also defective, as shown in Figure 2b. They contain a large number of charged defects (Zhang et al. 2019b) including iodine vacancies (Wu et al. 2020), X-terminating surfaces with nonstoichiometric compositions (Qiu et al. 2020), and improper bonding: (110)-X_2 halide surfaces with a large number of broken bonds (Jain et al. 2019; Kong et al. 2016), and Pb dangling bonds (Kong et al. 2016). SRH recombination at interfaces with the transport layers is the dominant loss mechanism in perovskite solar cells (Sherkar et al. 2017). With regard to stacking faults, they occur in crystals characterized by a periodic sequence of atomic planes due to an interruption in the typical regular arrangement. Song and co-workers (Song et al. 2015) showed that $MAPbI_3$ phases with I/Pb ratios ranging from 3.2 to 3.5 form stacked perovskite sheets with a large amount of stacking faults, whereas thin films with I/Pb ratios ranging from 2.9 to 3.1 form the conventional 3D perovskite with few stacking faults (alpha phase). First principle studies of the electronic properties of {111} twin boundaries in mixed HaP containing FA, Cs, Br, and I revealed that twin boundaries in these perovskites are nucleation sites for I-rich and Cs-rich formation, which are hole traps and can cause electron–hole recombination, leading to a loss in V_{oc} (Mckenna 2018). Direct imaging using TEM has revealed twin boundaries in a $MAPbI_3$ thin film range from 100 to 300 nm wide with twin boundaries parallel to $\{112\}_t$ (Rothmann et al. 2017). By varying the anti-solvent during deposition, Tan and co-workers (Tan et al. 2020) were able to change the defect density of the (111) twin boundary for $Cs_{0.05}FA_{0.81}MA_{0.14}PbI_{2.55}Br_{0.45}$ mixed perovskite to establish the relationship with PCE. It has been shown that recombination centers limiting charge carrier lifetimes in HaP are preferentially located close to the surface rather than in the bulk of the crystal (Stewart et al. 2016).

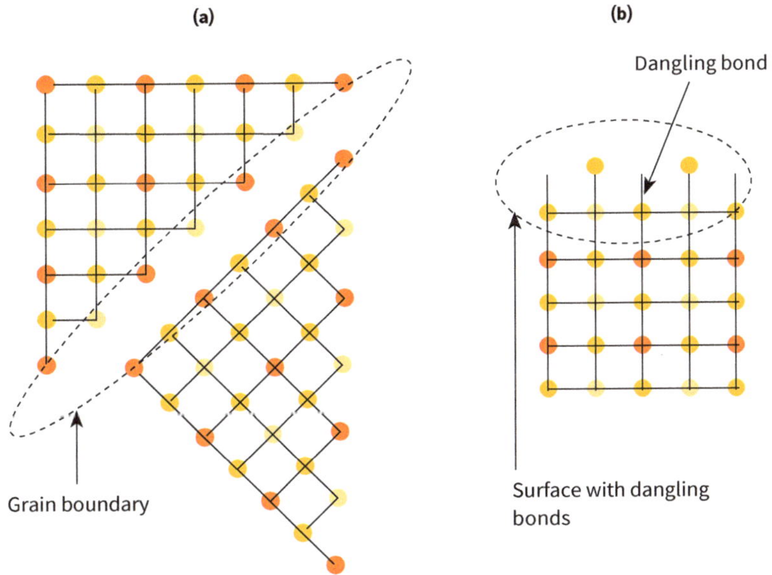

Figure 2. Schematic diagram showing grain boundaries (**a**) and surface defects (dangling bonds) (**b**) in HaP. Source: Graphic by authors.

2.1.4. The Reaction of HaP with Metal Back-Electrodes

Another primary source of instability is caused by the reaction of HaP with widely used metal electrodes when in direct or indirect contact. Gold (Au), silver (Ag), and copper (Cu) are widely preferred as back-electrodes in perovskite solar cells due to their high conductivity. Indirect contact occurs when the metal diffuses through the hole and electron transport layers into the active layer of the device and reacts with the perovskite to form insulating metal halide species or defect states in the bulk or at the surface (Domanski et al. 2016), reducing the thermal stability of the device (Boyd et al. 2018; Domanski et al. 2016). Halide can also diffuse out of the active layer and make contact with the electrodes. For instance, corrosion of a silver (Ag) electrode due to the reaction with diffused hydrogen iodide (HI), produced during the decomposition of a $MAPbI_3$ absorber of an encapsulated solar cell, has been shown to speed up the degradation of $MAPbI_3$ (Han et al. 2015). Wang and co-workers (Wang et al. 2018) showed that $MAPbI_3$ reacted rapidly with Ag electrodes, and the reaction was driven by diffused iodide (I^-) ions which caused corrosion of the electrode. Gold electrodes corrode rapidly due to the aggressive chemical interaction between gold and highly reactive iodine-containing by-products formed in the course of perovskite decomposition under illumination with intense

UV radiation (Shlenskaya et al. 2018). In addition, metal electrodes can diffuse into the HaP active layer through the hole transport layer or electron transport layer, leading to degradation of the perovskite (Ming et al. 2018; Zhao et al. 2016b). In $CH_3NH_3PbI_3$ devices with Al electrodes and the presence of moisture, Al rapidly reduces Pb^{2+} to Pb^0 and converts $CH_3NH_3PbI_3$ to $(CH_3NH_3)_4PbI_6 \cdot 2H_2O$ and then to CH_3NH_3I (Zhao et al. 2016b).

Electrodes are deposited directly on perovskite in hole conduction layer-free solar cells (Asad et al. 2019). The perovskites are deposited directly on metal electrodes in cells without charge transport layers (Lin et al. 2017). Metal electrodes also make direct contact with HaP in Schottky diodes, resistive switching devices, and photodetectors. The electrodes of Schottky diodes, resistive switching devices, and photodetectors also directly contact HaP (Li et al. 2019; Kang et al. 2019). Halide perovskite becomes unstable when in direct contact with metal electrodes. We have shown that methylammonium lead tri-bromide ($MAPbBr_3$) perovskite grains delaminate rapidly on Al-coated substrates as opposed to Au, Ag, Au-Zn, and Sn substrates (Fru et al. 2021). The direct contact of Ag with perovskites also speeds up their degradation, leading to a loss in organic cations (Svanström et al. 2020). These results are important in the selection of electrodes for stable charge transport layer-free solar cells.

2.2. HaP Panel/Module Lifetime

A sustainable transfer of the perovskite solar cell technology from the laboratory to market requires the module/panel lifetime to be greater than 20 years. A widely used definition of module lifetime, known as the T_{80} lifetime, is the time taken for its efficiency to decrease by 20% of the initial value (He et al. 2020). The failure of a module in photovoltaic technology is determined using the T_{80} lifetime. The T_{80} lifetime is calculated using Equation (2),

$$T_{80} = \frac{20\%}{Degrad_{rate}} \tag{2}$$

The median degradation rate ($Degrad_{rate}$) of commercially available solar modules ranges from 0.36%/year for monocrystalline silicon to 0.96%/year for copper indium gallium selenide (CIGS), providing T_{80} lifetimes of over 55.6 and 20.8 years, respectively. These degradation rates were determined from a field test with solar modules operating under normal working conditions. However, the average degradation rate of perovskite solar modules is 66%/year, corresponding to an average lifetime of 0.30 years (3.6 months). These results indicate that perovskite solar modules are very unstable under real operating conditions, and intensive research is needed to commercialize the technology. The lifetime of a solar panel/module is highly correlated with the stability of the constituent solar cells.

2.3. Improving the Stability of Perovskite Solar Cells/Panels

Various strategies have been employed to improve the stability of perovskite solar cells. These techniques can be grouped into stable materials synthesis, additives and passivation, alternative robust functional layers, encapsulation, and engineering of 2D and 2D/3D mixed-dimensional perovskites.

2.3.1. Towards Compositional Stability 3D Halide Perovskite Materials

Degradation due to light and heat is mitigated by improvement in the HaP material and interfaces of the solar cell. Careful selection of the organic cation in HaP is necessary to prevent irreversible degradation (decomposition) under the action of heat and light (Juarez-Perez et al. 2018). This decomposition is mainly due to the release of volatile components in MA-containing perovskites. Thus, it has been shown that going MA free will produce inherently stable perovskites (Turren-Cruz et al. 2018). An alternative organic cation for 3D HaP is formamidinum (FA). The high enthalpy and activation energy needed for its decomposition make FA more resistant to thermal decomposition and produce more thermally stable perovskites than MA (Juarez-Perez et al. 2019). However, FA-based perovskites lack phase stability under humidity and thermal stress (Chen et al. 2021). Much effort is being directed towards the stabilization of the black phase of FA-based perovskites through additives, doping, alloying, interfacial engineering, etc. (Chen et al. 2021). The use of a low-vapor pressure inorganic cation such as cesium (Cs) as a substitute for high-vapor pressure MA leads to a more stable completely inorganic HaP. By mixing the A-site cations, composition stability can also be achieved.

2.3.2. Additives and Passivation

Figure 3a,b show the schematic diagrams of defective and passivated surfaces. Most attempts to mitigate these surface defects involve using different additives that will either improve the film morphology by increasing the grain size to reduce the number of grain boundaries or cause surface passivation (Zhang et al. 2016a). Passivation can be conducted by using various additives including small molecules (Xu et al. 2016), polymers (Dunn et al. 2017), ligands (Zhang et al. 2019a), perovskite quantum dots (Zheng et al. 2019), and 2D perovskites (Rahmany and Etgar 2021). The effect of grain boundaries on the lifetime of charge carriers has been reduced by passivation of the perovskite surface with Lewis acid additives such as 1,2-ethanedithiol (Stewart et al. 2016) and Lewis base additives (Noel et al. 2014). These studies suggested that the Lewis bases donate electrons to surface traps, thus preventing them from capturing charge carriers, while the Lewis acids donate protons, as shown in Figure 3b. Surface treatment by post-deposition of a variety of Lewis bases (electron-donating molecules) and surface ligands passivates surface defects, thereby reducing non-radiative recombination. The presence of excess PbI_2 between grain boundaries also has a passivation effect (Chen et al. 2014). The addition of an optimum amount of potassium iodide (KI) in triple-cation $(Cs_{0.06}FA_{0.79}MA_{0.15})Pb(I_{0.85}Br_{0.15})_3$ perovskite reduces non-radiative losses and photoinduced halide ion migration by passivation of the perovskite film and interfaces (Abdi-jalebi et al. 2018). This is achieved by the excess iodide from KI compensating for any halide vacancies (trap states). At the same time, potassium ion selectively depletes bromide from the crystal, thereby reducing trap states that result from bromide-rich perovskites. The formation of benign (potassium-rich, halide-sequestering species) from excess halides at the grain boundaries and interfaces immobilizes halide ion migration. The addition of a strong electron acceptor of 2,3,5,6-tetrafluoro7,7,8,8-tetracyanoquinodimethane (F4TCNQ) into the perovskite functional layer fills grain boundaries, thus reducing metallic lead defects and iodide vacancies significantly (Liu et al. 2018). Excess MAI intrinsically passivates the surface of $MAPbI_3$ films, leading to a reduced surface recombination velocity and an improved total carrier lifetime (Yang et al. 2017a). Additives such as sulfonated carbon nanotubes (Zhang et al. 2016a), Lewis bases such as urea and thurea (Hsieh et al. 2018), and Lewis acid−base adducts (for example, the PbI_2 adduct with the O-donor DMSO is excellent for improving grain size in $MAPbI_3$ and PbI_2 adducts, while the S-donor thiourea is excellent for $FAPbI_3$) (Lee et al. 2015) mitigate defects by producing larger grains with fewer grain boundaries. The addition of sulfonated carbon nanotubes also passivates perovskite by filling grain boundaries (Zhang et al. 2016a). Other Lewis bases such as thiophene and pyridine passivate the perovskite surface by donating an electron to under-coordinated Pb atoms present in the crystal (Noel et al. 2014). Fullerenes (PCBM) deposited on the top of the perovskite have a passivation effect which reduces photocurrent hysteresis and the trap density (Shao et al. 2014). Fang and co-workers (Fang et al. 2020) showed that the 4-fluorophenylmethylammonium-trifluoroacetate additive passivates both uncoordinated lead and halide ions in the mixed-cation mixed HaP $FA_{0.33}Cs_{0.67}Pb(I_{0.7}Br_{0.3})_3$. This is possible because the trifluoroacetate anion binds with the lead cation, and the 4-fluorophenylmethylammonium cations bind with the halide ion. This dual passivation suppressed hysteresis, halide segregation, and ion migration, leading to an improvement in the operational lifetime of light-emitting diodes from 1.0 to 14.0 h. Qiao and co-workers (Qiao et al. 2019) showed that alkali metals mitigate I_i defects in two ways: by increasing their formation energy, thus reducing their concentration, and binding strongly to them, thereby eliminating mid-gap states that act as traps for electrons and holes, thus increasing the carrier density and extending the carrier lifetimes significantly.

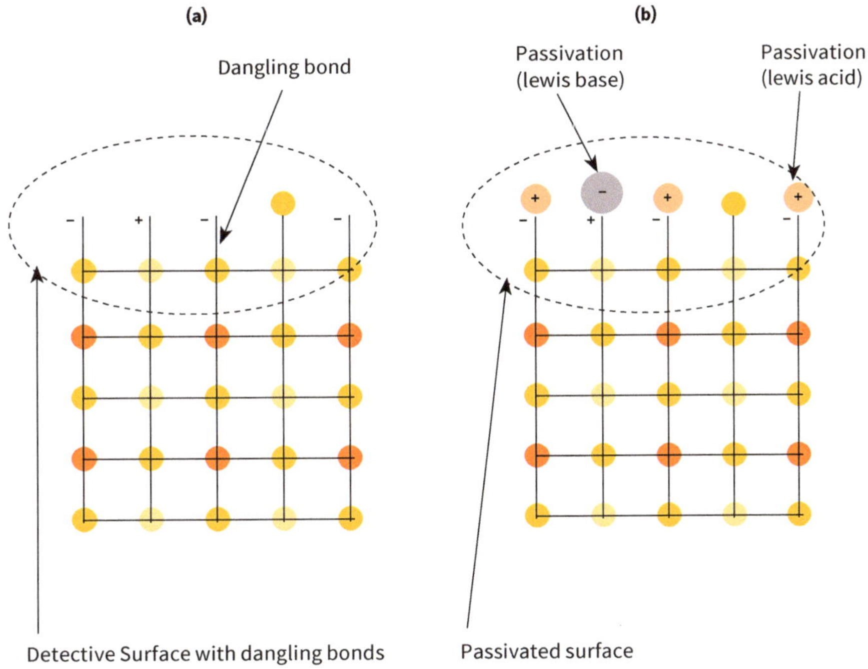

Figure 3. The schematic diagrams of defective (**a**) and passivated (**b**) surfaces. Source: Graphic by authors.

2.3.3. Encapsulation

Encapsulation is an important method to solve instability problems, prevent the leakage of toxic and water-soluble lead compounds to the environment, and help the perovskite solar module to pass the hail impact test (He et al. 2020). In addition, it prevents contact with ambient air, prevents leakage of volatile components, and reduces moisture and heat degradation. Table 1 describes various techniques and materials for encapsulating HaP solar cells. High-performance encapsulation materials should be easy to process and chemically inert and have a low oxygen transmission rate, low water vapor transmission rate, high dielectric constant, resistance to UV and thermal oxidation, high adhesion to perovskite solar modules, similar coefficients of thermal expansion to perovskite solar cell materials, and high mechanical impact strength (Griffini and Turri 2016; Aranda et al. 2021). The techniques used for perovskite solar cells include rigid glass–glass encapsulation, ultra-thin flexible glass sheet encapsulation, polymeric laminates, thin-film barrier-coated webs, and thin-film encapsulation (TFE).

Table 1. Summary of encapsulation methods and materials.

Encapsulation Method	Description	Materials
Glass–glass encapsulation (rigid, widely used, straightforward, very effective, very affordable, incompatible with flexible devices)	The device is sandwiched between two rigid glass sheets using thermo-curable adhesives or UV-curable sealants. Edges are sealed with sealants to prevent ingress of oxygen and moisture.	Examples of thermo-curable adhesives include ethylene-vinyl acetate (EVA) (Bush et al. 2017), surlyn ionomer (Cheacharoen et al. 2018), and polyisobutylene (PIB) (Shi et al. 2017). UV-curable adhesives include epoxy resin (Mansour Rezaei Fumani et al. 2020; Ierides et al. 2021). Edge sealants include butyl rubber and PIB (Vidal et al. 2021).
Ultra-thin flexible glass sheet encapsulation (most recent, flexible, high cost, effects on performance and long-term stability, needs investigation)	Flexible device sandwiched between ultra-thin flexible glass sheets.	Hermetic glass frit (Fantanas et al. 2018; Emami et al. 2020).
Polymeric laminates and thin-film barrier-coated webs (used for both flexible and rigid solar cells)	Can be used as substrates in flexible solar cells and as an encapsulating agent on various types of substrates.	Poly(methylmethacrylate) (PMMA), polyethylene terephthalate (PET), polydimethylsiloxane (PDMS), polyethylenenaphthalate (PEN).
Thin-film encapsulation (TFE) (emerging and promising, expensive, challenging)	Direct deposition of a single ultra thin-film flexible protective layer on the device using vacuum deposition methods including physical vapor deposition, chemical vapor deposition, plasma-enhanced chemical vapor deposition, and atomic layer deposition.	Metal oxides including Al_2O_3, SiO_x, TiO_2, and Zn_2SnO_4 (Aranda et al. 2021). Multilayer stacked organic/organic layers called dyads (Lee et al. 2018) and ultra-thin plasma polymeric films (Idígoras et al. 2018).

Source: Table by authors.

2.3.4. 2D and 2D/3D Mixed-Dimensional Perovskites

Two-dimensional HaP have layered structures that are similar to the Ruddlesden–Popper (RP) phases (Ruddlesden and Popper 1958), consisting of a nanoplatelet (nanosheet) perovskite that is separated by large spacer cations. The RP phase has the general formula $A_{n-1}L_2M_nX_{3n+1}$. In this form, A is a small-size monovalent cation (Cs^+, MA^+), L corresponds to a large-size aromatic or aliphatic alkylammonium spacer cation including phenyl-ethyl ammonium (PEA^+) and butylammonium (BA^+), M is a transition metal cation (such as Pb^{2+} and Sn^{2+}), X stands for a halide anion (such as I^-, Br^-, and Cl^-), and the integer n represents the number of metal halide octahedral $[MX_6]^{4-}$ layers between the two L-cations, determined by careful control of the stoichiometry (Shi et al. 2018). Two-dimensional perovskites have strong quantum confinement effects and large bandgaps (Zhang et al. 2020). In solar cells, 2D perovskites have been applied as primary light harvesters (Cao et al. 2015), capping layers (Chen et al. 2018), passivation layers (Jiang et al. 2019), and 2D/3D interfacial layers (Niu et al. 2019). Two-dimensional HaP solar cells are more stable than their 3D counterparts but less efficient. Moreover, their hydrophobicity and moisture resistance improve device stability under high humidity (Zheng et al. 2018).

Two-dimensional/three-dimensional mixed-dimensional perovskite solar cells combine the stability of 2D perovskites with the excellent light-harvesting properties of 3D perovskites to produce stable and efficient devices. When grown on 3D perovskites to form a 2D/3D mixed-dimensional perovskite, grain boundaries and surface charged defects are passivated to enhance stability (Wu et al. 2021). In 2017, Grancini et al. (2017) obtained a stable 10 cm × 10 cm perovskite solar cell that maintained its 11.6% PCE for more than 10,000 h under controlled standard conditions using a fully printable industrial process. Remarkable stability was achieved through 2D/3D interface engineering in which the 2D layer prevented moisture ingress.

2.3.5. Use of Stable Metal Electrodes and Very Thin Interlayers

As explained above, diffusion of the widely used Au, Al, Ag, and Cu electrodes into the HaP active layer is one of the leading causes of instability. Very thin barrier layers including chromium (Domanski et al. 2016), chromium oxide-chromium (Cr_2O_3/Cr) (Kaltenbrunner et al. 2015), MoO_x (Sanehira et al. 2016), bismuth (Bi) (Wu et al. 2019), and amine-mediated titanium suboxide ($AM\text{-}TiO_x$) (Back et al. 2016) have been employed between the perovskite and hole transport layers to protect metal top contacts from reaction with the halide perovskites. Domanski et al. (2016) showed that, at 70 °C, gold (Au) diffused through the HTL into the HaP layer. However, the diffusion was prevented by depositing a layer of chromium (Cr) between the HTL and the Au electrode. The Cr layer alleviated the severe degradation of the device performance at elevated temperatures. In comparison to Au and Ag, Cu electrodes do not diffuse into the perovskite active layer and produce more stable perovskite solar cells (Zhao et al. 2016a). Zhao et al. demonstrated that high-PCE Cu electrode-based solar cells with efficiency above 20% retain 98% of the initial PCE after 816 h of storage in an ambient environment without encapsulation (Zhao et al. 2016a). Cu and Ag do not form deep-level trap states in $MAPbI_3$-based solar cells (Ming et al. 2018). Additionally, the conventional noble metal electrodes are not sustainable because of the cost, scarcity, and complexity of metal ore extraction. To overcome these problems, carbon electrodes are gaining increased attention due to their low cost, excellent stability, and compatibility with up-scaling techniques. However, ultra-thin buffer layers of materials such as Cr are required between the electrode and the charge transport layers to ensure good electrical contact (Babu et al. 2020).

3. Solar Hydrogen Production

PEC water splitting, a technology for solar hydrogen production, is an attractive approach for numerous reasons. First, photocatalytic hydrogen production offers an attractive route for solar energy storage. This is because hydrogen energy storage has been considered as the most suitable means for storing excess off-peak power where long-term storage is a priority (Benato and Stoppato 2018). In addition, hydrogen can be easily transported via land, air, or sea, making it possible to transport solar energy (converted to hydrogen) from one geographical location to another. Additionally, hydrogen fuel already has a vast and established economy with numerous applications in homes and industries. Hydrogen can be converted directly into electricity for domestic consumption, use for the powering of automobiles, and as fuel in the aviation industry (Glanz 2010). The numerous applications of hydrogen make its production from solar energy more attractive considering the global need for clean energy production for a sustainable future.

The device used for harvesting solar energy for photocatalytic hydrogen production is often known as a PEC cell (Figure 4). The basic operation of a PEC device has been reported by many authors (Glanz 2010; Ihssen et al. 2014). Here, a summary of the operation of a PEC cell is explained using a device consisting of a photoanode and a metallic counter electrode immersed in an acidic electrolyte. Equation (3) presents an illustration of the basic operation of a PEC device for water splitting. First, the photoanode will absorb photons when irradiated with incident photon energy hv and become ionized, resulting in the generation of electron–hole pairs. If recombination does not occur, the hole (h^+) becomes separated from the electron (e^-), moves to the surface of the photoanode, and oxidizes water to produce oxygen gas and H^+ ions, as shown in Equation (1). The H^+ ions produced at the surface of the photoanode are transported to the cathode. Simultaneously, the electrons produced in the photoanode are driven to the cathode through the external circuit where they interact with the H^+ ions to produce H_2 gas, as shown in Equation (4). The chemical reaction for the decomposition of water into O_2 and H_2 via PEC water splitting is summarized in Equation (5). Examples of materials that could be used as a photoanode in PEC devices include n-type semiconductors such TiO_2, $BiVO_4$, and α-Fe_2O_3.

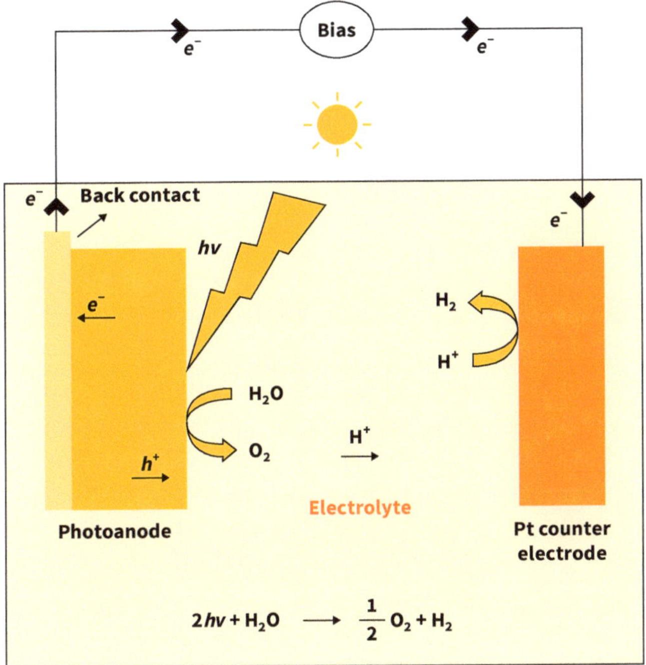

Figure 4. Schematic illustration of the basic operation of a PEC device. Source: Graphic by authors.

$$\text{At the photoanode}: 2h^+ + H_2O \;\rightarrow\; \frac{1}{2}O_2 + 2H^+ \tag{3}$$

$$\text{At the counter electrode: } 2H^+ + 2e^- \rightarrow H_2 \tag{4}$$

$$\text{Summary}: \ hv + H_2O \ \rightarrow \ \frac{1}{2} O_2 + H_2 \tag{5}$$

3.1. Hematite as Photocatalyst

Hardee and Bard were the first to use a hematite photoelectrode for water photolysis in 1976 (Cattarin and Decker 2009). The stability of hematite in an aqueous environment and its ability to absorb photons in the visible region are the major properties that have continued to attract increased research into its application in PEC water splitting. An increasing amount of research is still being channeled towards overcoming the major challenges inhibiting the use of hematite as a photoanode in solar hydrogen production. The main challenges are outlined in Section 1, which include its poor conductivity, high electron–hole recombination, and inefficient charge separation, among others. The strategies which have been developed over the years towards overcoming the problems that have been limiting the application of hematite-based photoanodes in solar hydrogen production are discussed in the following section.

3.2. Strategies for Enhancing the PEC Properties of α-Fe$_2$O$_3$ Films

3.2.1. Nanostructuring

Nanostructuring is the fabrication of materials consisting of structural features in the nanometer scale (Singh and Terasaki 2008). Nanostructured materials provide flexible space for ease of fabrication, enhanced mechanical stability, confinement effects, and a large surface area, making them suitable for photocatalytic applications (Rani et al. 2018). The nanostructuring approach has long been employed in the fabrication of α-Fe$_2$O$_3$ thin films to mitigate their poor charge transport property without compromising their photon absorption for PEC applications. α-Fe$_2$O$_3$ has a low absorption coefficient and, as a result, requires films of 400–500 nm thickness for complete light absorption. Because of the short hole diffusion length of 2–4 nm (Ahn et al. 2014), photogenerated charge carriers in bulk α-Fe$_2$O$_3$ films will likely recombine before reaching the surface of the films to perform water oxidation, which will result in a low photocurrent in the PEC device. Since thinner α-Fe$_2$O$_3$ films are not able to absorb sufficient photons for a significant photocatalytic activity, nanostructuring has been employed to help solve this paradox. Nanostructured α-Fe$_2$O$_3$ films that can absorb sufficient photons can also offer a large interfacial area for interaction with the electrolyte, making them suitable for promoting charge carrier transport during photocatalytic reactions (Tamirat et al. 2016; Annamalai et al. 2016).

The nanostructuring approach has been widely utilized in preparing α-Fe_2O_3 films of different morphologies and has been shown to help promote charge separation on the film's surfaces where water oxidation/reduction reactions occur during photocatalysis (Annamalai et al. 2016). Nanostructured α-Fe_2O_3 films with morphologies such as nanoparticles (Souza et al. 2009), nanorods (Ito et al. 2017), nanoflowers (Tsege et al. 2016), nanocones (Li et al. 2014), nanosheets (Peerakiatkhajohn et al. 2016), nanotubes (Kim et al. 2016a), and nanowires (Xie et al. 2018; Grigorescu et al. 2012) have been prepared for PEC water splitting, yielding an improved photocurrent density compared to the bulk films (Chou et al. 2013). Figure 5 presents a schematic illustration for some of the different morphologies of hematite films for PEC water splitting. One of the major limitations of nanostructuring is its inability to influence the intrinsic properties of hematite such as its low electrical conductivity of 10^{-14} Ω^{-1} cm^{-1} (Tamirat et al. 2016) and charge carrier lifetime of 3–10 ps (Grave et al. 2018).

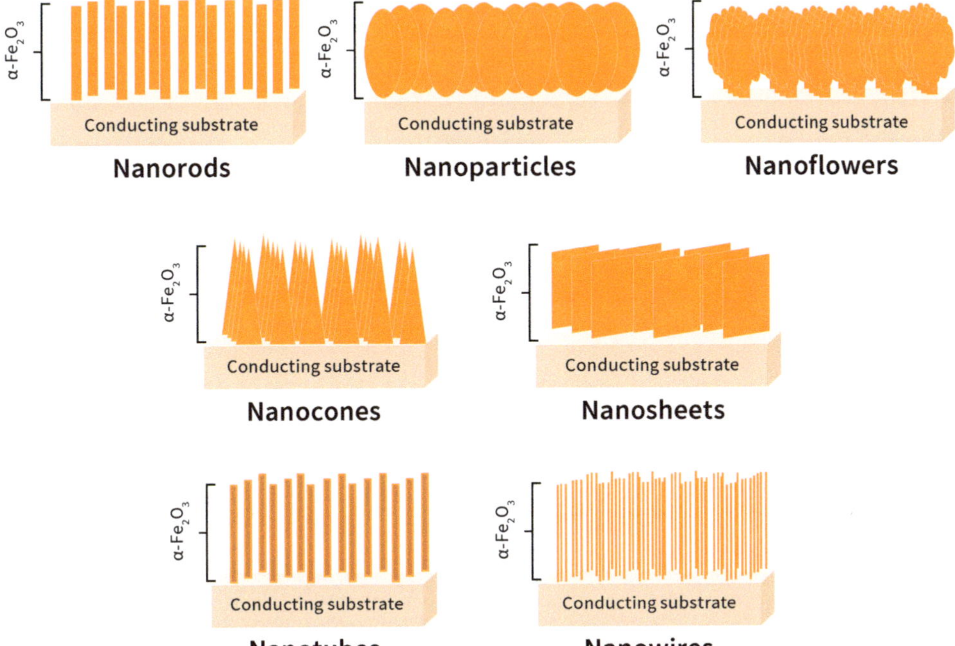

Figure 5. Schematic illustration of different morphologies of hematite films for PEC water splitting. Source: Graphic by authors.

3.2.2. Doping

The introduction of impurities into a semiconductor, termed doping (Grundmann 2010), can positively alter its intrinsic properties for PEC applications. Doping of semiconductor materials can help to narrow their optical bandgap and influence electrical properties, such as an increase in the charge carrier's concentration and mobility, thereby improving PEC performance (Yang et al. 2019). For α-Fe$_2$O$_3$, elemental doping involves replacing the lattice iron with foreign atoms, in order to influence its intrinsic properties for improved photocatalytic capability. The intrinsic properties of α-Fe$_2$O$_3$ which negatively affects its efficacy in PEC devices such as its low electrical conductivity of 10^{-14} Ω^{-1} cm^{-1}, charge carrier concentration on the order of 10^{18} cm^3, electron mobility of 10^{-2} cm^2 V^{-1} s^{-1}, hole mobility of 0.0001 cm^2 V^{-1} s^{-1}, and charge carrier lifetime of 3–10 ps have been improved through doping (Tamirat et al. 2016; Grave et al. 2018). Doping can significantly cause an increase in the charge carrier concentration in hematite films which directly improves their conductivity. Both experimental evidence (Gurudayal et al. 2014; Mao et al. 2011) and theoretical calculations (Zhang et al. 2016b) have confirmed the enhancement of the charge carrier concentration through doping. In addition, enhancement of the photocatalytic capabilities of hematite through doping has also been associated with the passivation of surface states and grain boundaries, shifting of band edge positions, and the distortion of its crystal structure which facilitates charge carrier hopping and transport (Grave et al. 2018).

The PEC performance of α-Fe$_2$O$_3$ films has been improved through doping with n-type dopants such as Ti (Feng et al. 2020; Peng et al. 2021), Pt (Mao et al. 2011), and Sn (Li et al. 2017), p-type dopants such as Mn^{2+} (Gurudayal et al. 2014), Cu^{2+} (Tsege et al. 2016), and Ag$^+$ (Shen et al. 2014) [4], and non-metals such as Si (Dias et al. 2014), S (Bemana and Rashid-Nadimi 2017), and P (Zhang et al. 2015). Feng et al. (2020) achieved an over 2-fold increment in the photocurrent density at 1.23 V vs. RHE and a negative onset potential shift of over 200 mV for α-Fe$_2$O$_3$ photoanodes through Ti doping. They attributed the improved PEC water splitting to an increase in the charge carrier density and enhanced charge separation. Elsewhere, a 3-fold increase in the photocurrent density was achieved for α-Fe$_2$O$_3$ nanorods through p-type doping with Mn, and the onset potential shifted by 30 mV to a more negative value. The boost in PEC water splitting was also associated with the increased charge carrier density as well as the reduced electron–hole recombination rate in Mn-doped α-Fe$_2$O$_3$ photoanodes (Gurudayal et al. 2014). In another study, a photocurrent density of 1.42 mA/cm^2 at 1.23 V vs. RHE was achieved for S-doped α-Fe$_2$O$_3$ nanorods, representing a 4-fold increase compared to the undoped films. The authors attributed the superior PEC activity to the improved charge carrier mobility of the S-doped α-Fe$_2$O$_3$ films (Zhang et al. 2017).

3.2.3. Heterojunction Formation

The heterojunction architecture involves the coupling of two semiconducting materials to improve PEC water splitting efficiency. Depending on the semiconductor materials used to form the heterostructure (*n*-type or *p*-type), *n–n*, *p–p*, or *p–n* junction structures could be formed. Heterojunction formation confers three major contributions: enhanced visible light absorption, improved charge separation, and increased lifetime of charge carriers (Tamirat et al. 2016). Heterojunction structures allow for the incorporation of materials of different bandgaps, broadening the photon absorption spectrum of the heterostructure for better photocatalysis (Mayer et al. 2012; Sharma et al. 2015; Kyesmen et al. 2021). Additionally, the formation of a heterojunction results in the development of an internal electric field at the space charge region between the heterostructures which helps in facilitating charge carrier transport. This will culminate in improving charge separation and increasing the carrier lifetime, leading to reduced electron–hole recombination and enhanced PEC efficiency during water splitting (Bai et al. 2018; Selim et al. 2019).

The charge transport mechanism and energy band diagram of hematite-based photoanodes during PEC water splitting can be explained using the *p–n* heterojunction structure presented in Figure 6. When a heterojunction is formed between two semiconductors, a space charge layer is created at the interface between them. For a *p–n* heterojunction with a hematite-based photoanode, the valence band (VB) and conduction band (CB) edges of the p-type semiconductor material both need to be more negative than those of α-Fe$_2$O$_3$ (Afroz et al. 2018). Additionally, the electrons from the CB of the p-type semiconductor are transferred to the CB of α-Fe$_2$O$_3$ and then to the fluorine-doped tin oxide (FTO) substrate, where they move onto the counter electrode through the back-contact to reduce H$^+$ to H$_2$. The movement of photogenerated charge carriers across the heterojunction is facilitated by the electric field formed at the interface between the composite materials, enhancing the effective charge separation and reducing the recombination rate of electron–hole pairs (Liu et al. 2015a). For an *n–n* heterojunction-structured hematite-based photoanode, a similar operation mechanism and energy band bending to those of the *p–n* junction apply. However, the semiconductor material is required to have more negative CB and VB band positions relative to those of hematite.

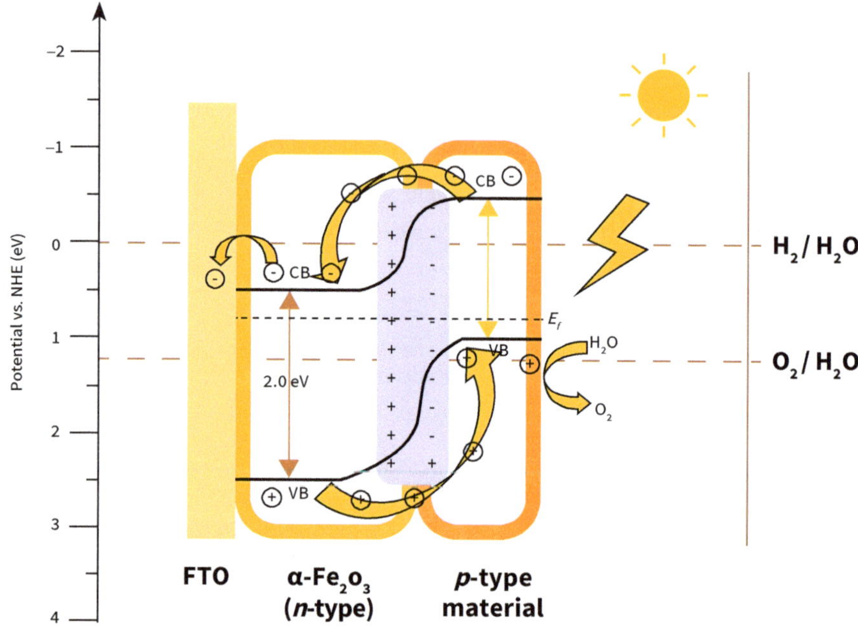

Figure 6. The charge transport mechanism and energy band diagram of the hematite-based *p–n* heterojunction structure during PEC water splitting. Source: Graphic by authors.

Furthermore, different composite materials have been employed in improving the PEC water splitting of α-Fe$_2$O$_3$. The formation of the α-Fe$_2$O$_3$/NiO heterojunction structure has been reported to improve the photocurrent density of α-Fe$_2$O$_3$ from 0.042 to 0.156 mA/cm^2 at 0.4 V vs. AgCl. The improvement was attributed to the enhanced charge transfer kinetics resulting from the formation of the α-Fe$_2$O$_3$/NiO heterostructure (Bemana and Rashid-Nadimi 2019). Natarajan et al. (2017) fabricated α-Fe$_2$O$_3$/CdS heterostructures and achieved a photocurrent density of 0.6 mA/cm^2 at 0.92 V vs. RHE and a 0.4 V negative shift in the onset potential compared to the value recorded for pristine α-Fe$_2$O$_3$ films. They attributed the enhancement in PEC water splitting to the improved photon absorption and facilitated charge transfer kinetics also resulting from the formation of the heterojunction structure (ibid. 2017). While different materials have been used to form the heterojunction structure with α-Fe$_2$O$_3$ for PEC applications, the choice of the composite material is important for achieving a notable improvement in water splitting efficiency. Materials that can enhance light absorption and promote charge transport can play a significant role in boosting the photocurrent density of α-Fe$_2$O$_3$-based heterojunction photocatalysts.

3.2.4. The Use of Co-Catalysts

One of the biggest challenges of PEC water splitting using α-Fe$_2$O$_3$ is the overpotential required to drive the water oxidation reaction due to its high activation energy barrier. The presence of a co-catalyst on photoanodes can improve PEC water splitting by facilitating water oxidation reactions and decreasing the overpotential and activation energy, thus shifting the onset potential to a more negative value (Tamirat et al. 2016).

Noble metal oxides (Badia-Bou et al. 2013), amorphous phosphates (Eftekharinia et al. 2017; Kwon et al. 2021), borates (Dang et al. 2017), and oxyhydroxides (Kim et al. 2016b) have been used as co-catalysts on α-Fe$_2$O$_3$ photoanodes. α-Fe$_2$O$_3$ has been modified with the iridium oxide (IrO$_2$) co-catalyst and used as a photoanode in PEC water splitting, yielding a photocurrent density of 200 μA/cm^2 at 1.29 V vs. RHE, while the pristine films required a positive shift of 300 mV to achieve the same photoresponse. The IrO$_2$ co-catalyst promoted charge separation and acted as a storage site for photogenerated holes, leading to an improvement in PEC water splitting achieved for modified α-Fe$_2$O$_3$ films (Badia-Bou et al. 2013). Additionally, the cobalt-phosphate (Co-Pi) co-catalyst has been used to modify α-Fe$_2$O$_3$ photoanodes and recorded a photocurrent density of 1.5 mA/cm^2 at 1.5 V vs. RHE, plus a negative shift of 185 mV in the onset potential. The improved performance was also attributed to the catalytic property of Co-Pi which can capture photogenerated holes, leading to suppressed charge recombination and facilitating water oxidation (Eftekharinia et al. 2017). Elsewhere, Kim et al. (2016b) used ultra-thin amorphous FeOOH as a co-catalyst on an α-Fe$_2$O$_3$ photoanode, recording a 2-fold increase in the photocurrent density, with an onset potential drop of about 120 mV, when applied towards PEC water splitting. The improved PEC behavior was attributed to the enhanced water oxidation kinetics and passivation of the surface states of the α-Fe$_2$O$_3$ photoanode due to the modification with the FeOOH co-catalyst (ibid. 2016b).

3.2.5. Plasmonic Enhancement Effects

Plasmonic metal nanostructures offer a promising route for improving the solar energy conversion efficiency of semiconductors (Li et al. 2013). Plasmonic metals can improve the performance of photoelectrodes in PEC water splitting via three major mechanisms. First, light scattering through localized surface plasmonic resonance (LSPR) absorption and re-emission can prolong the mean photon path in metal/semiconductor composites, resulting in an increased capture rate of incident photons. Second, hot electrons in the metal nanostructure generated through the decay of optically excited plasmons are transferred across the Schottky barrier to the nearby semiconductor, culminating in extra photoactivity. Finally, when metal/semiconductor composite nanostructures have overlapping LSPRs and energy band gaps, a large electric field enhancement occurs near the metal nanostructure's surface, leading to increased generation of electron–hole pairs in the nearby semiconductor, a concept known as the plasmonic near-field effect (Fan et al. 2016; Augustynski et al. 2016).

In efforts to improve the photocurrent density of α-Fe$_2$O$_3$ photoanodes during PEC water splitting, plasmonic metals such as Au (Archana et al. 2015; Shinde et al. 2017) and Ag (Liu et al. 2015a; Kwon et al. 2016) have been widely employed, showing great promise. Archana et al. (2015) deposited Au nanoparticles on α-Fe$_2$O$_3$ films and achieved a photocurrent enhancement that was three times higher than that of the pristine films at 0.6 V vs. Ag/AgCl. The photocurrent enhancement was attributed to a higher generation of charge carriers due to the plasmonic effects of Au nanoparticles on the α-Fe$_2$O$_3$ films (ibid. 2015). Additionally, Ag nanoparticles deposited on hydrothermally grown α-Fe$_2$O$_3$ nanowires produced a photocurrent density of about 0.18 mA/cm^2 at 1.23 vs. RHE when utilized as photoanodes in a PEC cell, representing a 10-fold enhancement relative to the value obtained for the pristine α-Fe$_2$O$_3$. The improvement was also associated with the surface plasmonic effects of Ag nanoparticles on the α-Fe$_2$O$_3$ nanowires (Kwon et al. 2016).

3.2.6. The Use of Multiple Approaches

The simultaneous use of multiple approaches to produce a single photoelectrode is a concept which harnesses the benefits of the different approaches to enhancing the PEC performance of hematite to produce a synergetic effect. The concurrent use of different approaches to produce a more efficient photocatalyst has been exploited by researchers with some significant successes recorded. Table 2 shows a list of hematite-based photoanodes in which multiple approaches to enhancing PEC performance were implemented, yielding a synergetic effect and an enhanced photocatalytic response.

Table 2. Hematite-based photoanodes in which multiple approaches to enhancing PEC performance were implemented.

Hematite-Based Photoelectrode	Strategies Engaged	Photocurrent Density Achieved Under 1 Sun	Photocurrent Density Increase Relative to That of Pristine α-Fe$_2$O$_3$	Reference
Ti-doped α-Fe$_2$O$_3$	Nanostructuring, doping	2.1 mA/cm^2 at 0.67 V vs. Ag/AgCl in 1 M NaOH electrolyte	2.8 times	(Lee et al. 2014)
α-Fe$_2$O$_3$/Co-Pi	Nanostructuring, co-catalyst loading	1.5 mA/cm^2 at 1.5 V vs. RHE 1 M NaOH electrolyte	1.39 times	(Eftekharinia et al. 2017)
α-Fe$_2$O$_3$/Au	Nanostrucring, plasmonic effects	1.0 mA/cm^2 at 1.23 V$_{RHE}$ in 1 M KOH electrolyte	2.86 times	(Wang et al. 2015)
α-Fe$_2$O$_3$/NiO	Nanostructing, heterojunction	1.55 mA/cm^2 at 1 V vs. RHE in 1M KOH electrolyte	19.37 times	(Rajendran et al. 2015)
α-Fe$_2$O$_3$/ BiVO$_4$/ NiFe-LDH	Nanostructuring, heterojunction, co-catalyst	1.7 mA/cm^2 at 1.8 V vs. RHE in 1 M NaOH electrolyte	4.25 times	(Bai et al. 2018)
Pt-doped α-Fe$_2$O$_3$/Co-Pi	Nanostructuring, doping, co-catalyst	4.32 mA/cm^2 at 1.23 V vs. RHE in 1 M NaOH electrolyte	3.43 times	(Kim et al. 2013)
Ti-doped α-Fe$_2$O$_3$/Cu$_2$O	Nanostructuring, doping, heterojunction	2.60 mA/cm^2 at 0.95 V vs. SCE in 1 M NaOH electrolyte	16.25 times	(Sharma et al. 2015)
α-Fe$_2$O$_3$/Au/ Co-Pi	Nanostructuring, plasmonic effects, co-catalyst	4.68 mA/cm^2 at 1.23 V vs. RHE in 1 M NaOH electrolyte	3 times	(Peerakiatkhajohn et al. 2016)
α-Fe$_2$O$_3$/ Nb-doped SnO$_2$/Co-Pi	Nanostructuring, heterojunction, doping, co-catalyst	3.16 mA/cm^2 at 1.23 V vs. RHE in … electrolyte under 1 sun	not given	(Yan et al. 2017)

Bai et al. (2018) in their work improved the performance of an α-Fe$_2$O$_3$ photoanode by combining the concepts of nanostructuring, heterojunction formation, and the use of co-catalysts. In their work, an α-Fe$_2$O$_3$/BiVO$_4$/NiFe-LDH photoanode was fabricated and applied towards PEC water splitting. A maximum photocurrent density of 1.7 mA/cm^2 was attained by the photoanode at 1.8 V vs. RHE, representing 1.3 and 4.25 times increases compared to the values obtained for α-Fe$_2$O$_3$/BiVO$_4$ and α-Fe$_2$O$_3$ films at the same potential, respectively (ibid. 2018). Elsewhere, Kim et al. (2013) prepared doped nanostructured α-Fe$_2$O$_3$ with the Pt dopant followed by surface modification with a Co-Pi co-catalyst-based photoanode when applied towards PEC water splitting. The doping of the pristine α-Fe$_2$O$_3$ photoanode with Pt increased its photocurrent density by 74% to 2.19 mA/cm^2 at 1.23 V vs. RHE, which was further enhanced to 4.32 mA/cm^2 at the same potential after loading with the Co-Pi co-catalyst (ibid. 2013). In a similar approach, Peerakiatkhajohn et al. (2016) demonstrated the synergetic effect of coating hematite nanosheets with Au nanoparticles for a plasmonic effect, followed by loading the surface with the Co-Pi co-catalyst, and achieved a photocurrent of 4.68 mA/cm (at 1.23 V vs. RHE), which is one of the highest performances reported in the literature for a modified hematite photoanode (ibid. 2016). The improved performances obtained for hematite-based photoanodes through the use of multiple approaches were achieved by harnessing the benefits of the different methods of boosting PEC performance via the systematic application of interface engineering.

4. Conclusion

In this chapter, promising materials for solar energy harnessing have been discussed with a special focus on HaP and α-Fe$_2$O$_3$ for direct conversion into electricity and hydrogen fuels, respectively. Long-term stability is an important requirement for the sustainable transfer of HaP solar cells from the laboratory to the market. The instability of perovskite solar cells depends on the Goldschmidt tolerance, chemical composition, and defects in halide perovskites. Other components of the solar cell architecture including the back-metal contact and the charge transport layers greatly contribute to the instability of the device. All these issues are responsible for the extremely low T_{80} (less than 2 years) for perovskite solar cells as opposed to the commercially available solar cells with T_{80} lifetimes exceeding 20 years. Protecting metal top contacts from reacting with halide perovskites, passivation of 2D perovskites to form 2D/3D mixed-dimensional perovskites, encapsulation of the devices and modules, and focusing on MA-free perovskites are credible strategies that, if well developed, will enhance the long-term stability and lifetime of perovskite solar cells. The intrinsic properties of α-Fe$_2$O$_3$ films such as their poor conductivity and short carrier lifetime have continued to limit their application for solar hydrogen production. Various strategies for improving the durability of HaP solar cells and the efficiency of α-Fe$_2$O$_3$ films in photocatalytic hydrogen production were discussed. The use of additives to achieve large grain sizes with few grain boundaries and to passivate the surface and boundaries of HaP is effective in improving the stability of HaP solar cells. Meanwhile, the concurrent use of multiple approaches such as nanostructuring, doping, the formation of heterostructures, the use of co-catalysts, and plasmonic enhancement effects has shown great promise in improving the photocatalytic efficiency of α-Fe$_2$O$_3$-based films for solar hydrogen production. Further research is still required for the eventual commercialization of solar hydrogen production and photovoltaic technologies using α-Fe$_2$O$_3$ and HaP, respectively.

Author Contributions: Conceptualization, J.N.F., and P.I.K.; methodology, J.N.F., and P.I.K.; software, J.N.F., and P.I.K.; validation, J.N.F., P.I.K., and M.D.; formal analysis, J.N.F., and P.I.K.; investigation, J.N.F., and P.I.K.; resources, J.N.F., P.I.K., and M.D.; data curation, J.N.F., and P.I.K.; writing—original draft preparation, J.N.F., and P.I.K.; writing—review and editing, J.N.F., and P.I.K.; visualization, J.N.F., and P.I.K.; supervision, M.D.; project administration, J.N.F., and P.I.K.; funding acquisition, M.D. substantially to the work reported.

Funding: This project was funded by the National Research Foundation, South Africa (grant number: N0115/115463) and the University of Pretoria (grant number: A0X816). APC funding is being sought externally.

Acknowledgments: The authors wish to thank the University of Pretoria, the National Research Foundation, UP postdoctoral fellowship program: grant DRI-cost center: A0X816, and the Externally Funded UP Post-Doctoral Fellowship Program: Grant Cost Centre N0115/115463, for the SARChI financial support.

Conflicts of Interest: The authors declare no conflict of interest.

References

Abdi-jalebi, Mojtaba, Zahra Andaji-garmaroudi, Stefania Cacovich, Camille Stavrakas, Bertrand Philippe, Eline M. Hutter, Andrew J. Pearson, Samuele Lilliu, Tom J. Savenije, Håkan Rensmo, and et al. 2018. Halide Perovskites with Potassium Passivation. *Nature Publishing Group* 555: 497–501. [CrossRef]

Afroz, Khurshida, Md Moniruddin, Nurlan Bakranov, Sarkyt Kudaibergenov, and Nurxat Nuraje. 2018. A Heterojunction Strategy to Improve the Visible Light Sensitive Water Splitting Performance of Photocatalytic Materials. *Journal of Materials Chemistry A* 6: 21696–718. [CrossRef]

Ahn, Hyo Jin, Myung Jun Kwak, Jung Soo Lee, Ki Yong Yoon, and Ji Hyun Jang. 2014. Nanoporous Hematite Structures to Overcome Short Diffusion Lengths in Water Splitting. *Journal of Materials Chemistry A* 2: 19999–20003. [CrossRef]

Annamalai, Alagappan, Pravin S. Shinde, Tae Hwa Jeon, Hyun Hwi Lee, Hyun Gyu Kim, Wonyong Choi, and Jum Suk Jang. 2016. Fabrication of Superior α-Fe_2O_3 Nanorod Photoanodes through Ex-Situ Sn-Doping for Solar Water Splitting. *Solar Energy Materials and Solar Cells* 144: 247–55. [CrossRef]

Aranda, Clara A., Laura Caliò, and Manuel Salado. 2021. Toward Commercialization of Stable Devices: An Overview on Encapsulation of Hybrid Organic-Inorganic Perovskite Solar Cells. *Crystals* 11: 1–16. [CrossRef]

Archana, Panikar Sathyaseelan, Neha Pachauri, Zhichao Shan, Shanlin Pan, and Arunava Gupta. 2015. Plasmonic Enhancement of Photoactivity by Gold Nanoparticles Embedded in Hematite Films. *Journal of Physical Chemistry C* 119: 15506–16. [CrossRef]

Augustynski, Jan, Krzysztof Bienkowski, and Renata Solarska. 2016. Plasmon Resonance-Enhanced Photoelectrodes and Photocatalysts. *Coordination Chemistry Reviews* 325: 116–24. [CrossRef]

Babu, Vivek, Rosinda Fuentes Pineda, Taimoor Ahmad, Agustin O. Alvarez, Luigi Angelo Castriotta, Aldo Di Carlo, Francisco Fabregat-Santiago, and Konrad Wojciechowski. 2020. Improved Stability of Inverted and Flexible Perovskite Solar Cells with Carbon Electrode. *ACS Applied Energy Materials* 3: 5126–34. [CrossRef]

Back, Hyungcheol, Geunjin Kim, Junghwan Kim, Jaemin Kong, Tae Kyun Kim, Hongkyu Kang, Heejoo Kim, Jinho Lee, Seongyu Lee, and Kwanghee Lee. 2016. Achieving Long-Term Stable Perovskite Solar Cells: Via Ion Neutralization. *Energy and Environmental Science* 9: 1258–63. [CrossRef]

Badia-Bou, Laura, Elena Mas-Marza, Pau Rodenas, Eva M. Barea, Francisco Fabregat-Santiago, Sixto Gimenez, Eduardo Peris, and Juan Bisquert. 2013. Water Oxidation at Hematite Photoelectrodes with an Iridium-Based Catalyst. *Journal of Physical Chemistry C* 117: 3826–33. [CrossRef]

Bai, Shouli, Haomiao Chu, Xu Xiang, Ruixian Luo, Jing He, and Aifan Chen. 2018. Fabricating of Fe_2O_3/$BiVO_4$ Heterojunction Based Photoanode Modified with NiFe-LDH Nanosheets for Efficient Solar Water Splitting. *Chemical Engineering Journal* 350: 148–56. [CrossRef]

Bemana, Hossein, and Sahar Rashid-Nadimi. 2017. Effect of Sulfur Doping on Photoelectrochemical Performance of Hematite. *Electrochimica Acta* 229: 396–403. [CrossRef]

Bemana, Hossein, and Sahar Rashid-Nadimi. 2019. Incorporation of NiO Electrocatalyst with α-Fe_2O_3 Photocatalyst for Enhanced and Stable Photoelectrochemical Water Splitting. *Surfaces and Interfaces* 14: 184–91. [CrossRef]

Benato, Alberto, and Anna Stoppato. 2018. Pumped Thermal Electricity Storage: A Technology Overview. *Thermal Science and Engineering Progress* 6: 301–15. [CrossRef]

Boyd, Caleb C., Rongrong Cheacharoen, Kevin A. Bush, Rohit Prasanna, Tomas Leijtens, and Michael D. McGehee. 2018. Barrier Design to Prevent Metal-Induced Degradation and Improve Thermal Stability in Perovskite Solar Cells. *ACS Energy Letters* 3: 1772–78. [CrossRef]

Bush, Kevin A., Axel F. Palmstrom, Zhengshan J. Yu, Mathieu Boccard, Rongrong Cheacharoen, Jonathan P. Mailoa, David P. McMeekin, Robert L. Z. Hoye, Colin D. Bailie, Tomas Leijtens, and et al. 2017. 23.6%-Efficient Monolithic Perovskite/Silicon Tandem Solar Cells With Improved Stability. *Nature Energy* 2: 1–7. [CrossRef]

Cao, Duyen H., Constantinos C. Stoumpos, Omar K. Farha, Joseph T. Hupp, and Kanatzidis G. Mercouri. 2015. 2D Homologous Perovskites as Light-Absorbing Materials for Solar Cell Applications. *Journal of the American Chemical Society* 137: 7843–50. [CrossRef]

Cattarin, Sandro, and Franco Decker. 2009. Electrodes | Semiconductor Electrodes. *Encyclopedia of Electrochemical Power Sources* 9: 121–33. [CrossRef]

Cheacharoen, Rongrong, Nicholas Rolston, Duncan Harwood, Kevin A. Bush, Reinhold H. Dauskardt, and Michael D. McGehee. 2018. Design and Understanding of Encapsulated Perovskite Solar Cells to Withstand Temperature Cycling. *Energy and Environmental Science* 11: 144–50. [CrossRef]

Chen, Haoran, Yuetian Chen, Taiyang Zhang, Xiaomin Liu, Xingtao Wang, and Yixin Zhao. 2021. Advances to High-Performance Black-Phase FAPbI$_3$ Perovskite for Efficient and Stable Photovoltaics. *Small Structures* 2: 2000130. [CrossRef]

Chen, Peng, Yang Bai, Songcan Wang, Miaoqiang Lyu, Jung-ho Yun, and Lianzhou Wang. 2018. In Situ Growth of 2D Perovskite Capping Layer for Stable and Efficient Perovskite Solar Cells. *Advanced Functional Materials* 28: 1706923. [CrossRef]

Chen, Qi, Huanping Zhou, Tze-bin Song, Song Luo, Ziruo Hong, Hsin-sheng Duan, Letian Dou, Yongsheng Liu, and Yang Yang. 2014. Controllable Self-Induced Passivation of Hybrid Lead Iodide Perovskites toward High-Performance Solar Cells. *Nano Letters* 14: 4158–63. [CrossRef] [PubMed]

Chou, Jen Chun, Szu An Lin, Chi Young Lee, and Jon Yiew Gan. 2013. Effect of Bulk Doping and Surface-Trapped States on Water Splitting with Hematite Photoanodes. *Journal of Materials Chemistry A* 1: 5908–14. [CrossRef]

Dang, Ke, Tuo Wang, Chengcheng Li, Jijie Zhang, Shanshan Liu, and Jinlong Gong. 2017. Improved Oxygen Evolution Kinetics and Surface States Passivation of Ni-Bi Co-Catalyst for a Hematite Photoanode. *Engineering* 3: 285–89. [CrossRef]

DeQuilettes, Dane W., Sarah M. Vorpahl, Samuel D. Stranks, Hirokazu Nagaoka, Giles E. Eperon, Mark E. Ziffer, Henry J. Snaith, and David S. Ginger. 2015. Impact of Microstructure on Local Carrier Lifetime in Perovskite Solar Cells. *Science* 348: 683–86. [CrossRef]

Dewinggih, Tanti, Shobih, Lia Muliani, Herman, and Rahmat Hidayat. 2017. The Temperature Effect on the Working Characteristics of Solar Cells Based on Organometal Halide Perovskite Crystals The Temperature Effect on the Working Characteristics of Solar Cells Based on Organometal Halide Perovskite Crystals. *Journal of Physics: Conference Series* 877: 012043. [CrossRef]

Dias, Paula, Tânia Lopes, Luísa Andrade, and Adélio Mendes. 2014. Temperature Effect on Water Splitting Using a Si-Doped Hematite Photoanode. *Journal of Power Sources* 272: 567–80. [CrossRef]

Domanski, Konrad, Juan Pablo Correa-Baena, Nicolas Mine, Mohammad Khaja Nazeeruddin, Antonio Abate, Michael Saliba, Wolfgang Tress, Anders Hagfeldt, and Michael Grätzel. 2016. Not All That Glitters Is Gold: Metal-Migration-Induced Degradation in Perovskite Solar Cells. *ACS Nano* 10: 6306–14. [CrossRef]

Dunn, Bruce, Yang Yang, Mingkui Wang, Lijian Zuo, Hexia Guo, Nicholas De Marco, David S. Ginger, Bruce Dunn, Mingkui Wang, and Yang Yang. 2017. Polymer-Modified Halide Perovskite Films for Efficient and Stable Planar Heterojunction Solar Cells. *Science Advances* 3: e1700106. [CrossRef]

Eftekhari, Ali, Veluru Jagadeesh Babu, and Seeram Ramakrishna. 2017. Photoelectrode Nanomaterials for Photoelectrochemical Water Splitting. *International Journal of Hydrogen Energy* 42: 11078–109. [CrossRef]

Eftekharinia, Behrooz, Ahmad Moshaii, Ali Dabirian, and Nader Sobhkhiz Vayghan. 2017. Optimization of Charge Transport in a Co-Pi Modified Hematite Thin Film Produced by Scalable Electron Beam Evaporation for Photoelectrochemical Water Oxidation. *Journal of Materials Chemistry A* 5: 3412–24. [CrossRef]

Eidsvåg, Håkon, Said Bentouba, Ponniah Vajeeston, Shivatharsiny Yohi, and Dhayalan Velauthapillai. 2021. TiO_2 as a Photocatalyst for Water Splitting—An Experimental and Theoretical Review. *Molecules* 26: 1–30. [CrossRef] [PubMed]

Emami, Seyedali, Jorge Martins, Dzmitry Ivanou, and Adélio Mendes. 2020. Advanced Hermetic Encapsulation of Perovskite Solar Cells: The Route to Commercialization. *Journal of Materials Chemistry A* 8: 2654–62. [CrossRef]

Fan, Wenguang, Michael K. H. Leung, Jimmy C. Yu, and Wing Kei Ho. 2016. Recent Development of Plasmonic Resonance-Based Photocatalysis and Photovoltaics for Solar Utilization. *Molecules* 21: 180. [CrossRef]

Fang, Zhibin, Wenjing Chen, Yongliang Shi, Jin Zhao, Shenglong Chu, and Ji Zhang. 2020. Dual Passivation of Perovskite Defects for Light-Emitting Diodes with External Quantum Efficiency Exceeding 20%. *Advanced Functional Materials* 30: 1909754. [CrossRef]

Fantanas, Dimitrios, Adam Brunton, Simon J. Henley, and Robert A. Dorey. 2018. Investigation of the mechanism for current induced network failure for spray deposited silver nanowires. *Nanotechnology* 29: 465705. [CrossRef]

Feng, Fan, Can Li, Jie Jian, Fan Li, Youxun Xu, Hongqiang Wang, and Lichao Jia. 2020. Gradient Ti-Doping in Hematite Photoanodes for Enhanced Photoelectrochemical Performance. *Journal of Power Sources* 449: 227473. [CrossRef]

Fru, Juvet N., Nolwazi Nombona, and Mmantsae Diale. 2021. Growth and Degradation of Methylammonium Lead Tri-Bromide Perovskite Thin Film at Metal/Perovskite Interfaces. *Thin Solid Films* 722: 138568. [CrossRef]

Glanz, Karen. 2010. Using Behavioral Theories to Guide Decisions of What to Measure, and Why. *International Journal of Hydrogen Energy* 27: 991–1022.

Grancini, Giulia, Cristina Roldán-Carmona, Ivan Zimmermann, Edoardo Mosconi, Xuhui Lee, Daniel Martineau, Stéphanie Narbey, F. Oswald, F. De Angelis, M. Graetzel, and et al. 2017. One-Year Stable Perovskite Solar Cells by 2D/3D Interface Engineering. *Nature Communications* 8: 1–8. [CrossRef]

Grave, Daniel A., Natav Yatom, David S. Ellis, Maytal Caspary Toroker, and Avner Rothschild. 2018. The 'Rust' Challenge: On the Correlations between Electronic Structure, Excited State Dynamics, and Photoelectrochemical Performance of Hematite Photoanodes for Solar Water Splitting. *Advanced Materials* 30: 1–10. [CrossRef] [PubMed]

Griffini, Gianmarco, and Stefano Turri. 2016. Polymeric Materials for Long-Term Durability of Photovoltaic Systems. *Journal of Applied Polymer Science* 133: 1–16. [CrossRef]

Grigorescu, Sabina, Chong-yong Lee, Kiyoung Lee, Sergiu Albu, Indhumati Paramasivam, Ioana Demetrescu, and Patrik Schmuki. 2012. Electrochemistry Communications Thermal Air Oxidation of Fe: Rapid Hematite Nanowire Growth and Photoelectrochemical Water Splitting Performance. *Electrochemistry Communications* 23: 59–62. [CrossRef]

Gurudayal, Sing Yang Chiam, Mulmudi Hemant Kumar, Prince Saurabh Bassi, Hwee Leng Seng, James Barber, and Lydia Helena Wong. 2014. Improving the Efficiency of Hematite Nanorods for Photoelectrochemical Water Splitting by Doping with Manganese. *ACS Applied Materials and Interfaces* 6: 5852–59. [CrossRef] [PubMed]

Han, Yu, Steffen Meyer, Yasmina Dkhissi, Karl Weber, Jennifer M. Pringle, Udo Bach, Leone Spiccia, and Yi-Bing Cheng. 2015. Degradation Observations of Encapsulated Planar $CH_3NH_3PbI_3$ Perovskite Solar Cells at High Temperatures and Humidity. *Journal of Materials Chemistry A* 3: 8139–47. [CrossRef]

He, Sisi, Longbin Qiu, Luis K. Ono, and Yabing Qi. 2020. How Far Are We from Attaining 10-Year Lifetime for Metal Halide Perovskite Solar Cells? *Materials Science and Engineering R: Reports* 140: 100545. [CrossRef]

Hoke, Eric T., Daniel J. Slotcavage, Emma R. Dohner, Andrea R. Bowring, Hemamala I. Karunadasa, and Michael D. McGehee. 2015. Reversible Photo-Induced Trap Formation in Mixed-Halide Hybrid Perovskites for Photovoltaics. *Chemical Science* 6: 613–17. [CrossRef]

Fujishima, Akira, and Kenichi Honda. 1972. Electrochemical Photolysis of Water at a Semiconductor Electrode. *Nature* 238: 737–40. [CrossRef]

Hsieh, Cheng-ming, Yung-sheng Liao, Yan-ru Lin, Chih-ping Chen, Cheng-min Tsai, Eric Wei-guang Diau, and Shih-ching Chuang. 2018. Of Perovskite Solar Cells Using Lewis Bases Urea and Thiourea as Additives: Stimulating Large Grain. *RSC Advances* 8: 19610–15. [CrossRef]

Idígoras, Jesús, Francisco J. Aparicio, Lidia Contreras-Bernal, Susana Ramos-Terrón, María Alcaire, Juan Ramón Sánchez-Valencia, Ana Borras, Ángel Barranco, and Juan A. Anta. 2018. Enhancing Moisture and Water Resistance in Perovskite Solar Cells by Encapsulation with Ultrathin Plasma Polymers. *ACS Applied Materials and Interfaces* 10: 11587–94. [CrossRef]

Ierides, Ioannis, Isaac Squires, Giulia Lucarelli, Thomas M. Brown, and Franco Cacialli. 2021. Inverted Organic Photovoltaics with a Solution-Processed ZnO/MgO Electron Transport Bilayer. *Journal of Materials Chemistry C* 9: 3901–10. [CrossRef]

Ihssen, Julian, Artur Braun, Greta Faccio, Krisztina Gajda-Schrantz, and Linda Thöny-Meyer. 2014. Light Harvesting Proteins for Solar Fuel Generation in Bioengineered Photoelectrochemical Cells. *Current Protein & Peptide Science* 15: 374–84. [CrossRef]

Ito, Nathalie Minko, Waldemir Moura Carvalho, Dereck Nills Ferreira Muche, Ricardo Hauch Ribeiro Castro, Gustavo Martini Dalpian, and Flavio Leandro Souza. 2017. High Temperature Activation of Hematite Nanorods for Sunlight Driven Water Oxidation Reaction. *Physical Chemistry Chemical Physics* 19: 25025–32. [CrossRef] [PubMed]

Jain, Deepak, Suryanaman Chaube, Prerna Khullar, Sriram Goverapet Srinivasan, and Beena Rai. 2019. Bulk and Surface DFT Investigations of Inorganic Halide Perovskites Screened Using Machine Learning and Materials Property Databases. *Physical Chemistry Chemical Physics* 21: 19423–36. [CrossRef] [PubMed]

Jeong, Jaeki, Minjin Kim, Jongdeuk Seo, Haizhou Lu, Paramvir Ahlawat, Aditya Mishra, Yingguo Yang, Michael A. Hope, Felix T. Eickemeyer, Maengsuk Kim, and et al. 2021. Pseudo-Halide Anion Engineering for α-FAPbI$_3$ Perovskite Solar Cells. *Nature* 592: 381–85. [CrossRef]

Jiang, Qi, Yang Zhao, Xingwang Zhang, Xiaolei Yang, Yong Chen, Zema Chu, Qiufeng Ye, Xingxing Li, Zhigang Yin, and Jingbi You. 2019. Surface Passivation of Perovskite Film for Efficient Solar Cells. *Nature Photonics* 13: 460–66. [CrossRef]

Jin, Handong, Elke Debroye, Masoumeh Keshavarz, Ivan G. Scheblykin, Maarten B. J. Roeffaers, and Julian A. Steele. 2020. It's a Trap ! On the Nature of Localised States and Charge Trapping in Lead Halide Perovskites. *Materials Horizons* 7: 397–410. [CrossRef]

Juarez-Perez, Emilio J., Luis K. Ono, Maki Maeda, Yan Jiang, Zafer Hawash, and Yabing Qi. 2018. Photodecomposition and Thermal Decomposition in Methylammonium Halide Lead Perovskites and Inferred Design Principles to Increase Photovoltaic Device Stability. *Journal of Materials Chemistry A* 6: 9604–12. [CrossRef]

Juarez-Perez, Emilio J., Luis K. Ono, and Yabing Qi. 2019. Thermal Degradation of Formamidinium Based Lead Halide Perovskites into *Sym*-Triazine and Hydrogen Cyanide Observed by Coupled Thermogravimetry-Mass Spectrometry Analysis. *Journal of Materials Chemistry A* 7: 16912–19. [CrossRef]

Kafizas, Andreas, Robert Godin, and James R. Durrant. 2017. Charge Carrier Dynamics in Metal Oxide Photoelectrodes for Water Oxidation. *Semiconductors and Semimetals* 97: 3–46. [CrossRef]

Kaltenbrunner, Martin, Getachew Adam, Eric Daniel Głowacki, Michael Drack, Reinhard Schwödiauer, Lucia Leonat, Dogukan Hazar Apaydin, Heiko Groiss, Markus Clark Scharber, Matthew Schuette White, and et al. 2015. Flexible High Power-per-Weight Perovskite Solar Cells with Chromium Oxide-Metal Contacts for Improved Stability in Air. *Nature Materials* 14: 1032–39. [CrossRef] [PubMed]

Kang, Keehoon, Heebeom Ahn, Younggul Song, Woocheol Lee, Junwoo Kim, Youngrok Kim, Daekyoung Yoo, and Takhee Lee. 2019. High-Performance Solution-Processed Organo-Metal Halide Perovskite Unipolar Resistive Memory Devices in a Cross-Bar Array Structure. *Advanced Materials* 31: 1804841. [CrossRef] [PubMed]

Kim, Do Hong, Dinsefa M. Andoshe, Young Seok Shim, Cheon Woo Moon, Woonbae Sohn, Seokhoon Choi, Taemin Ludvic Kim, Migyoung Lee, Hoonkee Park, Kootak Hong, and et al. 2016a. Toward High-Performance Hematite Nanotube Photoanodes: Charge-Transfer Engineering at Heterointerfaces. *ACS Applied Materials and Interfaces* 8: 23793–800. [CrossRef] [PubMed]

Kim, Jae Young, Duck Hyun Youn, Kyoungwoong Kang, and Jae Sung Lee. 2016b. Highly Conformal Deposition of an Ultrathin FeOOH Layer on a Hematite Nanostructure for Efficient Solar Water Splitting. *Angewandte Chemie - International Edition* 55: 10854–58. [CrossRef] [PubMed]

Kim, Gee Yeong, Alessandro Senocrate, Tae Youl Yang, Giuliano Gregori, Michael Grätzel, and Joachim Maier. 2018. Large Tunable Photoeffect on Ion Conduction in Halide Perovskites and Implications for Photodecomposition. *Nature Materials* 17: 445–49. [CrossRef]

Kim, Jae Young, Ganesan Magesh, Duck Hyun Youn, Ji Wook Jang, Jun Kubota, Kazunari Domen, and Jae Sung Lee. 2013. Single-Crystalline, Wormlike Hematite Photoanodes for Efficient Solar Water Splitting. *Scientific Reports* 3: 1–8. [CrossRef]

Kim, Jongseob, Sung-hoon Lee, Jung Hoon Lee, and Ki-ha Hong. 2014. The Role of Intrinsic Defects in Methylammonium Lead Iodide. *The journal of physical chemistry letters* 5: 1312–17. [CrossRef]

Kong, Weiguang, Tao Ding, Gang Bi, and Huizhen Wu. 2016. Optical Characterizations of the Surface States in Hybrid Lead–Halide Perovskites. *Physical Chemistry Chemical Physics* 18: 12626–32. [CrossRef]

Kwon, In, Mahadeo A. Mahadik, Jun Beom, Weon-sik Chae, Sun Hee, and Jum Suk. 2021. Lowering the Onset Potential of Zr-Doped Hematite Nanocoral Photoanodes by Al Co-Doping and Surface Modification with Electrodeposited Co–Pi. *Journal of Colloid And Interface Science* 581: 751–63. [CrossRef]

Kwon, Jinhyeong, Junyeob Yeo, Sukjoon Hong, Young D. Suh, Habeom Lee, Jun Ho Choi, Seung S. Lee, and Seung Hwan Ko. 2016. Photoreduction Synthesis of Hierarchical Hematite/Silver Nanostructures for Photoelectrochemical Water Splitting. *Energy Technology* 4: 271–77. [CrossRef]

Kyesmen, Pannan I., Nolwazi Nombona, and Mmantsae Diale. 2021. Heterojunction of Nanostructured α-Fe$_2$O$_3$/CuO for Enhancement of Photoelectrochemical Water Splitting. *Journal of Alloys and Compounds* 863: 158724. [CrossRef]

Lee, Jin-wook, Hui-seon Kim, and Nam-gyu Park. 2015. Lewis Acid − Base Adduct Approach for High Efficiency Perovskite Solar Cells. *Accounts of chemical research* 49: 311–19. [CrossRef] [PubMed]

Lee, Myeong Hwan, Jong Hoon Park, Hyun Soo Han, Hee Jo Song, In Sun Cho, Jun Hong Noh, and Kug Sun Hong. 2014. Nanostructured Ti-Doped Hematite (α-Fe$_2$O$_3$) Photoanodes for Efficient Photoelectrochemical Water Oxidation. *International Journal of Hydrogen Energy* 39: 17501–7. [CrossRef]

Lee, Young Il, Nam Joong Jeon, Bong Jun Kim, Hyunjeong Shim, Tae Youl Yang, Sang Il Seok, Jangwon Seo, and Sung Gap Im. 2018. A Low-Temperature Thin-Film Encapsulation for Enhanced Stability of a Highly Efficient Perovskite Solar Cell. *Advanced Energy Materials* 8: 1–8. [CrossRef]

Li, Jiangtian, Scott K. Cushing, Peng Zheng, Fanke Meng, Deryn Chu, and Nianqiang Wu. 2013. Plasmon-Induced Photonic and Energy-Transfer Enhancement of Solar Water Splitting by a Hematite Nanorod Array. *Nature Communications* 4: 1–8. [CrossRef]

Li, Jinkai, Yongcai Qiu, Zhanhua Wei, Qingfeng Lin, Qianpeng Zhang, Keyou Yan, Haining Chen, Shuang Xiao, Zhiyong Fan, and Shihe Yang. 2014. A Three-Dimensional Hexagonal Fluorine-Doped Tin Oxide Nanocone Array: A Superior Light Harvesting Electrode for High Performance Photoelectrochemical Water Splitting. *Energy and Environmental Science* 7: 3651–58. [CrossRef]

Li, Mingyang, Yi Yang, Yichuan Ling, Weitao Qiu, Fuxin Wang, Tianyu Liu, Yu Song, Xiaoxia Liu, Pingping Fang, Yexiang Tong, and et al. 2017. Morphology and Doping Engineering of Sn-Doped Hematite Nanowire Photoanodes. *Nano Letters* 17: 2490–95. [CrossRef]

Li, Bixin, Wei Hui, Xueqin Ran, Yingdong Xia, Fei Xia, Lingfeng Chao, Yonghua Chen, and Wei Huang. 2019. Memory Devices and Artificial Synapses. *Journal of Materials Chemistry C* 7: 7476–93. [CrossRef]

Lin, Xiongfeng, Askhat N. Jumabekov, Niraj N. Lal, Alexander R. Pascoe, Daniel E. Gómez, Noel W. Duffy, Anthony S.R. Chesman, Kallista Sears, Maxime Fournier, Yupeng Zhang, and et al. 2017. Dipole-Field-Assisted Charge Extraction in Metal-Perovskite-Metal Back-Contact Solar Cells. *Nature Communications* 8: 1–8. [CrossRef]

Liu, Cong, Zengqi Huang, Xiaotian Hu, Xiangchuan Meng, Liqiang Huang, and Jian Xiong. 2018. Grain Boundary Modi Fi Cation via F4TCNQ To Reduce Defects of Perovskite Solar Cells with Excellent Device Performance. *ACS Applied Materials & Interfaces* 10: 1909–16. [CrossRef]

Liu, Dong, David M. Bierman, Andrej Lenert, Hai-Tong Yu, Zhen Yang, Evelyn N. Wang, and Yuan-Yuan Duan. 2015a. Ultrathin Planar Hematite Film for Solar Photoelectrochemical Water Splitting. *Optics Express* 23: A1491. [CrossRef] [PubMed]

Liu, Xiangye, Wei Zhao, Houlei Cui, Yi Xie, Yaoming Wang, Tao Xu, and Fuqiang Huang. 2015b. Organic–Inorganic Halide Perovskite Based Solar Cells–Revolutionary Progress in Photovoltaics. *Inorganic Chemistry Frontiers* 2: 584–84. [CrossRef]

Grundmann, Marius. 2010. *Physics of Semiconductors*. Berlin: Springer, vol. 11, pp. 401–72. [CrossRef]

Mansour Rezaei Fumani, Nasibeh, Farzaneh Arabpour Roghabadi, Maryam Alidaei, Seyed Mojtaba Sadrameli, Vahid Ahmadi, and Farhood Najafi. 2020. Prolonged Lifetime of Perovskite Solar Cells Using a Moisture-Blocked and Temperature-Controlled Encapsulation System Comprising a Phase Change Material as a Cooling Agent. *ACS Omega* 5: 7106–14. [CrossRef] [PubMed]

Mao, Aiming, Nam Gyu Park, Gui Young Han, and Jong Hyeok Park. 2011. Controlled Growth of Vertically Oriented Hematite/Pt Composite Nanorod Arrays: Use for Photoelectrochemical Water Splitting. *Nanotechnology* 22: 175703. [CrossRef]

Mayer, Matthew T., Chun Du, and Dunwei Wang. 2012. Hematite/Si Nanowire Dual-Absorber System for Photoelectrochemical Water Splitting at Low Applied Potentials. *Journal of the American Chemical Society* 134: 12406–9. [CrossRef]

Mckenna, Keith P. 2018. Electronic Properties of {111} Twin Boundaries in a Mixed-Ion Lead Halide Perovskite Solar Absorber. *ACS Energy Letters* 3: 2663–68. [CrossRef]

Ming, Wenmei, Dongwen Yang, Tianshu Li, Lijun Zhang, and Mao Hua Du. 2018. Formation and Diffusion of Metal Impurities in Perovskite Solar Cell Material $CH_3NH_3PbI_3$: Implications on Solar Cell Degradation and Choice of Electrode. *Advanced Science* 5: 1700662. [CrossRef]

Motti, Silvia G., Daniele Meggiolaro, Samuele Martani, Roberto Sorrentino, Alex J. Barker, Filippo De Angelis, and Annamaria Petrozza. 2019. Defect Activity in Lead Halide Perovskites. *Advanced Materials* 31: 1901183. [CrossRef]

Murphy, Anthony B., Piers R. F. Barnes, Lakshman K. Randeniya, Ian C. Plumb, Ian E. Grey, Mike D. Horne, and Julie A. Glasscock. 2006. Efficiency of Solar Water Splitting Using Semiconductor Electrodes. *International Journal of Hydrogen Energy* 31: 1999–2017. [CrossRef]

Natarajan, Kaushik, Mohit Saraf, and Shaikh M. Mobin. 2017. Visible-Light-Induced Water Splitting Based on a Novel α-Fe$_2$O$_3$/CdS Heterostructure. *ACS Omega* 2: 3447–56. [CrossRef]

Niu, Tianqi, Jing Lu, Xuguang Jia, Zhuo Xu, Ming-chun Tang, Dounya Barrit, Ningyi Yuan, Jianning Ding, Xu Zhang, Yuanyuan Fan, and et al. 2019. Interfacial Engineering at the 2D/3D Heterojunction for High-Performance Perovskite Solar Cells. *Nano Letters* 19: 7181–90. [CrossRef] [PubMed]

Noel, Nakita Kimberly, Antonio Abate, Samuel David Stranks, Elizabeth Parrott, Victor Burlakov, Alain Goriely, and Henry J. Snaith. 2014. Enhanced Photoluminescence and Solar Cell Performance via Lewis Base Passivation of Organic-Inorganic Lead Halide Perovskites. *ACS nano* 8: 9815–21. [CrossRef] [PubMed]

Rothmann, Mathias Uller, Wei Li, Ye Zhu, Udo Bach, Leone Spiccia, Joanne Etheridge, and Yi-Bing Cheng. 2017. Direct observation of intrinsic twin domains in tetragonal CH$_3$NH$_3$PbI$_3$. *Nature Communications* 8: 6–13. [CrossRef] [PubMed]

Peerakiatkhajohn, Piangjai, Jung Ho Yun, Hongjun Chen, Miaoqiang Lyu, Teera Butburee, and Lianzhou Wang. 2016. Stable Hematite Nanosheet Photoanodes for Enhanced Photoelectrochemical Water Splitting. *Advanced Materials* 28: 6405–10. [CrossRef] [PubMed]

Peng, Yong, Qingdong Ruan, Chun Ho Lam, Fanxu Meng, Chung-yu Guan, Shella Permatasari Santoso, Xingli Zou, Edward T. Yu, Paul K. Chu, and Hsien-yi Hsu. 2021. Plasma-Implanted Ti-Doped Hematite Photoanodes with Enhanced Photoelectrochemical Water Oxidation Performance. *Journal of Alloys and Compounds* 870: 159376. [CrossRef]

Qiao, Lu, Wei-hai Fang, and Run Long. 2019. Forschungsartikel Extending Carrier Lifetimes in Lead Halide Perovskites with Alkali Metals by Passivating and Eliminating Halide Interstitial Defects Forschungsartikel. *Angewandte Chemie* 132: 4714–20. [CrossRef]

Qiu, Longbin, Sisi He, Luis K. Ono, and Yabing Qi. 2020. Progress of Surface Science Studies on ABX$_3$-Based Metal Halide Perovskite Solar Cells. *Advanced Energy Materials* 10: 1902726. [CrossRef]

Rahmany, Stav, and Lioz Etgar. 2021. Two-Dimensional or Passivation Treatment: The Effect of Hexylammonium Post Deposition Treatment on 3D Halide Perovskite-Based Solar Cells. *Materials Advances* 2: 2617–25. [CrossRef]

Rajendran, Kumar, Vithiya Karunagaran, Biswanath Mahanty, and Shampa Sen. 2015. Biosynthesis of Hematite Nanoparticles and Its Cytotoxic Effect on HepG2 Cancer Cells. *International Journal of Biological Macromolecules* 74: 376–81. [CrossRef]

Rani, Ankita, Rajesh Reddy, Uttkarshni Sharma, Priya Mukherjee, Priyanka Mishra, Aneek Kuila, Lan Ching Sim, and Pichiah Saravanan. 2018. A Review on the Progress of Nanostructure Materials for Energy Harnessing and Environmental Remediation. *Journal of Nanostructure in Chemistry* 8: 255–91. [CrossRef]

Ruddlesden, S. N., and Paul Popper. 1958. The Compound $Sr_3Ti_2O_7$ and Its Structure. *Acta Crystallographica* 11: 54–55. [CrossRef]

Sanehira, Erin M., Bertrand J. Tremolet De Villers, Philip Schulz, Matthew O. Reese, Suzanne Ferrere, Kai Zhu, Lih Y. Lin, Joseph J. Berry, and Joseph M. Luther. 2016. Influence of Electrode Interfaces on the Stability of Perovskite Solar Cells: Reduced Degradation Using MoO_x/Al for Hole Collection. *ACS Energy Letters* 1: 38–45. [CrossRef]

Selim, Shababa, Laia Francàs, Miguel García-Tecedor, Sacha Corby, Chris Blackman, Sixto Gimenez, James R. Durrant, and Andreas Kafizas. 2019. WO_3/$BiVO_4$: Impact of Charge Separation at the Timescale of Water Oxidation. *Chemical Science* 10: 2643–52. [CrossRef]

Asad, Jihad, Samy K. K. Shaat, Hussam Musleh, Nabil Shurrab, Ahmed Issa, Abelilah Lahmar, Amal Al-Kahlout, and Naji Al Dahoudi. 2019. Perovskite Solar Cells Free of Hole Transport Layer. *Journal of Sol-Gel Science and Technology*, 443–49. [CrossRef]

Shao, Yuchuan, Yanjun Fang, Tao Li, Qi Wang, Qingfeng Dong, Yehao Deng, Yongbo Yuan, Haotong Wei, Meiyu Wang, Alexei Gruverman, and et al. 2016. Environmental Science Grain Boundary Dominated Ion Migration in Polycrystalline Organic – Inorganic Halide Perovskite Films. *Energy & Environmental Science* 9: 1752–59. [CrossRef]

Shao, Yuchuan, Zhengguo Xiao, Cheng Bi, Yongbo Yuan, and Jinsong Huang. 2014. Origin and Elimination of Photocurrent Hysteresis by Fullerene Passivation in $CH_3NH_3PbI_3$ Planar Heterojunction Solar Cells. *Nature Communications* 5: 1–7. [CrossRef] [PubMed]

Sharma, Dipika, Sumant Upadhyay, Anuradha Verma, Vibha R. Satsangi, Rohit Shrivastav, and Sahab Dass. 2015. Nanostructured Ti-Fe_2O_3/Cu_2O Heterojunction Photoelectrode for Efficient Hydrogen Production. *Thin Solid Films* 574: 125–31. [CrossRef]

Shen, Shaohua, Jigang Zhou, Chung Li Dong, Yongfeng Hu, Eric Nestor Tseng, Penghui Guo, Liejin Guo, and Samuel S. Mao. 2014. Surface Engineered Doping of Hematite Nanorod Arrays for Improved Photoelectrochemical Water Splitting. *Scientific Reports* 4: 1–9. [CrossRef]

Sherkar, Tejas S., Cristina Momblona, A. Jorge, Michele Sessolo, Henk J. Bolink, and L. Jan Anton Koster. 2017. Recombination in Perovskite Solar Cells: Significance of Grain Boundaries, Interface Traps, and Defect Ions. *ACS Energy Letters* 2: 1214–22. [CrossRef]

Shi, Enzheng, Gao Yao, Blake P. Finkenauer, Akriti, Aidan H. Coffey, and Letian Dou. 2018. Two-Dimensional Halide Perovskite Nanomaterials and Heterostructures. *Chemical Society Reviews* 47: 6001–446. [CrossRef]

Shi, Lei, Trevor L. Young, Jincheol Kim, Yun Sheng, Lei Wang, Yifeng Chen, Zhiqiang Feng, Mark J. Keevers, Xiaojing Hao, Pierre J. Verlinden, and et al. 2017. Accelerated Lifetime Testing of Organic-Inorganic Perovskite Solar Cells Encapsulated by Polyisobutylene. *ACS Applied Materials and Interfaces* 9: 25073–81. [CrossRef] [PubMed]

Shinde, Pravin S., Su Yong Lee, Jungho Ryu, Sun Hee Choi, and Jum Suk Jang. 2017. Enhanced Photoelectrochemical Performance of Internally Porous Au-Embedded α-Fe_2O_3 Photoanodes for Water Oxidation. *Chemical Communications* 53: 4278–81. [CrossRef] [PubMed]

Shlenskaya, Natalia N., Nikolai A. Belich, Michael Grätzel, Eugene A. Goodilin, and Alexey B. Tarasov. 2018. Light-Induced Reactivity of Gold and Hybrid Perovskite as a New Possible Degradation Mechanism in Perovskite Solar Cells. *Journal of Materials Chemistry A* 6: 1780–86. [CrossRef]

Singh, David J., and Ichiro Terasaki. 2008. Thermoelectrics: Nanostructuring and More. *Nature Materials* 7: 616–17. [CrossRef]

Sivula, Kevin, Florian Le Formal, and Michael Grätzel. 2011. Solar Water Splitting: Progress Using Hematite (α-Fe_2O_3) Photoelectrodes. *ChemSusChem* 4: 432–49. [CrossRef]

Song, Zhaoning, Suneth C. Watthage, Adam B. Phillips, Brandon L. Tompkins, Randy J. Ellingson, and Michael J. Heben. 2015. Impact of Processing Temperature and Composition on the Formation of Methylammonium Lead Iodide Perovskites. *Chemistry of Materials* 27: 4612–19. [CrossRef]

Souza, Flavio L., Kirian P. Lopes, Pedro A. P. Nascente, and Edson R. Leite. 2009. Nanostructured Hematite Thin Films Produced by Spin-Coating Deposition Solution: Application in Water Splitting. *Solar Energy Materials and Solar Cells* 93: 362–68. [CrossRef]

Stewart, Robert J., Christopher Grieco, Alec V. Larsen, Joshua J. Maier, and John B. Asbury. 2016. Approaching Bulk Carrier Dynamics in Organo-Halide Perovskite Nanocrystalline Films by Surface Passivation. *The Journal of Physical Chemistry Letters* 7: 1148–53. [CrossRef]

Svanström, Sebastian, T. Jesper Jacobsson, Gerrit Boschloo, Erik M. J. Johansson, Håkan Rensmo, and Ute B. Cappel. 2020. Degradation Mechanism of Silver Metal Deposited on Lead Halide Perovskites. *ACS Applied Materials and Interfaces* 12: 7212–21. [CrossRef]

Tamirat, Andebet Gedamu, John Rick, Amare Aregahegn Dubale, Wei Nien Su, and Bing Joe Hwang. 2016. Using Hematite for Photoelectrochemical Water Splitting: A Review of Current Progress and Challenges. *Nanoscale Horizons* 1: 243–67. [CrossRef]

Tan, Chih Shan, Yi Hou, Makhsud I. Saidaminov, Andrew Proppe, Yu Sheng Huang, Yicheng Zhao, Mingyang Wei, Grant Walters, Ziyun Wang, Yongbiao Zhao, and et al. 2020. Heterogeneous Supersaturation in Mixed Perovskites. *Advanced Science* 7: 1903166. [CrossRef] [PubMed]

Tee, Si Yin, Khin Yin Win, Wee Siang Teo, Leng Duei Koh, Shuhua Liu, Choon Peng Teng, and Ming Yong Han. 2017. Recent Progress in Energy-Driven Water Splitting. *Advanced Science* 4: 1600337. [CrossRef] [PubMed]

Thomson, Stuart. 2018. *Observing Phase Transitions in a Halide Perovskite Using Temperature Dependent Photoluminescence Spectroscopy*. Livingston: Edinburgh Instruments, AN_P45.

Tsege, Ermias Libnedengel, Timur Sh Atabaev, Md Ashraf Hossain, Dongyun Lee, Hyung Kook Kim, and Yoon Hwae Hwang. 2016. Cu-Doped Flower-like Hematite Nanostructures for Efficient Water Splitting Applications. *Journal of Physics and Chemistry of Solids* 98: 283–89. [CrossRef]

Turren-Cruz, Silver Hamill, Anders Hagfeldt, and Michael Saliba. 2018. Methylammonium-Free, High-Performance, and Stable Perovskite Solar Cells on a Planar Architecture. *Science* 362: 449–53. [CrossRef] [PubMed]

Uratani, Hiroki, and Koichi Yamashita. 2017. Charge Carrier Trapping at Surface Defects of Perovskite Solar Cell Absorbers: A First-Principles Study. *The Journal of Physical Chemistry Letters* 8: 742–46. [CrossRef]

Vidal, Rosario, Jaume-Adrià Alberola-Borràs, Núria Sánchez-Pantoja, and Iván Mora-Seró. 2021. Comparison of Perovskite Solar Cells with Other Photovoltaics Technologies from the Point of View of Life Cycle Assessment. *Advanced Energy and Sustainability Research* 2: 2000088. [CrossRef]

Wang, Jie, Xiaolian Chen, Fangyuan Jiang, Qun Luo, Lianping Zhang, Mingxi Tan, Menglan Xie, Yan-Qing Li, Yinhua Zhou, Wenming Su, and et al. 2018. Electrochemical Corrosion of Ag Electrode in the Silver Grid Electrode-Based Flexible Perovskite Solar Cells and the Suppression Method. *Solar RRL* 2: 1800118. [CrossRef]

Wang, Lei, Xuemei Zhou, Nhat Truong Nguyen, and Patrik Schmuki. 2015. Plasmon-Enhanced Photoelectrochemical Water Splitting Using Au Nanoparticles Decorated on Hematite Nanoflake Arrays. *ChemSusChem* 8: 618–22. [CrossRef]

Wang, Songcan, Jung Ho Yun, and Lianzhou Wang. 2017. Nanostructured Semiconductors for Bifunctional Photocatalytic and Photoelectrochemical Energy Conversion. *Semiconductors and Semimetals* 97: 315–47. [CrossRef]

Wu, Shaohang, Rui Chen, Shasha Zhang, B. Hari Babu, Youfeng Yue, Hongmei Zhu, Zhichun Yang, Chuanliang Chen, Weitao Chen, Yuqian Huang, and et al. 2019. A Chemically Inert Bismuth Interlayer Enhances Long-Term Stability of Inverted Perovskite Solar Cells. *Nature Communications* 10: 1–11. [CrossRef]

Wu, Wu-qiang, Peter N. Rudd, Zhenyi Ni, Charles Henry Van Brackle, Haotong Wei, Qi Wang, Benjamin R. Ecker, Yongli Gao, and Jinsong Huang. 2020. Reducing Surface Halide Deficiency for Efficient and Stable Iodide- Based Perovskite Solar Cells. *Journal of the American Chemical Society* 142: 3989–96. [CrossRef] [PubMed]

Wu, Yinghui, Dong Wang, Jinyuan Liu, and Houzhi Cai. 2021. Review of Interface Passivation of Perovskite Layer. *Nanomaterials* 11: 1–19. [CrossRef] [PubMed]

Xi, Lifei, and Kathrin M. Lange. 2018. Surface Modification of Hematite Photoanodes for Improvement of Photoelectrochemical Performance. *Catalysts* 8: 497. [CrossRef]

Xie, Yiyuan, Yang Ju, Yuhki Toku, and Yasuyuki Morita. 2018. Synthesis of a Single-Crystal Fe_2O_3 Nanowire Array Based on Stress-Induced Atomic Diffusion Used for Solar Water Splitting. *Royal Society Open Science* 5: 172126. [CrossRef]

Xu, Ming, Jing Feng, Xia Li Ou, Zhen Yu Zhang, Yi Fan Zhang, Hai Yu Wang, and Hong Bo Sun. 2016. Surface Passivation of Perovskite Film by Small Molecule Infiltration for Improved Efficiency of Perovskite Solar Cells. *IEEE Photonics Journal* 8: 1–7. [CrossRef]

Yan, Keyou, Yongcai Qiu, Shuang Xiao, Junbo Gong, Shenghe Zhao, Jiantie Xu, Xiangyue Meng, Shihe Yang, and Jianbin Xu. 2017. Self-Driven Hematite-Based Photoelectrochemical Water Splitting Cells with Three-Dimensional Nanobowl Heterojunction and High-Photovoltage Perovskite Solar Cells. *Materials Today Energy* 6: 128–35. [CrossRef]

Yang, Dongwen, Wenmei Ming, Hongliang Shi, Lijun Zhang, and Mao-Hua Du. 2016. Fast Diffusion of Native Defects and Impurities in Perovskite Solar Cell Material $CH_3NH_3PbI_3$. *Chemistry of Materials* 28: 4349–57. [CrossRef]

Yang, Woon Seok, Byung-Wook Park, Eui Hyuk Jung, and Nam Joong Jeon. 2017a. Iodide Management in Formamidinium-Lead-Halide–Based Perovskite Layers for Efficient Solar Cells. *Science* 356: 1376–791. [CrossRef]

Yang, Wooseok, Rajiv Ramanujam Prabhakar, Jeiwan Tan, S. David Tilley, and Jooho Moon. 2019. Strategies for Enhancing the Photocurrent, Photovoltage, and Stability of Photoelectrodes for Photoelectrochemical Water Splitting. *Chemical Society Reviews* 48: 4979–5015. [CrossRef]

Yang, Ye, Mengjin Yang, David T. Moore, Yong Yan, Elisa M. Miller, Kai Zhu, and Matthew C. Beard. 2017b. Top and bottom surfaces limit carrier lifetime in lead iodide perovskite films. *Nature Energy* 2: 1–7. [CrossRef]

Yilmaz, Ceren, and Ugur Unal. 2016. Morphology and Crystal Structure Control of α-Fe_2O_3 Films by Hydrothermal-Electrochemical Deposition in the Presence of Ce^{3+} and/or Acetate, F^- Ions. *RSC Advances* 6: 8517–27. [CrossRef]

Yun, Jae Sung, Jincheol Kim, Trevor Young, Robert J. Patterson, Dohyung Kim, Jan Seidel, Sean Lim, Martin A. Green, Shujuan Huang, and Anita Ho-Baillie. 2018. Humidity-Induced Degradation via Grain Boundaries of $HC(NH_2)_2PbI_3$ Planar Perovskite Solar Cells. *Advanced Functional Materials* 28: 1705363. [CrossRef]

Zhang, Chengxi, Dai-bin Kuang, and Wu-Qiang Wu. 2020. A Review of Diverse Halide Perovskite Morphologies for Efficient Optoelectronic Applications. *Small Methods* 4: 1900662. [CrossRef]

Zhang, Hong, Mohammad Khaja Nazeeruddin, and Wallace C. H. Choy. 2019a. Perovskite Photovoltaics: The Significant Role of Ligands in Film Formation, Passivation, and Stability. *Advanced Materials* 31: 1–29. [CrossRef] [PubMed]

Zhang, Moyao, Qi Chen, Rongming Xue, Yu Zhan, Cheng Wang, Junqi Lai, Jin Yang, Hongzhen Lin, Jianlin Yao, Yaowen Li, and et al. 2019b. Reconfiguration of Interfacial Energy Band Structure for High-Performance Inverted Structure Perovskite Solar Cells. *Nature Communications* 10: 1–9. [CrossRef] [PubMed]

Zhang, Rong, Yiyu Fang, Tao Chen, Fengli Qu, Zhiang Liu, Gu Du, Abdullah M. Asiri, Tao Gao, and Xuping Sun. 2017. Enhanced Photoelectrochemical Water Oxidation Performance of Fe_2O_3 Nanorods Array by S Doping. *ACS Sustainable Chemistry and Engineering* 5: 7502–6. [CrossRef]

Zhang, Yong, Licheng Tan, Qingxia Fu, Lie Chen, Ting Ji, and Yiwang Chen. 2016a. Enhancing the grain size of organic halide perovskites by sulfonate-carbon nanotube incorporation in high performance perovskite solar cells. *Chemical Communications* 52: 5674–77. [CrossRef]

Zhang, Yuchao, Hongwei Ji, Wanhong Ma, Chuncheng Chen, Wenjing Song, and Jincai Zhao. 2016b. Doping-Promoted Solar Water Oxidation on Hematite Photoanodes. *Molecules* 21: 868. [CrossRef]

Zhang, Yuchao, Shiqi Jiang, Wenjing Song, Peng Zhou, Hongwei Ji, Wanhong Ma, Weichang Hao, Chuncheng Chen, and Jincai Zhao. 2015. Nonmetal P-Doped Hematite Photoanode with Enhanced Electron Mobility and High Water Oxidation Activity. *Energy and Environmental Science* 8: 1231–36. [CrossRef]

Zhao, Jingjing, Xiaopeng Zheng, Yehao Deng, Tao Li, Yuchuan Shao, Alexei Gruverman, Jeffrey Shield, and Jinsong Huang. 2016a. Is Cu a Stable Electrode Material in Hybrid Perovskite Solar Cells for a 30-Year Lifetime? *Energy and Environmental Science* 9: 3650–56. [CrossRef]

Zhao, Lianfeng, Ross A. Kerner, Zhengguo Xiao, Yunhui Lisa, Kyung Min Lee, Jeffrey Schwartz, Barry P. Rand, and Barry P. Rand. 2016b. Redox Chemistry Dominates the Degradation and Decomposition of Metal Halide Perovskite Optoelectronic Devices Redox Chemistry Dominates the Degradation and Decomposition of Metal Halide Perovskite Optoelectronic Devices. *ACS Energy Letters* 1: 595–602. [CrossRef]

Zhao, Yi Cheng, Wen Ke Zhou, Xu Zhou, Kai Hui Liu, Da Peng Yu, and Qing Zhao. 2017. Quantification of Light-Enhanced Ionic Transport in Lead Iodide Perovskite Thin Films and Its Solar Cell Applications. *Light: Science and Applications* 6: e16243. [CrossRef] [PubMed]

Zheng, Haiying, Guozhen Liu, Liangzheng Zhu, Jiajiu Ye, Xuhui Zhang, Ahmed Alsaedi, Tasawar Hayat, Xu Pan, and Songyuan Dai. 2018. The Effect of Hydrophobicity of Ammonium Salts on Stability of Quasi-2D Perovskite Materials in Moist Condition. *Advanced Energy Materials* 8: 1–8. [CrossRef]

Zheng, Kaibo, and Tönu Pullerits. 2019. Two Dimensions Are Better for Perovskites. *Journal of Physical Chemistry Letters* 10: 5881–85. [CrossRef]

Zheng, Xiaopeng, Joel Troughton, Nicola Gasparini, Yuanbao Lin, Mingyang Wei, Yi Hou, Jiakai Liu, Kepeng Song, Zhaolai Chen, Chen Yang, and et al. 2019. Quantum Dots Supply Bulk- and Surface-Passivation Agents for Efficient and Stable Perovskite Solar Cells. *Joule* 3: 1963–76. [CrossRef]

Advanced Energy Management Systems and Demand-Side Measures for Buildings towards the Decarbonisation of Our Society

Fabiano Pallonetto

1. Introduction

The synchronised power system is one of the top human engineering achievements of the twentieth century (Almassalkhi and Hiskens 2015). Throughout the early part of the twentieth century, the number of electrified cities increased, leading to a connected system identified as the synchronised power grid (Hughes 1993). Since then, the power system has evolved, presenting new hurdles for system operators, both at transmission (TSO) and distribution (DNO) level. Environmental impact, increased penetration of renewable energies, the continued growth in demand, and the uncertainty of fuel reserves are just a tiny part of a set of new challenges that the power systems research community is addressing (Nolan and O'Malley 2015). Grid reinforcement can be part of the solution to these challenges; however, it is costly and does not always improve the system's robustness. Recent blackouts in Germany, Texas and Italy caused by a domino effect of small evaluation mistakes are the empirical evidence of a more significant research problem (Boemer et al. 2011; Gimon and Fellow 2021). Assessing these issues requires complex modelling and extensive computational capabilities and can lead to counterintuitive results.

In 2012, researchers at the Max Planck Institute for Dynamics and Self Organisation in Göttingen, discovered that the power grid is affected by Braess' paradox. This phenomenon was discovered by the German mathematician Dietrich Braess in 1968 while undertaking studies on road network models (Pas and Principio 1997). The definition of the paradox as stated in Braess (1968, p. 1) is as follows:

> "For each point of a road network, let there be given the number of cars starting from it and the destination of the cars. Under these conditions, one wishes to estimate the distribution of traffic flow. Whether one street is preferable to another depends not only on the quality of the road but also on the density of the flow. If every driver takes the path that looks

most favourable to him, the resultant running times need not be minimal. Furthermore, it is indicated by an example that an extension of the road network may cause a redistribution of the traffic that results in longer individual running times."

The same principle applies to the power grid, where adding one or more links to the power grid could degrade the overall efficiency of the system (Witthaut and Timme 2012). The increasing energy demand, environmental concerns and the installation of interconnected Renewable Energy Systems (RES) add to the underlying complexity of the problem. RESs are associated with low carbon emissions; however, for the general public, the threats caused by their intermittent nature are underrated and not well understood (Vargas et al. 2015). Despite these technical challenges, post-Kyoto regulations endorsed by the European Union have established the target of full decarbonisation by 2050 (International Energy Agency 2016).

Historically, system operators owned most of the system, and they planned the generation mix a day-ahead while tuning the daily electricity production to compensate for unplanned generator outages or unexpected load oscillations. High penetration of renewable energy increases the complexity of this process to an unexplored level (Bozalakov et al. 2014). Furthermore, the increasing electricity consumption caused by a larger adoption of low-carbon technologies in end-use sectors represents another influencing factor on the demand side of the network. The increasing percentage of electric vehicles and heat pumps can strain the network capacity and ultimately lead to blackouts (Veldman et al. 2011).

Extending the control to the demand-side of the system can become part of the solution (Cecati et al. 2011; Fuller et al. 2011; McKenna and Keane 2016; Nolan and O'Malley 2015; Torriti et al. 2010). The adaptability of demand is not new to the dynamics of the power grid infrastructure. These measures have been promoted in various countries across the world to clip winter or summer peaks and defer grid reinforcement (Paterakis et al. 2017).

Following these measures, power grids have gradually adapted to the increased demand and are adopting a higher percentage of new renewable energy generators such as photovoltaics and wind turbines. System operators did not embrace the penetration of RES until it started to affect the supply/demand balance of the whole system, altering the system frequency beyond safety thresholds (Ulbig et al. 2014). At that point, system operators had to take into account not only the unscheduled load demand but also the variability of power generation caused by weather conditions (Bozalakov et al. 2014). These open challenges cannot be addressed within the boundaries of the existing power system (Farhangi 2010). The integration of information technology, communications, and circuit infrastructure could lead to disruptive technological innovations for the integration of higher penetration of RES, increasing assets efficiency and reducing overall carbon emissions (Yan et al. 2013).

In this chapter, relevant research on the topic of the built environment, Demand Response (DR) and optimisation algorithms for DR are critically reviewed, and the key results and advancements in the area are contextualised. Section 2 assesses the impact of buildings on the energy system while Section 3 introduces the concept of demand-side management and DR, assessing the advantages and disadvantages of automatic versus manual DR. Section 4 assesses how an energy management systems (EMS) can be used to implement demand response strategies. Section 4.1 examines the idea of home area network (HAN) or local area network (LAN) and how technological innovation. is changing the interaction between users and buildings. This part discusses the effect of technological advancements in developing interconnected appliances and communication protocols. It also focuses on the definition and characterisation of EMS in a smart-grid scenario. This section discloses several research gaps on the communication infrastructure between buildings and the power grid. In this part, an extensive analysis of optimisation algorithms for DR is presented. Section 5 analyses how advanced controllers can foster the transition to a lower-carbon economy, reducing the energy costs and facilitating the integration of renewable in the system. In Section 6, an overall contribution of buildings towards the full decarbonisation is analysed. The chapter concludes with Section 7 by identifying a path towards the decarbonisation of our society through advanced energy management systems.

2. Buildings as a Fundamental Asset for the Decarbonisation

Generally, the building stock can be divided into residential and commercial buildings. Census data or building surveys can be used to collect relevant information to characterise the building stock at the country level (Mata et al. 2014). In recent years, building energy certificates and other geographical information systems have enriched existing databases and increased data accuracy (Võsa et al. 2021). Moreover, some European and national projects have compiled available information for a country or group of countries or new methodologies to certify energy rating for buildings such as Active Building Research Programme (2013); ePANACEA (2020); Episcope EU (2013); U-CERT (2019).

The built environment accounts at least for 40% of the total electricity consumption (Pérez-Lombard et al. 2008). Seasonal peaks are caused by increased lighting, cooling or heating demands, and such profiles are also peaking wholesale electricity prices and reduced reliability due to tight generation reserve margins. Higher penetration of electric vehicles and heat pumps have caused an increased demand which is being modulated by RES and the smart grid rollout (Arteconi et al. 2013; Smith 2010). In these transitional circumstances, where the global target of 2050 is looming closer, the motivations of massively employing demand response programs using buildings have never been so compelling.

As illustrated in Figure 1, the European Union has a large and old building stock that requires retrofitting and upgrading. Despite a significant variation in the EU members' built environment, full decarbonisation would not be possible without a massive retrofitting plan across the EU. Furthermore, within the EU, there is a large variance in the energy consumption required to heat or cool the buildings. Figure 2 illustrates how more than half of the European countries have an average consumption per square meter above the average and far from NZEB or passive building standards.

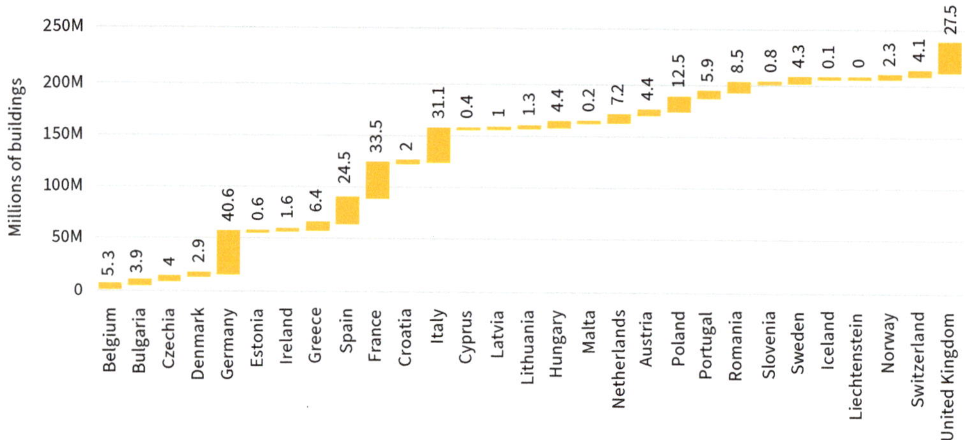

Figure 1. Building older than 15 years old of retrofit potential for Europe. Source: Graphic by author, adapted from Pallonetto et al. (2022).

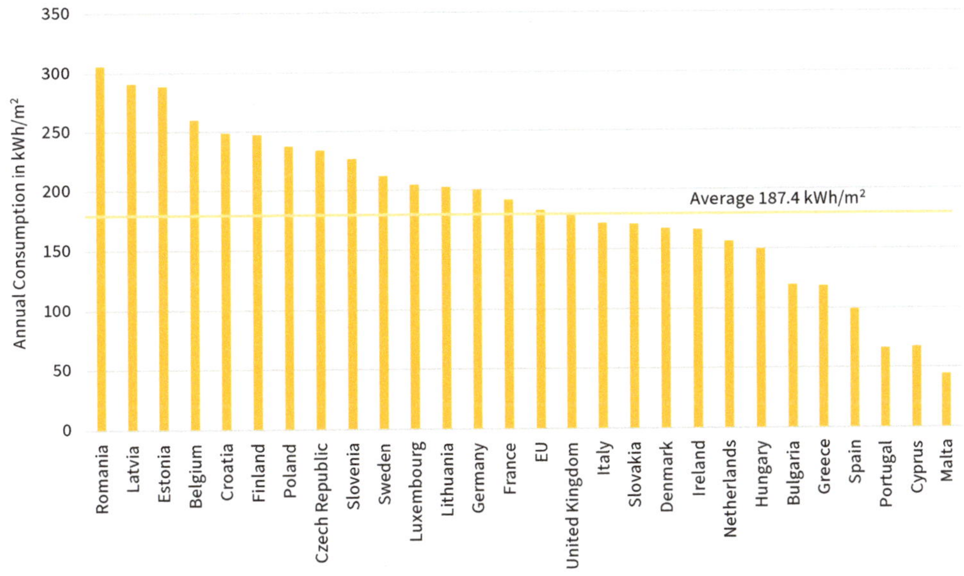

Figure 2. Average building consumption for square meter across Europe. Source: Graphic by author, data from Eurostat (2020).

Among low carbon technologies worth mentioning for the built environment, there is the heat pump. Heat pumps can reduce the energy consumption of the building and can be installed and used in tandem with renewable energy sources such as photovoltaic systems. Heat pumps deployment can also support the evolution of the power system and contribute to the high penetration of renewable sources through end-users participation in demand-response markets. Such a combination makes heat pumps more attractive. Such an advanced technology can meet the heat demand of the building while reducing carbon emissions by a factor of three (Boemer et al. 2011; Eriksen et al. 2005; Paterakis et al. 2017). From the power system perspective, innovations such as building home automation, smart grid rollout, diffusion of intelligent appliances and EMS integration are necessary prerequisites to boost the efficiency of the power system while increasing the RES penetration to meet the emissions target (European Commission 2010). These features will enable electricity end-users to modulate their electricity consumption by dynamically responding to fluctuations in the power generation caused by RES (Mohsenian-Rad et al. 2010b; Pedersen et al. 2017). End-users can manually or automatically alter their consumption patterns via home automation or EMS controllers in a smart built environment. Grid reliability and evolution in the regulations to enable DR in the electricity market are the primary reasons for the intense interest to develop intelligent EMS that can reduce energy costs and dynamically adapt to grid constraints (Conka et al. 2014; Pereira and da Silva 2017).

3. Demand-Side Measures towards the Full Decarbonisation of Our Society

DR is one of the Demand Side Measures (DSM) measures promoted as a mechanism to increase the percentage of renewable energies in the system (Albadi and El-Saadany 2008). It is defined as "changes in electricity use by demand-side resources from their normal consumption patterns in response to changes in the price of electricity or to incentive payments designed to induce lower electricity use at times of high wholesale market prices or when system reliability is jeopardised" (Gils 2014, p. 1).

This measure is being implemented worldwide by various TSOs through the remuneration of DR aggregators in the electricity market. In some cases, the aggregators were the TSOs or a related entity. Aggregators can control the energy demand of residential and commercial buildings, representing 40% of the total primary energy consumption.

The widespread adoption of DR programs leads to a paradigm shift in how TSOs manage the grid. Such changes require a bi-directional communication link between buildings and operators occurring with the smart grid rollout (Silva et al. 2012).

3.1. DR Objective and Programs

A DR signal by an aggregator or TSO, triggers the intentional reshape of the electricity demand profile. The variation can be measured as the level of instantaneous demand or total electricity consumption deferred. DR assets can dynamically change the electricity demand curve, providing peak shaving, frequency control, load shifting and forcing measures (Nolan and O'Malley 2015).

DR programs can be classified by financial schemes. DR aggregators can remunerate DR events to end-users with Incentive Based Programs (IBP) or Price Based Programs (PBP) (Aghamohamadi et al. 2018; Albadi and El-Saadany 2008). The difference between PBP and IBP is that in the latter one, end users get a financial benefit or a price reduction due to their affiliation to the scheme.

Among IBP programs, there are market-centred schemes. Market-centred DR is for medium size users or demand response aggregators. The schemer requires market access and equipment to connect to the TSOs communication infrastructure. In these cases, the financial benefit for the user is correlated with the flexibility provided. As described in Qdr (2006), the Market-centred DR programs include the following categories:

1. Emergency DR. These incentives are proportional to load reduction during emergency reserve shortfalls events. When utilised, a demand reduction signal is sent to all large users enrolled.
2. Capacity Market Program. This program is for users who can precisely estimate a determined load reduction when system contingencies happen. The users involved have a day-ahead notice, and if they do not answer, they are penalised. The payment is based on the declared peak load reduction achievable by the asset.
3. Ancillary Service Program. Operating reserve is bid in terms of curtailment capacity. If the bid is accepted after the measure is implemented, customers are paid the spot market price.
4. Demand Bidding (also called Buyback). In this program, consumers bid the load reduction in the wholesale market, where a bid is accepted if it is less than the market price.

In the PBP, the electricity price is directly correlated with the market price (Albadi and El-Saadany 2008). The objective of these schemes is to flatten the electricity profile to lower peak demand. Typical PBPs may encompass some or all of the following features:

1. Time of Use Tariffs (TOU) tariffs where there are two or more time blocks such as night, peak and off-peak electricity prices.
2. Critical Peak Price (CPP) is often utilised during high contingencies or higher electricity usage for a few days or hours or months.
3. Extreme Day Price (EDP) is a specific subset of the CPP program. In this case, the electricity tariff increases during a specific time of the day. During the rest of the day, a flat tariff is used. In this case, the DR event is set for one or more days.
4. Real Time Price (RTP), where the electricity tariff is synchronised to the market time resolution, which typically changes every hour.

3.1.1. Lessons Learnt from DR Pilot Programs

The development and testing of demand response programs have shown benefits and challenges yet to be addressed. China started piloting DR programs in 1990, but energy shortages during 2003–2008 reinforced the implementation of DR pilot projects. Since then, the established DR programs were based on TOU rates, Curtailable/Interruptible loads, the use of off-peak storage devices such as heat storage boilers and ice-storage air conditioners. These programs highlighted challenges related to human behaviour, absence of competitive electricity markets, customers unawareness of prices and absence of recovery mechanism for users and utility investments (Tahir et al. 2020). It should be noted that shifting from manual load-shifting in response to network stress at predictable times of day to dynamic programs price- or quantity-based requires additional Information and Communication Technology (ICT) support. The uncertainties associated with human behaviour are the main challenges in the implementation of these programs. The RealValue project included a test bed of more than 800 households across three different EU nations (Darby et al. 2018). Customers who were used to paying for a service from the power system became prosumers through distributed generation, storage and demand response. As a consequence, the connections across the resources dynamically managed by users and the established actors such as utilities required innovative ideas and additional user and ICT support to provide a fraction of potential theoretical flexibility estimated (Darby 2020).

Table 1 summarises and compares the experience of different demand response trials deployed worldwide and highlights different types of barriers, benefits and technology enablers (Lu et al. 2020). The table shows how critical is the use and identification of shiftable/curtailable loads coupled with storage to enable the deployment of DR programs. Switching to automated direct load control is complex and requires a reliable and trustworthy IT infrastructure and data exchange mechanisms. Additional, similar works highlight the importance of user acceptance and how occupants behaviour is the primary barrier to the success of these programs (Anaya and Pollitt 2021).

Table 1. Demand response measures and limitations Legend: ✓ Support it, (✓) Partially supported.

Type of Measure	Distributed Generation	Shiftable/Curtailable Load	Storage	Complexity	Signal Type	Signal Volatility	Privacy Risk	Price Risk	Human Behaviour Uncertainty Risk	Notes
TOU pricing	(✓)	✓	✓	Low	Price	Static	Low	Low	High	Distributed generation can support it if is controllable or synchronise with high prices periods
Dynamic pricing	(✓)	✓	✓	High	Price	Dynamic	Medium	High	High	It requires a complex IT infrastructure, high risk
Fixed load capping	✓	✓	✓	Low	Volume	Static	Low	None	Low	If hardware controlled is simple to deploy
Dynamic load capping	✓	✓	✓	High	Volume	Dynamic	Low	None	High	Can be complex and the signal can be hacked
Direct load control	(✓)	✓	✓	High	Control	Predefined	High	None	Low	Complex and high privacy risk, high potential, it requires detailed data
Critical Peak Price	(✓)	✓	✓	Low	Price	Static	Low	None	Low	Simple but limited benefits, depends on occupant behaviour
Extreme Day Price	(✓)	✓	✓	Low	Price	Static	Low	None	Low	Simple but limited benefits, depends on occupant behaviour

Source: Table by author, data from Lu et al. (2020).

3.1.2. Summary of DR Benefits

Figure 3 shows the correlation between stakeholders and DR schemes. The benefit for end-users is typically a reduction in the electricity bill or financial remuneration. On the other side, operators such as TSOs can increase the efficiency of the market, reducing the volatility and the use of peak generators (Albadi and El-Saadany 2008). Moreover, from a broad market perspective, DR programs can reduce the electricity price increasing the capacity of the system (Aghamohamadi et al. 2018; Braithwait and Eakin 2002). Such benefits also defer grid reinforcement, reducing the running costs and improving the market efficiency (Paterakis et al. 2017;

Qdr 2006). The overall reliability of the grid increases thanks to the use of DR schemes because the dynamic demand curtailment reduces the outage and transmission strains risks. Furthermore, reducing the contribution of peak generators and reducing the curtailment operation caused by a surplus of RES generation (EDP Consortium 2016; Hamidi et al. 2009) reduces the carbon emission of the system.

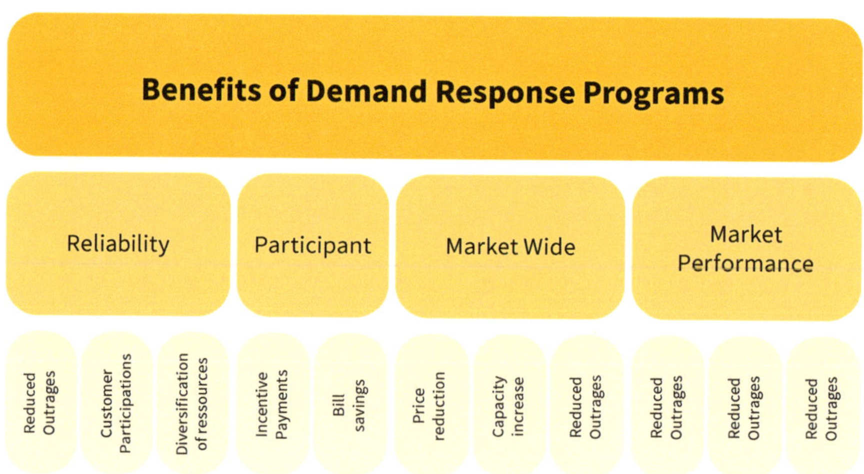

Figure 3. Overview of demand response schemes for different stakeholders and from the market perspective. Source: Graphic by author, data from Qdr (2006).

3.1.3. DR Operational Challenges

DR programs can also be classified by the level of automation. Manual DR measures rely on human actions to reduce or increase loads or alter the demand profile. For semi-automated DR measures, the user controls a digital system to trigger a demand response action. When using automated DR strategies, an external signal operates a programmed method and consequently does no human intervention is required. However, in this case, users must always be able to override the system (Piette et al. 2006; Rothleder and Loutan 2017).

Within the many challenges to DR schemes, a key factor is the availability of reliable resources. In some cases the system cannot respond to the DR signal; therefore capturing the available flexibility and capacity of the resources using flexibility metrics and metering equipment is a fundamental requirement to implement such programs. Additionally, the stochastic consumer behaviour could reduce the benefit of DR schemes if not considered. The high variability of the DR resources could be smoothed by aggregation. In fact, in Nolan and O'Malley (2015), when stochastic behaviour is accounted for, the aggregation of few thousand households represents a stable DR asset. Hence, the domestic sector electricity demand has the potential to provide services such as spinning reserves, frequency controls and Short Term Operating Reserve (STOR). In the building sector, STOR can be exploited using automation or increasing the energy awareness of the occupants.

4. The Role of Advanced Energy Management Systems

The smart grid is the next generation of the power system that enables a two-way communication channel from the end-user to the TSO, with the objective of monitoring and controlling end-user electricity demand in a power system with high RES penetration (Farhangi 2010).

The employment of smart grid technologies will support European countries to reach their CO_2 emissions reduction target and renewable generation increase (European Union 2017). In particular, the EU climate-neutral goal by 2050 is ambitious; increase penetration of RES to meet annual maximum generation. To reach the target, RES generators with their variable and uncertain electricity supply have been connected to the grid, thus increasing the operational challenges for system operators. The increasing wind and solar penetration impose significant technical difficulties such as large frequency variations, which require strict voltage control. These requirements have led to the utilisation of the smart grid for automatic DR projects in commercial, industrial and residential buildings (O'Sullivan et al. 2014).

Automatic DR control of heating, cooling and light systems requires the presence of one or more interconnected sensors and one or more corresponding controller devices. The sensors are usually connected to the cloud with a HAN or LAN. A DR controller device, called EMS, can read the sensor data and reshape the energy demand of the building according to a price or an interrupt signal from a TSO. Some of the systems are defined as intelligent. In this context, an intelligent appliance, algorithm or control indicates a system that uses various artificial intelligence computing approaches like neural networks, Bayesian network or optimisation techniques (Antsaklis and Passino 1993).

4.1. HAN/LAN, Definition and Developments

A HAN is a dedicated data network infrastructure within buildings built for data transfer and device communication. In the late 1990s, HANs became the emerging gateway to connect devices to the Internet. The availability of Internet access in buildings has boosted the diffusion of HAN systems since the early 2000s (Clements et al. 2011).

The de facto standard for the first period of HAN development was the Ethernet and 802.11 Wi-Fi standard (Huq and Islam 2010).

In the coming decades, the rate of diffusion of HANs across buildings is set to increase exponentially. In fact, the total number of connected devices is expected to reach 50 billion by 2030 (Ahmed et al. 2016). A good percentage of such devices will exchange sensor data in real-time and require low bandwidth, meaning that there is less stress on the overall network infrastructure. The phenomenon of connecting any device to the Internet is covered by the all-encompassing phrase the IoT (Ahmed et al. 2016; Atzori et al. 2010).

The use IoT devices connected to the HAN to perform actions that could reduce carbon emissions or peak power consumption has different requirements than the standard use of HAN (Darby 2006). In fact, only a small percentage of HAN connected devices can be classified as smart-grid enabled.

A smart-grid enabled device provides a two-way communication system to utility companies and could be remotely controlled to increase the overall efficiency of the power grid (Balakrishnan 2012; Bazydło and Wermiński 2018). It is necessary with these devices to have a stable link and low bandwidth allocation (Gungor et al. 2011). The devices affected by these changes include thermostats, HVAC systems, major appliances, home automation systems, EMS, lighting, gas meters, water meters, and electricity meters.

In recent years, the increased installation of local renewable energy systems such as PV and solar panels has raised additional challenges for the HAN research community (Liserre et al. 2010). Controlling in real time a RES at the building level with smart-grid enabled devices increases the complexity of the problem.

Additionally, inhibitors to the adoption of HAN as part of the smart-grid infrastructure are categorised in Eustis et al. (2007):

- Energy pricing that provides financial benefits to control energy use more efficiently and enable consumers to reduce their costs.
- Open, flexible, secure and efficient communication protocol established and accessible.
- Compliance of the services with consumer choice and privacy; wherein the consumer, ultimately, is the decision-maker.

Despite a general awareness that an interconnected system can enable utilities to more effectively balance energy demands and integrate with renewable energies systems, the above challenges raise additional questions about the architecture of HAN, different communication protocols (Huq and Islam 2010) and the strategy and algorithms that can be implemented by an EMS connected to HANs.

As previously noted, there is a trade-off between keeping the HAN architecture simple and efficient versus safeguarding the privacy and security of the users. In a smart grid scenario, it is essential to guarantee the safety of the network against cyberattacks that could compromise the entire power grid.

The interconnection between the various devices and the HAN should consider the bandwidth allocation and the potential vulnerabilities that each device could expose. As illustrated in Figure 4, installing a Home Energy Gateway or EMS that can separate the smart-grid-enabled devices from other devices could increase the security of the system. This architecture design utilises a connected gateway as a demilitarised zone, enhancing the security of the data transfer and controls. The presence of an EMS or smart gateway is also mentioned in Clements et al. (2011), where they draw a clear distinction between the two different layouts.

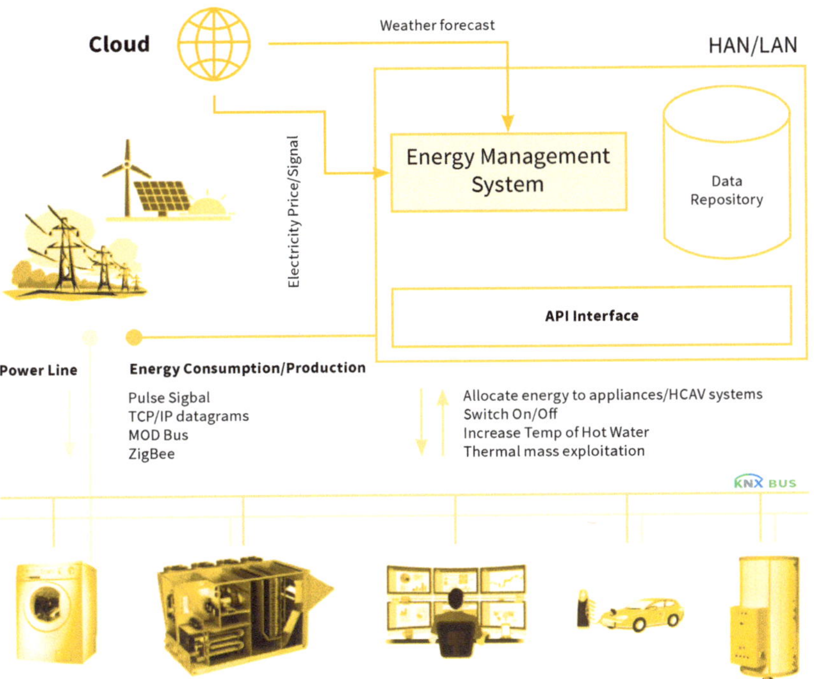

Figure 4. Overview of connected devices and local renewable energies systems to a HAN infrastructure via a home energy gateway devices. Source: Adapted from Pallonetto et al. (2021).

The layout in Figure 5 assumes the presence of a dedicated network with the TSO connected to the appliances. In this example, the home gateway is a proprietary stand-alone device. This layout keeps the data transmission physically separated through a virtual private network. The main advantage of this configuration is the increased data security and reliability. The dedicated network can also provide a minimum band allocation to ensure a sufficient data throughput. The main disadvantage of this layout is the high infrastructural cost. Hybrid designs require less network infrastructure because the appliances are connected to the utility using a shared network such as the Internet. In this scenario, the band allocation may represent a challenge, and several security risks have been identified (Huq and Islam 2010).

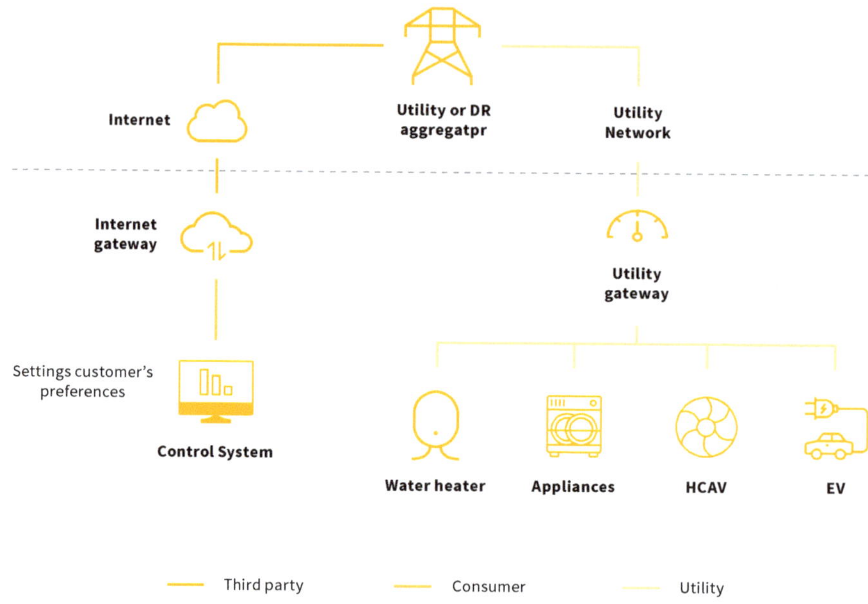

Figure 5. Overview of HAN infrastructure on a dedicated network. Source: Graphic by author, data from Pedreiras et al. (2002).

In contrast, in rural areas which lack sufficient network infrastructure, the layout in Figure 6 represents the only viable solution for data exchange between the HAN and TSO. Different communication protocols can also affect the layout of the interconnection between appliances and EMS or between the EMS and TSO. The following section examines the required band allocation, common communication protocols and their characteristics.

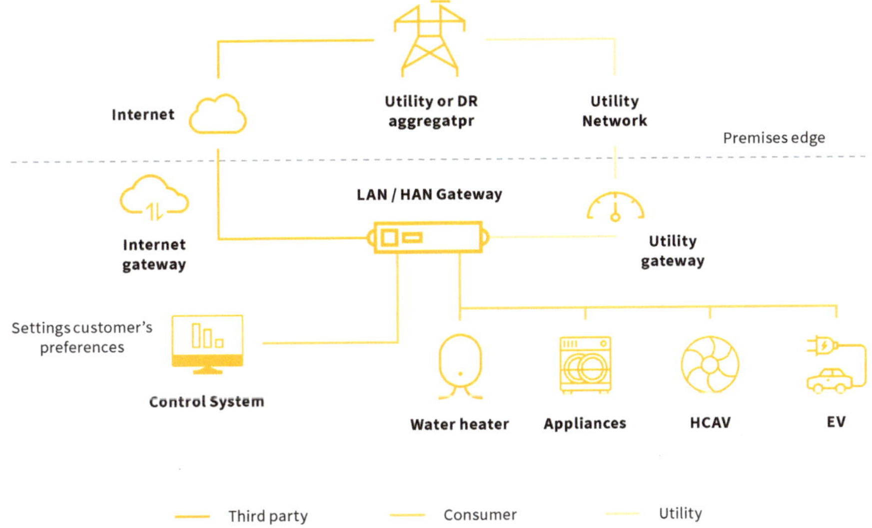

Figure 6. Overview of HAN infrastructure on a shared network. Source: Graphic by author, data from Pedreiras et al. (2002).

4.3. Communication Protocols for HAN

Communication protocols are used for data transmission between sensors and the HAN. During the last decade, various attempts at protocol standardisation have been made, each with limited success (Eustis et al. 2007). Different protocols can be categorised into three broad categories: new wires, no new wires and wireless, as illustrated in Table 2.

Table 2. Overview of different commercial communication protocols and evaluation of their suitability for smart grid applications.

Technology	Category	Frequency	Data Rate	Range	Power Consumption	Application
Bluetooth	wireless	2.4 GHz	25 MB/s	10 m	Low	Smart health devices, close range communication
DASH7	wireless	433 MHz	up to 200 KB/s	1000 m wireless	Low	Smart cities, smart buildings, smart transport, smart health
ZigBee	wireless	2.4 GHz, 0.915 GHz, 868 MHz	250 KB/s	up to 100 m wireless	Low	Smart homes, smart health
WiFi	wireless	2.4 GHz or 5 GHz	6.75 Gb/s	up to 1 km wireless	Medium	Smart homes, smart health, smart buildings
3G	wireless	0.85 GHz	24.8 Mb/s	1–8 km	High	Smart cities, smart transport, smart industries, smart grid
4G	wireless	up to 2.5 GHz	800 Mb/s	1–10 km	High	Smart homes, smart industry, smart grid
Ethernet	wired	up to 100 GHz	100 Gb/s	up to 500 m	Medium	Smart homes, smart industry, smart grid
Power-line	no new wires	up to 250 MHz	1.3 Gb/s	300 m	Low	Smart homes, broadband, smart grid

Source: Adapted from Ahmed et al. (2016).

In the new wires category, the de facto standard is ethernet (Pedreiras et al. 2002). The ethernet protocol is widely established and reliable. This protocol allows for greater integration with modern security mechanisms and procedures. The data transfer performance, from 10 Mb/s to more than 100 Gb/s, is sufficient for the throughput required. The main disadvantage of this technology is that the cable cannot be shared among different devices, so it requires a star design with one link for each device. Consequently, it is not extensively used in residential but mostly in commercial buildings (Ahmed et al. 2016).

In the wireless category, there are different protocols such as Bluetooth, ZigBee, Wifi and DASH7. A shared feature between the ZigBee, DASH7 and Bluetooth protocols are low energy consumption. However, Bluetooth has a lower communication range compared to ZigBee and DASH7 (Hayajneh et al. 2014). This limited range makes it suitable for smart health devices connected with phones or accessories paired to a central device. Although the hardware complexity of Bluetooth is lower than ZigBee or DASH7, it is not reliable for smart buildings or mobile devices that require longer ranges. Comparing DASH7 and ZigBee, the main differential is the trade-off between range and data rate transmission. ZigBee has a higher data transmission rate while it has a lower distance range.

In the no new wire category, one of the most common protocols is the powerline. Powerline protocols allow for communication via electricity sockets (Pavlidou et al. 2003). The disadvantages of this protocol are the limited distance range and the interference in the power supply, which can lead to fluctuations in the quality of the communication.

4.4. Overview of EMS Features and Objectives

During the last few years, the research and development on EMS have increased (Beaudin and Zareipour 2015). An EMS can be defined as a group of technologies used to manage the energy profile of a building, reducing the overall energy expenditure. Among these technologies, it is possible to include sensors, smart thermostats, electronic displays and smartphone apps that increase energy consumption awareness and offer remote or automatic control.

As suggested by Aman et al. (2013), an EMS should exhibit the following characteristics:

1. Monitor the energy consumption. The system provides energy consumption information at various time resolutions.
2. Disaggregation of the energy consumption. End-users can benefit from information about the real-time impact of appliances over a period of time.
3. Data availability and accessibility. The system makes the information available to the end-user via an interface. The interface is deployed as a physical device or through a web or mobile portal.
4. Appliances control. The EMS should provide programmable, remote and automatic control of devices
5. Data Integration. Integration of different types of information such as indoor temperature, humidity, acoustics, and light; and consumer historical data.
6. Ensuring cyber-security and data privacy. The system must restrict unauthorised access to third parties.
7. Intelligent controls and insights of data analytics. A requirement is to trigger smart actions that optimise energy consumption, maintaining consumer comfort.

As Paradiso et al. (2011) highlighted, EMS should perform intelligent actions that balance energy consumption and comfort. Specific algorithm techniques such as machine learning, data analysis or predictive control can be used. From the power system perspective, an EMS must be used more extensively for DR or to draw up a house profile o target energy improvement measures.

In Table 3, several studies have been examined based on some of features suggested by Aman et al. (2013) and Paradiso et al. (2011). All the EMSs have energy monitoring capabilities, and five have data disaggregation capabilities. The feature absent in the studies was the possibility to control load using intelligent algorithms.

In general, consumers are not aware of how an electrical system inside the building works, and, due to a low electricity price, they are not motivated to use their time to make energy-related decisions (Bartram et al. 2010). Therefore, to reach the objective of the EMS, the algorithm must perform intelligent actions to balance consumer comfort and energy consumption. Moreover, in a DR scenario, a smart EMS can adjust the power consumption to reduce the cost by exploiting the price signals, such as RTP, CPP or TOU. The action reduces the responsibility of the consumer to control and manipulate all their appliances all the time, while also providing flexibility to the power system for the integration of RES. The load controllability and the use of intelligent algorithms represent a research gap in the current literature that should be addressed.

Table 3. Features and characteristics of EMS technologies.

Evaluation Criteria	Metering and Analytics	Dis-Aggregation	Availability	Interoperability	Scalability	Actuators	Cybersecurity	Smart Controls and Intelligence
Totu (Totu et al. 2013)	Yes	No	Not discussed	Yes for large scale infrastructures	Yes	Yes	No	Yes advanced algorithms
PERSON (Yang and Li 2010)	Yes (API)	Yes	Decentralised at user premise; no web or mobile interface	Not discussed	High scalability Low cost and low power consumption	Manual remote control of the switches and dimmers in the home.	No	Context-aware intelligent algorithm
Bess (Mahfuz-Ur-Rahman et al. 2021)	Yes	No	Not available	Yes	Not available	Yes	No	Yes, smart AI control
WattDepot (Brewer and Johnson 2010)	Yes	No	Web based interface	No	Open source; freely available not scalable as it is	No	Limited	No
Viridiskope (Kim et al. 2009)	Yes (discontinued)	Yes	Not discussed	No API present	Requires indirect sensors; no inline installation requires	No	No	No
Mobile feedback (Weiss et al. 2009)	Yes	Yes	Interactive; readily available feedback on smartphone	Not discussed	Low scalability because of the mobile app	No	No	No
DEHEMS (Liu et al. 2013)	Yes	Yes	Web based UI, real-time display unit	No API but possibility to have integrated sensors, electric supply, gas supply line	Medium, it requires third party sensors	No	No	No
EnergyWiz (Petkov and Foth 2011)	Yes	No	Mobile phone app	Integrated historical usage and user info from peers	Requires mobile app installation	No	No	No
Nobel (Karnouskos 2011)	Yes	Yes	Mobile phone app	No	Low, requires mobile app installation and sensor integration not present	No	No	Yes, smart algorithm but requires human interaction
Simapi (Pallonetto et al. 2021)	Yes	No	Web and mobile app	API	Yes, high scalability	Yes	No	Yes
Alis (Rodgers and Bartram 2010)	Yes	No	Web, smart phone app, touch panel	Integrated API, based on community usage	No, requires extensive installation; less affordable	Yes (limited)	No	No

Source: Adapted from Aman et al. (2013).

4.5. EMS in Smart and Active Buildings

As illustrated in Figure 7, one of the key features of the smart grid is to enhance the communication capabilities between building systems and the power grid. Such communication includes a network infrastructure inside buildings that could monitor and control the electric systems connected to the HAN.

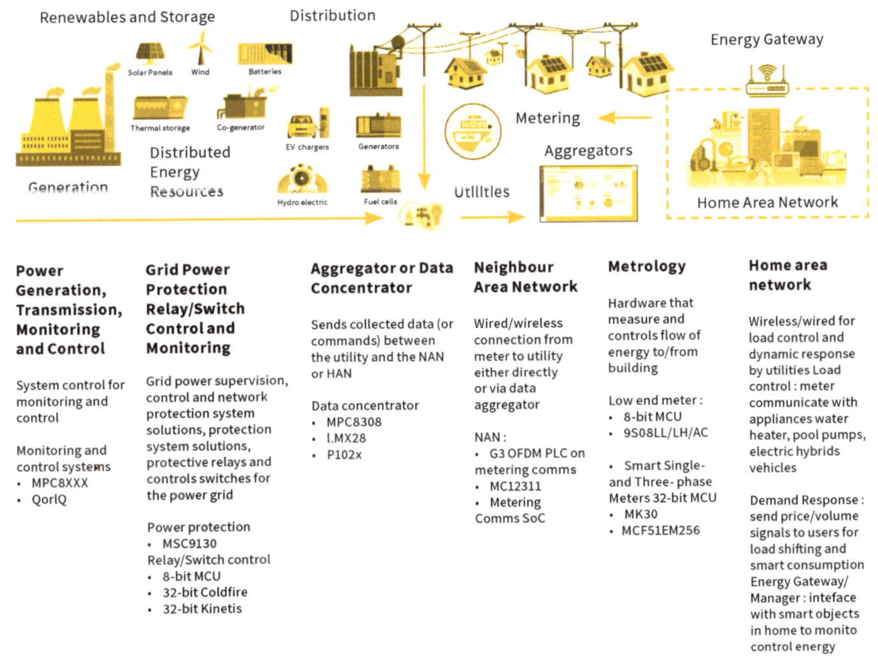

Power Generation, Transmission, Monitoring and Control	Grid Power Protection Relay/Switch Control and Monitoring	Aggregator or Data Concentrator	Neighbour Area Network	Metrology	Home area network
System control for monitoring and control	Grid power supervision, control and network protection system solutions, protection system solutions, protective relays and controls switches for the power grid	Sends collected data (or commands) between the utility and the NAN or HAN	Wired/wireless connection from meter to utility either directly or via data aggregator	Hardware that measure and controls flow of energy to/from building	Wireless/wired for load control and dynamic response by utilities Load control: meter communicate with appliances water heater, pool pumps, electric hybrids vehicles
Monitoring and control systems • MPC8XXX • QorIQ		Data concentrator • MPC8308 • I.MX28 • P102x	NAN: • G3 OFDM PLC on metering comms • MC12311 • Metering Comms SoC	Low end meter: • 8-bit MCU • 9S08LL/LH/AC • Smart Single- and Three- phase Meters 32-bit MCU • MK30 • MCF51EM256	Demand Response: send price/volume signals to users for load shifting and smart consumption Energy Gateway/ Manager: inteface with smart objects in home to monito control energy
	Power protection • MSC9130 Relay/Switch control • 8-bit MCU • 32-bit Coldfire • 32-bit Kinetis				

Figure 7. Example of a HAN/LAN and EMS communication in a smart grid context. Source: Adapted from Balakrishnan (2012).

Alam et al. (2012) define a smart dwelling as the end node of the smart grid that provides services in ambient intelligence, remote home control or home automation. Furthermore, each dwelling or node of the smart grid has the possibility to broadcast information about its electricity consumption profile and status. In a smart house, the EMS adapts the house energy consumption to the overall grid requirements without affecting the comfort of the occupants. However, to provide flexibility to the grid and reduce carbon emissions, buildings require communication capabilities (smart buildings) and advanced energy features. Therefore, in the last few years, researchers have pushed for a standardisation effort to formally define and categorise buildings with the capability of being integrated into the power system as active buildings. An active building is a building that can generate, store and modulate energy to adapt to their own demand or to the needs of the local grid (Fosas et al. 2021). The massive deployment of EMS in active and smart buildings could boost the decarbonisation both at the end-user and system level.

5. Control Algorithms for Implementing Demand Response in EMS

An EMS is defined as a system that can access information on energy consumption and generation at the building level and can implement DR measures to control heating and cooling systems, appliances or other devices connected to optimise the power usage and respond to grid signals. As illustrated in Pallonetto et al. (2020), the EMS in a DR scheme aims to different objectives:

- Reducing the overall energy consumption by increasing awareness providing data analytics and decomposition of energy use. Additionally, the EMS can control systems and appliances i.e., operating an HP at maximum COP, or optimising the inverter efficiency in a PV system with an MPPT algorithm.
- Shifting energy demand. Reducing peak consumption by exploiting TES or electric storage is a common DR measure. A signal can also trigger the measure and so that the control can shift the load to off-peak hours.
- Forcing loads. Forcing the use of high-load systems can be facilitated by storage and can be triggered by a DR signal during high penetration of RES in the system or locally generated electricity/energy.

The consumption reduction method is implementable in many different ways, whereas, the implementations for shifting and forcing is challenging. The main issue is the lack of standard flexibility metrics and the slow adoption of domotics systems. Additionally, time-dependent electricity tariffs provided by utilities or the market are not necessarily aligned to end-user demand profiles, hence storage is required to implement DR measures (Gottwalt et al. 2017).

Encoding a smart algorithm in the EMS can potentially minimise the overall energy consumption and cost while ensuring the expected service level and thermal comfort.

Optimisation Problems and Solution Methods

The use of control algorithms for building management systems is a recurrent theme in the literature. (Gatsis and Giannakis 2011b; Mariano-Hernández et al. 2021; McKenna and Keane 2016; Yoon et al. 2014). The control algorithms are characterised by a specific objective function and a set of constraints. In a DR program, the objective function aims to cost or energy minimisation or welfare maximisation. Welfare is defined as the utility profit minus the generation cost and system losses (Dong et al. 2012).

As described in Pallonetto et al. (2020), Table 4 summarised an extensive literature on DR optimisation algorithms. Optimisation methods are reported on the columns while rows indicate the objective functions. The control algorithms assessed in this table have been tested to enable buildings to participate in DR programs. Nevertheless, other perspectives for intelligent EMS can include the market, the distribution grid and the buildings. Thus, 4 of the 38 papers assessed (3L, 7U, 20W, 36B) embed multiple optimisation strategies, such as mixed linear integer, continuous integer and quadratic programming, to reduce power flow overloads caused by variable renewable energy generation or load variations. In 7U and 36B, the EMS use electric storage to provide flexibility to the power system. The paper 20W illustrates a distributed algorithm with a minimal communication overhead. The system force loading in proportion to high uncertainty loads or generation such as renewable. The paper 3L include specialised constraints for balancing the distribution network. These two papers, despite the different approaches, top-down and bottom-up, respectively, aim to maximise the welfare in a smart grid system. One of the limitations of the control systems analysed is that none of these papers provides a comparable optimal solution.

Table 4. Optimisation problems and solution methods in for DR in the literature (see Table 5 for legend).

Objective Function / Optimisation Method	Linear Integer Programming	Mixed Integer Programming	Mixed Integer Linear Programming	Mixed Integer Non-Linear Programming	Mixed Discrete/Continuous Programming	Convex Optimisation Problem	Non Convex Optimisation Problem	Non Linear Programming
Min. Cost	23U 38U	1U	22B	24B		36U	13Y	
Min. Consumption	29Q	7U	22T 9B	8B		19U 23U	26U	
Max. Welfare				25B	3L	2U 18U		10M
Min. Cost and Min. Consumption		30T	27B					
Max. Welfare Min. Consumption					20W			

Objective Function / Optimisation Method	Heuristics	Particle Swarm Optimisation	Binary Partical Swarm Optimisation	Stochastic Optimisation	Markov Decision Problem	Robust Optimisation	Other Methods	Game Theory
Min. Cost	21D 28E 35U	14U 13B	14U	15F			31T 32T 33T	28A
Min. Consumption		16B		11U	26U	34T	23U	
Max. Welfare							8E 8O	24O 6U
Min. Cost and Min. Consumption	39U						4U 17B	5U
Max. Welfare Min. Consumption						12A 37U		

Source: Reused from Pallonetto et al. (2020).

Table 5. Legend for Table 4.

Position	Reference	Reference	Algorithm
1	Behrangrad et al. (2010)	A	Interior point method
2	Cao et al. (2012)	B	Commercial software
3	Cecati et al. (2011)	C	Multiple-looping algorithm
4	Chang et al. (2012)	D	Evolutionary algorithm
5	Chen et al. (2011)	E	Greedy search algorithm
6	Chen et al. (2012)	F	Lyapunov optimisation technique
7	Choi et al. (2011)	G	Relaxed convex programming
8	Cui et al. (2012)	H	Simulated annealing
9	Zhang et al. (2011)	I	Lagrange–Newton method
10	Doostizadeh and Ghasemi (2012)	L	Sequential Quadratic Programming
11	Ferreira et al. (2012)	M	Benders decomposition
12	Gatsis and Giannakis (2011a)	N	Q-Learning algorithm
13	Gatsis and Giannakis (2011b)	O	Filling method
14	Gudi et al. (2012)	P	Co-Evolutionary PSO algorithm
15	Guo et al. (2012)	Q	Branch and bound method
16	Jiang and Fei (2011)	R	Parallel distribution computation
17	Hedegaard et al. (2017)	S	Signaled particle swarm optimisation
18	Alibabaei et al. (2016)	T	MPC (Model Predictive Control)
19	Joe-Wong et al. (2012)	U	Author's software
20	Kallitsis et al. (2012)	V	Distributed subgradient algorithm
21	Logenthiran et al. (2012)	W	Iterative decentralised algorithm
22	Mohsenian-Rad et al. (2010a)	Y	Lagrangian dual algorithm
23	Molderink et al. (2009)		
24	Soares et al. (2011)		
25	Sortomme and El-Sharkawi (2012)		
26	Totu et al. (2013)		
27	Wang et al. (2012)		
28	Xiao et al. (2010)		

Table 5. *Cont.*

Position	Reference	Reference Algorithm
29	Zhu et al. (2012)	
30	Yoon et al. (2014)	
31	Ma et al. (2011)	
32	Cole et al. (2014)	
33	Bianchini et al. (2016)	
34	Kircher and Zhang (2015)	
35	Schibuola et al. (2015)	
36	Knudsen and Petersen (2016)	
37	Park et al. (2017)	
38	Alimohammadisagvand et al. (2016)	
39	Pallonetto et al. (2019)	

Furthermore, the analysis elicits a trade-off between optimisation at a single building and power grid level. It is a requirement for a smart grid DR algorithm to ensure the optimisation of the resources at an isolated building level while contributing to the power grid stability and reduction of the environmental impact via two-way communication to aggregators or TSOs.

This double aim can be reached if the optimisation algorithm objective function minimises both cost and consumption. As also demonstrated (Cole et al. 2014; Hedegaard et al. 2017) (32T, 17T), in the case of merely cost minimisation, the energy consumption and associated emissions can increase.

As illustrated in Figure 9, the majority of the optimisation algorithms which were analysed have a single objective function that minimises costs. Nevertheless, various studies used a double objective function (4U, 5U, 27B, 30T). In this category, different techniques were utilised such as heuristics, analytical solutions and game theory. Only the heuristic controller (30T) was able to reduce the consumption by

9.2% and the costs by 14.4%, using a threshold limit to operate the controllable loads under RTP prices.

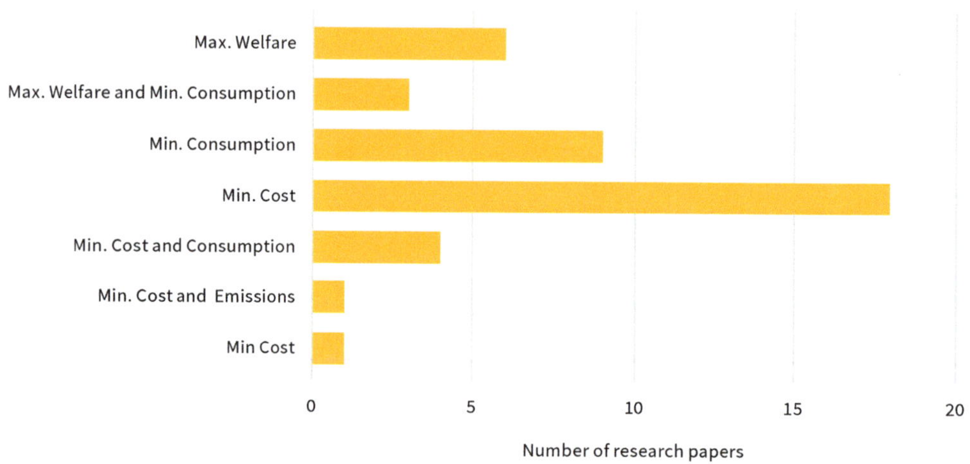

Figure 8. Classification of the most common algorithm objective functions in the literature. Source: Graphic by author, data from Pallonetto et al. (2020).

Two works (4U, 5U) used a cluster of residential buildings (10 and 60, respectively) to assess the results of the algorithm. When tested on the test load profiles, 5U showed a demand reduction of 13.5% and cost savings of 3.6%, while 4U used a randomly generated problem, and the approach cannot be compared with equivalent works. The remaining two works (27B, 30T) utilised a model of a single building to assess the benefits of the double objective function algorithm. Wang et al. (2012) (27B) reached an overall cost savings of 9% and a load reduction of 6%.

Although these results were significant, the MPC approach outperformed the others. Among the literature examined which aimed to minimise the cost of the energy expenditure, the MPC systems (17T, 18T, 31T, 32T, 33T, 38U) reached savings up to 28%. Above all, the papers analysed used a white box model such as EnergyPlus. The literature includes both residential (17T, 18T, 31T, 32T, 38U) and commercial buildings (33T). The predictive models used for the forecast were linear models (38U), autoregressive statistical models (31T, 33T), reduced-order model (32T) or grey-box model (17T) while other used machine learning algorithms (39U). It should be also noted that none of the MPC systems used a white box as a predictive model but as a testbed.

In Knudsen and Petersen (2016) (38U), the authors developed an MPC with two objective functions (emissions and electricity price). The MPC was a state-space model which is similar to a reduced-order model (Dehkordi and Candanedo 2016). Such an EMS reduced the carbon footprint by 5 to 10 per cent.

Moreover, as illustrated in Figure 10, of all the works analysed, the majority of them was tested on a single residential building. However, none of the papers mentioned any calibration of the building model despite, as illustrated in Figure 8, the majority of the works used a BES model for testing.

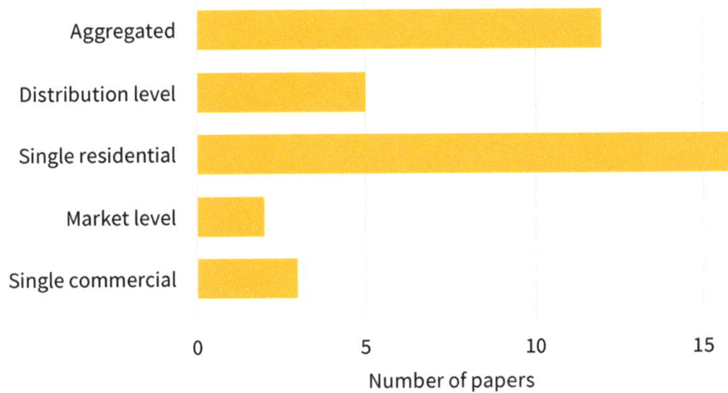

Figure 9. Classification of testing methodology in the literature. Source: Graphic by author, data from Pallonetto et al. (2020).

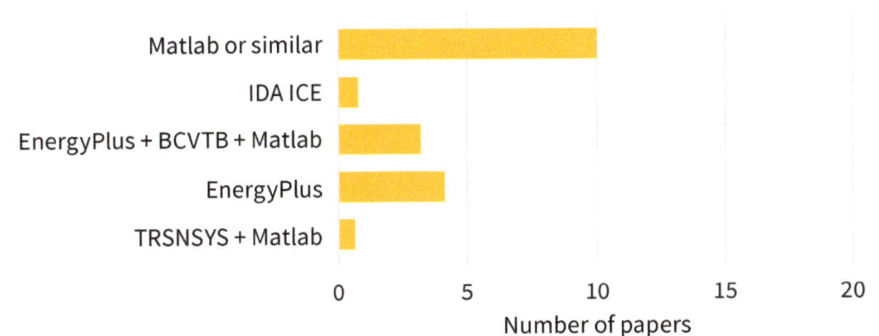

Figure 10. Software platform utilised for control algorithm. Source: Graphic by author, data from Pallonetto et al. (2020).

Figure 11 shows that the RTP price was used in the majority of the assessments. The RTP price is proportional to the market price but requires a fully automated EMS and could incur in low acceptance among end-users. Moreover, using RTP price, the assessment of control algorithms is more complex from considering the electricity profile perspective.

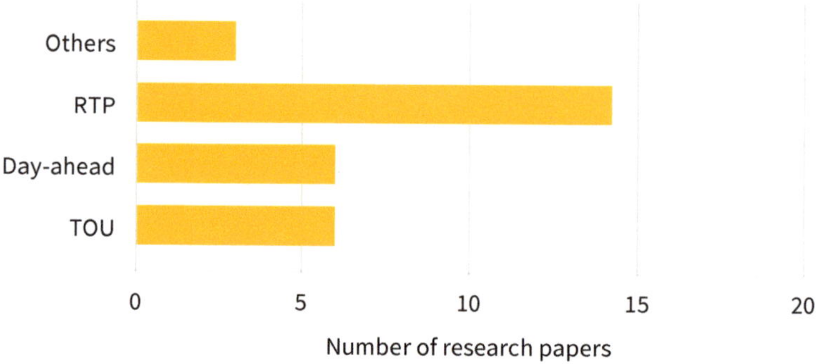

Figure 11. Most common electricity price schemes for algorithm testing. Source: Graphic by author, data from Pallonetto et al. (2020).

6. A Path towards the Full Decarbonisation

In the past decade, the distinction between energy efficiency measures and DR has become less stratified (Goldman et al. 2010). High penetration of variable RES generation and storage has widened the use of DR beyond peak hours to be applicable throughout the day (Calvillo et al. 2017; Jiang and Low 2011). From an end-user perspective, energy efficiency measures have been enhanced by controlling technologies in buildings (EDP Consortium 2016) that allow the exploitation of thermal storage and local renewable energy system as dynamic controllable load.

The effectiveness of full decarbonisation of our building stock using advanced energy management systems is dependable on energy efficiency regulations and policies that generally target the national building stock from an isolated (or single) building point of view. The building retrofit measures are designed to reduce energy consumption by deploying energy-efficient and low-carbon building technologies as illustrated by the IEA (2013) roadmap towards 2050. These methodologies do not take into account that the building, as a responsive leaf element of a smart grid system, could dynamically provide flexibility, both at distribution and transmission level, exploiting thermal and electric storage as well as deferrable loads within end-user thermal comfort constraints. Therefore, there is a need to deploy a more holistic approach to designing energy efficiency regulations that recognise the building as a dynamic energy asset. The implications of evaluating deferrable loads, storage and end-user comfort and constraints, can provide a comprehensive assessment of the impact of these new technologies and the influence of human behaviour on the energy profiles of residential buildings.

Additionally, the assessment of a DR resource must be calculated based on the quantity of energy that could be altered compared to a baseline use. The identification of a baseline and the quantification of the flexibility as a deviation from a baseline consumption are still open challenges in the research community (Jazaeri et al. 2016; Mathieu et al. 2011; Wijaya et al. 2014). The temporal quantification of available DR resources is a critical research need to enable DR at a household level (Gils 2014; Herter et al. 2007; Hurtado et al. 2017; McKenna and Keane 2016). The use of high-resolution simulation models defined as digital twins for commercial and residential buildings, embedding occupancy, consumer behaviour profiles and comfort constraints, is a new research frontier that could lead to a better understanding of the benefit of DR measures and the impact of large scale trials on the power system.

From an integration perspective, the implementation process of control algorithms that ensure occupant thermal comfort while increasing the energy flexibility of the system by providing DR capabilities is a critical research need for the smart grid rollout (CER 2011; Farhangi 2010; Gottwalt et al. 2017; Nolan and O'Malley 2015). Although in the research community, the design of new algorithms for DR optimisation is a subject undergoing intense study, new machine learning techniques and technology advancements need to be assessed as suitable for the development of intelligent residential controllers.

The contribution of the built environment is paramount for the full decarbonisation of our society. Five different perspectives have been the narrative threads of this chapter:

- Energy perspective: the implications of evaluating retrofit measures, load shifting, storage and end-user comfort and constraints provided a comprehensive assessment of the impact of new technologies as well as the influence of human behaviour on the energy profiles for buildings. The methodology for an energy perspective assessment was further applied to the evaluation of control algorithms for estimating the impact of EMS and RES at a building level upon the energy consumption and profiles.
- End-user perspective: evaluating the research findings for the benefit of end-users in terms of reduction of energy expenditure, adherence to thermal

comfort constraints and accessibility to energy data is an essential instrument for new technology adoption. The end-user perspective also concerns the transferability of the technology benefit across all the building categories.

- Utility perspective: where the evaluation of the results of the research benefits the utility in terms of reducing the system contingencies caused by the penetration of RES via demand-side measures and improving the predictability of building electricity demand is a fundamental criterion for the deployment of these new technologies. The utility perspective concerns the reliability, resilience and stability of the power grid with the objective of improving the overall efficiency hence leading to a generation cost reduction.
- Integration perspective: where the interaction between buildings, electromobility and their control system, the end-user, utilities and a bi-directional communication infrastructure defined within the smart grid needs to be evaluated from the point of view of the accessibility, interoperability, availability and controllability of the assets. The integration side of the research also allows the convergence of objectives for both the utility and the end-user.
- Environmental perspective: through an online data-driven assessment of the carbon emissions associated with the electricity consumption of the building at system and building level can increase awareness and drive the change. The implications of the retrofit measures and the use of control algorithms on the carbon footprint of the building are of interest to policy-makers and local government authorities.

7. Conclusions

Besides demand-side management and the installation of advanced energy management systems, several future lines of intervention are critical towards the full decarbonisation of the built environment: monitoring equipment, IoT data collection infrastructures, user engagement, distributed energy generation, storage and energy efficiency management measures. Data gathering and sensors installation can provide useful insights into the occupancy profile patterns of the building. They could also help to locate areas where the temperature constraints need to be precisely defined because of ongoing activities at low metabolic rates or areas where the setpoints can be dynamic. Sensor installations and data collection can also support a more accurate calibration of digital twins and therefore provide insight even on fine-tuned energy efficiency measures such as detailed multi-zone air ventilation rate and thermostatic set points. Occupants and buildings users are at the core of the transition to a low carbon economy. They can be empowered as smart energy users of the building. A smart energy user should not only consume energy but also be a conscious actor in energy savings, building energy policy and even in future energy measures either as an individual or as part of an energy community. Smart energy users are well aware of the consequences of their choices and way of life for energy consumption and the environment. Smart energy users are also actors in climate policy not only with their individual and collective decisions but also by creating awareness among fellow users and organising community events and movements to accelerate the transition to a post-carbon world. To facilitate user empowerment, the energy manager could support the engagement with screens and dynamic visualisation of the building energy consumption. Interactive screens with carbon emissions, zone energy performances, temperature and weather indicators can also facilitate gamification among users and promote a virtuous cycle where individuals gain awareness and lead energy efficiency measures. Another line of action is the installation of renewable energy generators and storage. Different technologies such as photovoltaic, solar thermal systems, heat pumps or biomass cogenerator (CHP) can represent viable solutions for further reducing the emissions of buildings. A biomass cogenerator can provide both electricity and thermal energy, making buildings a central element of a green circular ecosystem. Additionally, a CHP can contribute to heating and cooling loads through a trigeneration system. Other options to consider are power purchase agreements with wind and solar farms or heat recovery systems from nearby processing plants. Electric and thermal storage are the main enabler of all these technologies and can support the mass deployment of advanced DR programs. It should also be noted that the electrification of the mobility sector will allow a capillary diffusion of EVs. The latest charging technologies for EV batteries support a bidirectional electricity transfer with the grid (V2G). Therefore, EVs batteries can also provide distributed electricity storage to further support the increased penetration of renewable energies in the power system if the flexibility of these devices can be dynamically controlled and dispatched. Energy efficiency and management are at the heart of the decarbonisation process. The current work has highlighted several measures for reducing the energy consumption of the building and the provision of energy system services. However, the suggested line of actions can be further developed to depict a comprehensive roadmap for the full decarbonisation of the built environment. For instance, multi-zone thermostatic set points or zone air ventilation controllers can contribute to adapting the energy demand to the current occupancy profile of the building. Areas of the building that are not used or with a reduced occupancy profile can have a dynamic schedule. Additionally, a digital twin can be used to further design an innovative insulation layer with advanced materials such as phase change materials and similar. These lines of intervention coupled with advanced energy management systems and demand-side measures will allow reaching the ambitious emission target for 2050 and beyond reducing the overall impact of climate changes.

Funding: SFI Grant Number for NexSys is 21/SPP/3756.

Data Availability Statement: The data that support the findings of this study are available from the corresponding author, F.P., upon reasonable request.

Acknowledgments: This work was conducted in the Innovation Value Institute Institute within the Sustainability Cluster and the School of Business in the National University of Ireland, Maynooth (NUIM) in collaboration with the Energy Institute at the University College Dublin, Ireland. This publication has emanated from research conducted with the financial support of Science Foundation Ireland under Grant Number 21/SPP/3756.

Conflicts of Interest: The author declare no conflict of interest.

References

Active Building Research Programme. 2013. The Active Building Centre Research Programme. Available online: https://abc-rp.com/ (accessed on 25 October 2021).

Aghamohamadi, Mehrdad, Mohammad Ebrahim Hajiabadi, and Mahdi Samadi. 2018. A novel approach to multi energy system operation in response to dr programs; an application to incentive-based and time-based schemes. *Energy* 156: 534–47. doi:10.1016/j.energy.2018.05.034.

Ahmed, Ejaz, Ibrar Yaqoob, Abdullah Gani, Muhammad Imran, and Mohsen Guizani. 2016. Internet-of-things-based smart environments: State of the art, taxonomy, and open research challenges. *IEEE Wireless Communications* 23: 10–16. doi:10.1109/MWC.2016.7721736.

Alam, Muhammad Raisul, Mamun Bin Ibne Reaz, and Mohd Alauddin Mohd Ali. 2012. A review of smart homes—Past, present, and future. *IEEE Transactions on Systems, Man, and Cybernetics, Part C (Applications and Reviews)* 42: 1190–203. doi:10.1109/TSMCC.2012.2189204.

Albadi, Mohamed H., and Ehab. F. El-Saadany. 2008. A summary of demand response in electricity markets. *Electric Power Systems Research* 78: 1989–996. doi:10.1016/j.epsr.2008.04.002.

Alibabaei, Nima, Alan S. Fung, and Kaamran Raahemifar. 2016. Development of Matlab-TRNSYS co-simulator for applying predictive strategy planning models on residential house HVAC system. *Energy and Buildings* 128: 81–98. doi:10.1016/j.enbuild.2016.05.084.

Alimohammadisagvand, Behrang, Juha Jokisalo, Simo Kilpeläinen, Mubbashir Ali, and Kai Sirén. 2016. Cost-optimal thermal energy storage system for a residential building with heat pump heating and demand response control. *Applied Energy* 174: 275–87. doi:10.1016/j.apenergy.2016.04.013.

Almassalkhi, Mads R., and Ian A. Hiskens. 2015. Model-Predictive Cascade Mitigation in Electric Power Systems With Storage and Renewables—Part I: Theory and Implementation. *IEEE Transactions on Power Systems* 30: 67–77. doi:10.1109/TPWRS.2014.2320982.

Aman, Saima, Yogesh Simmhan, and Viktor K. Prasanna. 2013. Energy management systems: State of the art and emerging trends. *IEEE Communications Magazine* 51: 114–19. doi:10.1109/MCOM.2013.6400447.

Anaya, Karim L., and Michael G. Pollitt. 2021. How to Procure Flexibility Services within the Electricity Distribution System: Lessons from an International Review of Innovation Projects. *Energies* 14: 4475. doi:10.3390/en14154475.

Antsaklis, Panos J., and Kevin M. Passino. 1993. *An Introduction to Intelligent and Autonomous Control*. Dordrecht: Kluwer Academic Publishers.

Arteconi, Alessia, Neil J. Hewitt, and Fabio Polonara. 2013. Domestic demand-side management (DSM): Role of heat pumps and thermal energy storage (TES) systems. *Applied Thermal Engineering* 51: 155–65. doi:10.1016/j.applthermaleng.2012.09.023.

Atzori, Luigi, Antonio Iera, and Giacomo Morabito. 2010. The Internet of Things: A survey. *Computer Networks* 54: 2787–805. doi:10.1016/j.comnet.2010.05.010.

Balakrishnan, Meera. 2012. Smart Energy Solutions for Home Area Networks and Grid-End Applications. Available online: https://www.nxp.com/docs/en/brochure/PWRARBYNDBITSSES.pdf (accessed on 6 April 2022).

Bartram, Lyn, Johnny Rodgers, and Kevin Muise. 2010. Chasing the Negawatt: Visualization for Sustainable Living. *IEEE Computer Graphics and Applications* 30: 8–14. doi:10.1109/MCG.2010.50.

Bazydło, Grzegorz, and Szymon Wermiński. 2018. Demand side management through home area network systems. *International Journal of Electrical Power & Energy Systems* 97: 174–85. doi:10.1016/j.ijepes.2017.10.026.

Beaudin, Marc, and Hamidreza Zareipour. 2015. Home energy management systems: A review of modelling and complexity. *Renewable and Sustainable Energy Reviews* 45: 318–35. doi:10.1016/j.rser.2015.01.046.

Behrangrad, Mahdi, Hideharu Sugihara, and Tsuyoshi Funaki. 2010. Analyzing the system effects of optimal demand response utilization for reserve procurement and peak clipping. Paper presented at Power and Energy Society General Meeting, Minneapolis, MN, USA, 25–29 July 2010. doi:10.1109/PES.2010.5589597.

Bianchini, Gianni, Marco Casini, Antonio Vicino, and Donato Zarrilli. 2016. Demand-response in building heating systems: A Model Predictive Control approach. *Applied Energy* 168: 159–70. doi:10.1016/j.apenergy.2016.01.088.

Boemer, Jens C., Karsten Burges, Pavel Zolotarev, Joachim Lehner, Patrick Wajant, Markus Fürst, Rainer Brohm, and Thomas Kumm. 2011. Overview of German grid issues and retrofit of photovoltaic power plants in Germany for the prevention of frequency stability problems in abnormal system conditions of the ENTSO-E region Continental Europe. Paper presented at 1st International Workshop on Integration of Solar Power into Power Systems, Aarhus, Denmark, 24 October 2011.

Bozalakov, Dimitar, Tine Vandoorn, Bart Meersman, and Lieven Vandevelde. 2014. Overview of increasing the penetration of renewable energy sources in the distribution grid by developing control strategies and using ancillary services. Paper presented at IEEE Young Researchers Symposium, Ghent, Belgium, 24–25 April 2014.

Braess, Dietrich. 1968. Über ein paradoxon aus der verkehrsplanung. *Unternehmensforschung* 12: 258–68.

Braithwait, Steven D., and Kelly Eakin. 2002. *The Role of Demand Response in Electric Power Market Design*. Madison: Edison Electric Institute, pp. 1–57.

Brewer, Robert S., and Philip M. Johnson. 2010. WattDepot: An Open Source Software Ecosystem for Enterprise-Scale Energy Data Collection, Storage, Analysis, and Visualization. Paper presented at 2010 First IEEE International Conference on Smart Grid Communications, Gaithersburg, MD, USA, 4–6 October 2010. doi:10.1109/SMARTGRID.2010.5622023.

Calvillo, Christian F., Karen Turner, Keith Bell, and Peter McGregor. 2017. Impacts of residential energy efficiency and electrification of heating on energy market prices. Paper presented at 15th IAEE European Conference 2017, Vienna, Austria, 3–6 September 2017.

Cao, Jun, Bo Yang, Cailian Chen, and Xinping Guan. 2012. Optimal demand response using mechanism design in the smart grid. Paper presented at Chinese Control Conference (CCC), Hefei, China, 25–27 July 2012.

Cecati, Carlo, Costantino Citro, and Pierluigi Siano. 2011. Combined operations of renewable energy systems and responsive demand in a smart grid. *IEEE Transactions on Sustainable Energy* 2: 468–76. doi:10.1109/TSTE.2011.2161624.

CER. 2011. Demand Side Vision for 2020. Available online: https://www.semcommittee.com/sites/semcommittee.com/files/media-files/SEM-11-022%20Demand%20Side%20Vision%20for%202020.pdf (accessed on 6 April 2022).

Chang, Tsung-Hui, Mahnoosh Alizadeh, and Anna Scaglione. 2012. Coordinated home energy management for real-time power balancing. Paper presented at IEEE General Meeting Power & Energy Society, San Diego, CA, USA, 22–26 July 2012. doi:10.1109/PESGM.2012.6345639.

Chen, Chen, Shalinee Kishore, and Lawrence V. Snyder. 2011. An innovative RTP-based residential power scheduling scheme for smart grids. Paper presented at International Conference on Acoustics, Speech, and Signal Processing (ICASSP), Prague, Czech Republic, 22–27 May 2011. doi:10.1109/ICASSP.2011.5947718.

Chen, Yan, W. Sabrina Lin, Feng Han, Yu-Han Yang, Zoltan Safar, and K. J. Ray Liu. 2012. A cheat-proof game theoretic demand response scheme for smart grids. Paper presented at IEEE International Conference on Communications (ICC), Ottawa, ON, Canada, 10–15 June 2012. doi:10.1109/ICC.2012.6364397.

Choi, Soojeong, Sunju Park, Dong-Joo Kang, Seung-jae Han, and Hak-Man Kim. 2011. A microgrid energy management system for inducing optimal demand response. Paper presented at IEEE International Conference on Smart Grid Communications (SmartGridComm), Brussels, Belgium, 17–20 October 2011. doi:10.1109/SmartGridComm.2011.6102317.

Clements, Samuel L., Thomas E. Carroll, and Mark D. Hadley. 2011. *Home Area Networks and the Smart Grid*. Technical report No. PNNL-20374; Richland: Pacific Northwest National Laboratory.

Cole, Wesley J., Kody M. Powell, Elaine T. Hale, and Thomas F. Edgar. 2014. Reduced-order residential home modeling for model predictive control. *Energy and Buildings* 74: 6–7. doi:10.1016/j.enbuild.2014.01.033.

Conka, Zsolt, Pavol Hocko, Matuš Novák, Michal Kolcun, and Morva György. 2014. Impact of renewable energy sources on stability of EWIS transmision system. Paper presented at International Conference on Environment and Electrical Engineering (EEEIC), Krakow, Poland, 10–12 May 2014. doi:10.1109/EEEIC.2014.6835840.

Cui, Tiansong, Hadi Goudarzi, Safar Hatami, Shahin Nazarian, and Massoud Pedram. 2012. Concurrent optimization of consumer's electrical energy bill and producer's power generation cost under a dynamic pricing model. Paper presented at Innovative Smart Grid Technologies (ISGT), Washington, DC, USA, 16–20 January 2012. doi:10.1109/ISGT.2012.6175810.

Darby, Sarah. 2006. The Effectiveness of Feedback on Energy Consumption. A Review for DEFRA of the Literature on Metering, Billing and Direct Displays. Available online: https://copperalliance.starringjane.com/uploads/2018/04/smart-metering-report.pdf (accessed on 6 April 2022).

Darby, Sarah J. 2020. Demand response and smart technology in theory and practice: Customer experiences and system actors. *Energy Policy* 143: 111573. doi:10.1016/j.enpol.2020.111573.

Darby, Sarah, Sarah Higginson, Marina Topouzi, Julie Goodhew, and Stefanie Reiss. 2018. Getting the Balance Right: Can Smart Thermal Storage Work for Both Customers and Grids? Available online: https://ec.europa.eu/research/participants/documents/downloadPublic?documentIds=080166e5bb1417aa&appId=PPGMS (accessed on 6 April 2022).

Dehkordi, Vahid Raissi, and José Agustín Candanedo. 2016. State-Space Modeling of Thermal Spaces in a Multi-Zone Building. Paper presented at 4th International High Performance Buildings Conference, West Lafayette, IN, USA, 1–14 July 2016.

Dong, Qifen, Li Yu, Wen-Zhan Song, Lang Tong, and Shaojie Tang. 2012. Distributed Demand and Response Algorithm for Optimizing Social-Welfare in Smart Grid. Paper presented at International Symposium on Parallel and Distributed Processing (IPDPS), Shanghai, China, 21–25 May 2012. doi:10.1109/IPDPS.2012.112.

Doostizadeh, Meysam, and Hassan Ghasemi. 2012. A day-ahead electricity pricing model based on smart metering and demand-side management. *Energy* 46: 221–30. doi:10.1016/j.energy.2012.08.029.

EDP Consortium. 2016. Demand Response. The Road Ahead. Available online: http://www.wedgemere.com/wp-content/uploads/2013/03/Evolution-of-DR-Dialogue-Project-Report.pdf (accessed on 6 April 2022).

ePANACEA. 2020. ePANACEA Horizon 2020 Project. Available online: https://epanacea.eu/ (accessed on 25 October 2021).

Episcope EU. 2013. Project Episcope. Available online: https://episcope.eu/ (accessed on 25 October 2021).

Eriksen, Peter Borre, Thomas Ackermann, Hans Abildgaard, Paul Smith, Wilhelm Winter, and JuanMa Rodríguez García. 2005. System operation with high wind penetration. *IEEE Power and Energy Magazine* 3: 65–74. doi:10.1109/MPAE.2005.1524622.

European Commission. 2010. Energy Roadmap 2050. Available online: https://www.roadmap2050.eu/attachments/files/Volume1_fullreport_PressPack.pdf (accessed on 21 March 2022).

European Union. 2017. 2050 Climate and Energy Package. Available online: https://ec.europa.eu/clima/policies/strategies/2050_en (accessed on 25 October 2021).

Eurostat. 2020. Final Energy Consumption in Households. Available online: https://ec.europa.eu/eurostat/web/products-datasets/-/t2020_rk200 (accessed on 21 March 2022).

Eustis, Conrad, Gale Horst, and Donald J. Hammerstrom. 2007. Appliance Interface for Grid Responses. Paper presented at Grid-Interop 2007 Forum, Albuquerque, NM, USA, 7–9 November 2007.

Farhangi, Hassan. 2010. The path of the smart grid. *IEEE Power and Energy Magazine* 8: 18–28. doi:10.1109/MPE.2009.934876.

Ferreira, Rafael de Sa, Luiz Augusto N. Barroso, and Martha Martins Carvalho. 2012. Demand response models with correlated price data: A robust optimization approach. *Applied Energy* 96: 133–49. doi:10.1016/j.apenergy.2012.01.016.

Fosas, Daniel, Elli Nikolaidou, Matthew Roberts, Stephen Allen, Ian Walker, and David Coley. 2021. Towards active buildings: Rating grid-servicing buildings. *Building Services Engineering Research and Technology* 42: 129–155. doi:10.1177/0143624420974647.

Fuller, Jason C., Kevin P. Schneider, and David Chassin. 2011. Analysis of Residential Demand Response and double-auction markets. Paper presented at IEEE General Meeting Power & Energy Society, Detroit, MI, USA, 24–28 July 2011. doi:10.1109/PES.2011.6039827.

Gatsis, Nikolaos, and Georgios B. Giannakis. 2011a. Cooperative multi-residence demand response scheduling. Paper presented at Annual Conference on Information Sciences and Systems (CISS), Baltimore, MD, USA, 23–25 March 2011. doi:10.1109/CISS.2011.5766245.

Gatsis, Nikolaos, and Georgios B. Giannakis. 2011b. Residential demand response with interruptible tasks: Duality and algorithms. Paper presented at IEEE Conference on Decision and Control, Orlando, FL, USA, 12–15 December 2011. doi:10.1109/CDC.2011.6161103.

Gils, Hans Christian. 2014. Assessment of the theoretical demand response potential in Europe. *Energy* 67: 1–18. doi:10.1016/j.energy.2014.02.019.

Gimon, Eric, and Senior Fellow. 2021. Lessons from the Texas Big Freeze. Available online: https://energyinnovation.org/wp-content/uploads/2021/05/Lessons-from-the-Texas-Big-Freeze.pdf (accessed on 21 March 2022).

Goldman, Charles, Michael Reid, Roger Levy, and Alison Silverstein. 2010. *Coordination of Energy Efficiency and Demand Response.* Technical report No. LBNL-3044E. Berkeley: Lawrence Berkeley National Lab (LBNL). doi:10.2172/981732.

Gottwalt, Sebastian, Johannes Gärttner, Hartmut Schmeck, and Christof Weinhardt. 2017. Modeling and Valuation of Residential Demand Flexibility for Renewable Energy Integration. *IEEE Transactions on Smart Grid* 8: 2565–74. doi:10.1109/TSG.2016.2529424.

Gudi, Nikhil, Lingfeng Wang, and Vijay Devabhaktuni. 2012. A demand side management based simulation platform incorporating heuristic optimization for management of household appliances. *International Journal of Electrical Power & Energy Systems* 4: 185–93. doi:10.1016/j.ijepes.2012.05.023.

Gungor, Vehbi C., Dilan Sahin, Taskin Kocak, Salih Ergut, Concettina Buccella, Carlo Cecati, and Gerhard P. Hancke. 2011. Smart Grid Technologies: Communication Technologies and Standards. *IEEE Transactions on Industrial Informatics* 7: 529–39. doi:10.1109/TII.2011.2166794.

Guo, Yuanxiong, Miao Pan, and Yuguang Fang. 2012. Optimal Power Management of Residential Customers in the Smart Grid. *IEEE Transactions on Parallel and Distributed Systems* 23: 1593–606. doi:10.1109/TPDS.2012.25.

Hamidi, Vandad, Furong Li, and Francis Robinson. 2009. Demand response in the UK's domestic sector. *Electric Power Systems Research* 79: 1722–26. doi:10.1016/j.epsr.2009.07.013.

Hayajneh, Thaier, Ghada Almashaqbeh, Sana Ullah, and Athanasios V. Vasilakos. 2014. A survey of wireless technologies coexistence in WBAN: Analysis and open research issues. *Wireless Networks* 20: 2165–99. doi:10.1007/s11276-014-0736-8.

Hedegaard, Rasmus Elbæk, Theis Heidmann Pedersen, and Steffen Petersen. 2017. Multi-market demand response using economic model predictive control of space heating in residential buildings. *Energy and Buildings* 150: 25–61. doi:10.1016/j.enbuild.2017.05.059.

Herter, Karen, Patrick McAuliffe, and Arthur Rosenfeld. 2007. An exploratory analysis of California residential customer response to critical peak pricing of electricity. *Energy* 32: 25–34. doi:10.1016/j.energy.2006.01.014.

Hughes, Thomas Parke. 1993. *Networks of Power: Electrification in Western Society, 1880–1930.* Baltimore: JHU Press.

Huq, Md. Zahurul, and Syed Islam. 2010. Home Area Network technology assessment for demand response in smart grid environment. Paper presented at Australasian Universities Power Engineering Conference, AUPEC, Christchurch, New Zealand, 5–8 December 2010.

Hurtado, L. A., J. D. Rhodes, P. H. Nguyen, I. G. Kamphuis, and M. E. Webber. 2017. Quantifying demand flexibility based on structural thermal storage and comfort management of non-residential buildings: A comparison between hot and cold climate zones. *Applied Energy* 195: 1047–54. doi:10.1016/j.apenergy.2017.03.004.

IEA. 2013. *Transition to Sustainable Buildings: Strategies and Opportunities to 2050.* Technical report. Paris: Internationl Energy Agency.

International Energy Agency. 2016. *Renewables and Energy Market Report Information 2016.* Paris: IEA Publications. Available online: https://www.iea.org/reports/medium-term-renewable-energy-market-report-2016 (accessed on 6 April 2022).

Jazaeri, Javad, Tansu Alpcan, Robert Gordon, Miguel Brandao, Tim Hoban, and Chris Seeling. 2016. Baseline methodologies for small scale residential demand response. Paper presented at IEEE Innovative Smart Grid Technologies-Asia (ISGT Asia), Melbourne, Australia, 28 November–1 December 2016. doi:10.1109/ISGT-Asia.2016.7796478.

Jiang, Bingnan, and Yunsi Fei. 2011. Dynamic Residential Demand Response and Distributed Generation Management in Smart Microgrid with Hierarchical Agents. *Energy Procedia* 12: 76–90. doi:10.1016/j.egypro.2011.10.012.

Jiang, Libin, and Steven Low. 2011. Multi-period optimal energy procurement and demand response in smart grid with uncertain supply. Paper presented at IEEE Conference on Decision and Control, Orlando, FL, USA, 12–15 December 2011. doi:10.1109/CDC.2011.6161320.

Joe-Wong, Carlee, Soumya Sen, Sangtae Ha, and Mung Chiang. 2012. Optimized Day-Ahead Pricing for Smart Grids with Device-Specific Scheduling Flexibility. *IEEE Journal on Selected Areas in Communications* 30: 1075–85. doi:10.1109/JSAC.2012.120706.

Kallitsis, Michael G., George Michailidis, and Michael Devetsikiotis. 2012. Optimal Power Allocation Under Communication Network Externalities. *IEEE Transactions on Smart Grid* 3: 162–73. doi:10.1109/TSG.2011.2169995.

Karnouskos, Stamatis. 2011. Demand Side Management via prosumer interactions in a smart city energy marketplace. Paper presented at IEEE PES Innovative Smart Grid Technologies Conference Europe (ISGT Europe), Manchester, UK, 5–7 December 2011. doi:10.1109/ISGTEurope.2011.6162818.

Kim, Younghun, Thomas Schmid, Zainul M. Charbiwala, and Mani B. Srivastava. 2009. ViridiScope: Design and implementation of a fine grained power monitoring system for homes. Paper presented at Ubicomp '09: The 11th International Conference on Ubiquitous Computing, Orlando, FL, USA, 30 September–3 October 2009.

Kircher, Kevin J., and K. Max Zhang. 2015. Model predictive control of thermal storage for demand response. Paper presented at American Control Conference (ACC), Chicago, IL, USA, 1–3 July 2015. doi:10.1109/ACC.2015.7170857.

Knudsen, Michael Dahl, and Steffen Petersen. 2016. Demand response potential of model predictive control of space heating based on price and carbon dioxide intensity signals. *Energy and Buildings* 125: 196–204. doi:10.1016/j.enbuild.2016.04.053.

Liserre, Marco, Thilo Sauter, and John Y. Hung. 2010. Future Energy Systems: Integrating Renewable Energy Sources into the Smart Power Grid Through Industrial Electronics. *IEEE Industrial Electronics Magazine* 4: 18–37. doi:10.1109/MIE.2010.935861.

Liu, Qi, Grahame Cooper, Nigel Linge, Haifa Takruri, and Richard Sowden. 2013. DEHEMS: Creating a digital environment for large-scale energy management at homes. *IEEE Transactions on Consumer Electronics* 59: 62–69. doi:10.1109/TCE.2013.6490242.

Logenthiran, Thillainathan, Dipti Srinivasan, and Tan Zong Shun. 2012. Demand Side Management in Smart Grid Using Heuristic Optimization. *IEEE Transactions on Smart Grid* 3: 1244–52. doi:10.1109/TSG.2012.2195686.

Lu, Xiaoxing, Kangping Li, Hanchen Xu, Fei Wang, Zhenyu Zhou, and Yagang Zhang. 2020. Fundamentals and business model for resource aggregator of demand response in electricity markets. *Energy* 204: 117885. doi:10.1016/j.energy.2020.117885.

Ma, Jingran, S. Joe Qin, Bo Li, and Tim Salsbury. 2011. Economic model predictive control for building energy systems. Paper presented at Innovative Smart Grid Technologies (ISGT), Anaheim, CA, USA, 17–19 January 2011. doi:10.1109/ISGT.2011.5759140.

Mahfuz-Ur-Rahman, A. M., Md. Rabiul Islam, Kashem M. Muttaqi, and Danny Sutanto. 2021. An Effective Energy Management With Advanced Converter and Control for a PV-Battery Storage Based Microgrid to Improve Energy Resiliency. *IEEE Transactions on Industry Applications* 57: 6659–68. doi:10.1109/TIA.2021.3115085.

Mariano-Hernández, D., L. Hernández-Callejo, A. Zorita-Lamadrid, O. Duque-Pérez, and F. Santos García. 2021. A review of strategies for building energy management system: Model predictive control, demand side management, optimization, and fault detect & diagnosis. *Journal of Building Engineering* 33: 101692. doi:10.1016/j.jobe.2020.101692.

Mata, Érika, A. Sasic Kalagasidis, and Filip Johnsson. 2014. Building-stock aggregation through archetype buildings: France, Germany, Spain and the UK. *Building and Environment* 81: 270–82. doi:10.1016/j.buildenv.2014.06.013.

Mathieu, Johanna L., Duncan S. Callaway, and Sila Kiliccote. 2011. Examining uncertainty in demand response baseline models and variability in automated responses to dynamic pricing. Paper presented at IEEE Conference on Decision and Control, Orlando, FL, USA, 12–15 December 2011. doi:10.1109/CDC.2011.6160628.

McKenna, Killian, and Andrew Keane. 2016. Residential Load Modeling of Price-Based Demand Response for Network Impact Studies. *IEEE Transactions on Smart Grid* 7: 2285–94. doi:10.1109/TSG.2015.2437451.

Mohsenian-Rad, Amir-Hamed, and Alberto Leon-Garcia. 2010a. Optimal Residential Load Control With Price Prediction in Real-Time Electricity Pricing Environments. *IEEE Transactions on Smart Grid* 1: 120–33. doi:10.1109/TSG.2010.2055903.

Mohsenian-Rad, Amir-Hamed, Vincent W. S. Wong, Juri Jatskevich, and Robert Schober. 2010b. Optimal and autonomous incentive-based energy consumption scheduling algorithm for smart grid. Paper presented at Innovative Smart Grid Technologies (ISGT), Gaithersburg, MD, USA, 19–21 January 2010. doi:10.1109/ISGT.2010.5434752.

Molderink, Albert, Vincent Bakker, Maurice G.C. Bosman, Johann L. Hurink, and Gerard J.M. Smit. 2009. Domestic energy management methodology for optimizing efficiency in Smart Grids. Paper presented at Power Tech Conference, Bucharest, Romania, 28 June–2 July 2009. doi:10.1109/PTC.2009.5281849.

Nolan, Sheila, and Mark O'Malley. 2015. Challenges and barriers to demand response deployment and evaluation. *Applied Energy* 152: 1–10. doi:10.1016/j.apenergy.2015.04.083.

O'Sullivan, Jon, Alan Rogers, Damian Flynn, Paul Smith, Alan Mullane, and Mark O'Malley. 2014. Studying the Maximum Instantaneous Non-Synchronous Generation in an Island System—Frequency Stability Challenges in Ireland. *IEEE Transactions on Power Systems* 29: 2943–51. doi:10.1109/TPWRS.2014.2316974.

Pallonetto, Fabiano, Mattia De Rosa, Federico Milano, and Donal P. Finn. 2019. Demand response algorithms for smart-grid ready residential buildings using machine learning models. *Applied Energy* 239: 1265–82. doi:10.1016/j.apenergy.2019.02.020.

Pallonetto, Fabiano, Mattia De Rosa, Francesco D'Ettorre, and Donal P. Finn. 2020. On the assessment and control optimisation of demand response programs in residential buildings. *Renewable and Sustainable Energy Reviews* 127: 109861. doi:10.1016/j.rser.2020.109861.

Pallonetto, Fabiano, Mattia De Rosa, and Donal P. Finn. 2021. Impact of intelligent control algorithms on demand response flexibility and thermal comfort in a smart grid ready residential building. *Smart Energy* 2: 100017. doi:10.1016/j.segy.2021.100017.

Pallonetto, Fabiano, Mattia De Rosa, and Donal P. Finn. 2022. Environmental and economic benefits of building retrofit measures for the residential sector by utilizing sensor data and advanced calibrated models. *Advances in Building Energy Research* 16: 89–117. doi:10.1080/17512549.2020.1801504.

Paradiso, Joseph, Prabal Dutta, Hans Gellersen, and Eve Schooler. 2011. Guest Editors' Introduction: Smart Energy Eystems. *IEEE Pervasive Computing* 10: 11–12. doi:10.1109/MPRV.2011.4.

Park, Laihyuk, Yongwoon Jang, Sungrae Cho, and Joongheon Kim. 2017. Residential Demand Response for Renewable Energy Resources in Smart Grid Systems. *IEEE Transactions on Industrial Informatics* 13: 3165–73 doi:10.1109/TII.2017.2704282.

Pas, Eric I., and Shari L. Principio. 1997. Braess' paradox: Some new insights. *Transportation Research Part B: Methodological* 31: 265–76. doi:10.1016/S0191-2615(96)00024-0.

Paterakis, Nikolaos G., Ozan Erdinc, and JoalP.S. Catalao. 2017. An overview of Demand Response: Key-elements and international experience. *Renewable and Sustainable Energy Reviews* 69: 871–91. doi:10.1016/j.rser.2016.11.167.

Pavlidou, Niovi, A.J. Han Vinck, Javad Yazdani, and Bahram Honary. 2003. Power line communications: State of the art and future trends. *IEEE Communications magazine* 41: 34–40. doi:10.1109/MCOM.2003.1193972.

Pedersen, Theis Heidmann, Rasmus Elbæk Hedegaard, and Steffen Petersen. 2017. Space heating demand response potential of retrofitted residential apartment blocks. *Energy and Buildings* 141: 158–66. doi:10.1016/j.enbuild.2017.02.035.

Pedreiras, Paulo, Luis Almeida, and Paolo Gai. 2002. The FTT-Ethernet Protocol: Merging Flexibility, Timeliness and Efficiency. Paper presented at Euromicro RTS 2002, Vienna, Austria, 19–21 June 2002 doi:10.1109/EMRTS.2002.1019195.

Pereira, Guillermo Ivan, and Patrícia Pereira da Silva. 2017. Energy efficiency governance in the EU-28: Analysis of institutional, human, financial, and political dimensions. *Energy Efficiency* 10: 1279–97. doi:10.1007/s12053-017-9520-9.

Pérez-Lombard, Luis, José Ortiz, and Christine Pout. 2008. A review on buildings energy consumption information. *Energy and Buildings* 40: 394–98. doi:10.1016/j.enbuild.2007.03.007.

Petkov, Petromil, and Marcus Foth. 2011. EnergyWiz. Available online: https://eprints.qut.edu.au/80773/ (accessed on 6 April 2022).

Piette, Mary Ann, David Watson, Naoya Motegi, Sila Kiliccote, and Eric Linkugel. 2006. Automated Demand Response Strategies and Commissioning Commercial Building Controls. Paper presented at National Conference on Building Commissioning, San Francisco, CA, USA, 19–21 April 2006.

Qdr, Q. 2006. *Benefits of Demand Response in Electricity Markets and Recommendations for Achieving Them*. Technical Report. Washington, DC: U.S. Department of Energy.

Rodgers, Johnny, and Lyn Bartram. 2010. ALIS: An interactive ecosystem for sustainable living. Paper presented at Ubicomp '10: The 2010 ACM Conference on Ubiquitous Computing, Copenhagen, Denmark, 26–29 September 2010. doi:10.1145/1864431.1864467.

Rothleder, Mark, and Clyde Loutan. 2017. Case Study–Renewable Integration: Flexibility Requirement, Potential Overgeneration, and Frequency Response Challenges. In *Renewable Energy Integration (Second Edition)*. Edited by Lawrence E. Jones. Amsterdam: Elsevier B.V., pp. 69–81.

Schibuola, Luigi, Massimiliano Scarpa, and Chiara Tambani. 2015. Demand response management by means of heat pumps controlled via real time pricing. *Energy and Buildings* 90: 15–28 doi:10.1016/j.enbuild.2014.12.047.

Silva, Marisa, Hugo Morais, and Zita Vale. 2012. An integrated approach for distributed energy resource short-term scheduling in smart grids considering realistic power system simulation. *Energy Conversion and Management* 64: 273–88. doi:10.1016/j.enconman.2012.04.021.

Smith, William J. 2010. Can EV (electric vehicles) address Ireland's CO_2 emissions from transport? *Energy* 35: 4514–21. doi:10.1016/j.energy.2010.07.029.

Soares, João, Tiago Sousa, Hugo Morais, Zita Vale, and Pedro Faria. 2011. An optimal scheduling problem in distribution networks considering V2G. Paper presented at IEEE Symposium on Computational Intelligence Applications in Smart Grid (CIASG), Paris, France, 11–15 April 2011. doi:10.1109/CIASG.2011.5953342.

Sortomme, Eric, and Mohamed A. El-Sharkawi. 2012. Optimal Scheduling of Vehicle-to-Grid Energy and Ancillary Services. *IEEE Transactions on Smart Grid* 3: 351–59. doi:10.1109/TSG.2011.2164099.

Tahir, Muhammad Faizan, Haoyong Chen, Asad Khan, Muhammad Sufyan Javed, Khalid Mehmood Cheema, and Noman Ali Laraik. 2020. Significance of demand response in light of current pilot projects in China and devising a problem solution for future advancements. *Technology in Society* 63: 101374. doi:10.1016/j.techsoc.2020.101374.

Torriti, Jacopo, Mohamed G. Hassan, and Matthew Leach. 2010. Demand response experience in Europe: Policies, programmes and implementation. *Energy* 35: 1575–83. doi:10.1016/j.energy.2009.05.021.

Totu, Luminita C., John Leth, and Rafael Wisniewski. 2013. Control for large scale demand response of thermostatic loads. Paper presented at American Control Conference (ACC), Washington, DC, USA, 17–19 June 2013. doi:10.1109/ACC.2013.6580618.

U-CERT. 2019. ePANACEA Horizon 2020 Project. Available online: https://u-certproject.eu (accessed on 25 October 2021).

Ulbig, Andreas, Theodor S. Borsche, and Göran Andersson. 2014. Impact of Low Rotational Inertia on Power System Stability and Operation. *IFAC Proceedings Volumes* 47: 7290–97. doi:10.3182/20140824-6-ZA-1003.02615.

Vargas, Luis S., Gonzalo Bustos-Turu, and Felipe Larrain. 2015. Wind Power Curtailment and Energy Storage in Transmission Congestion Management Considering Power Plants Ramp Rates. *IEEE Transactions on Power Systems* 30: 2498–506. doi:10.1109/TPWRS.2014.2362922.

Veldman, Else, Madeleine Gibescu, Han Slootweg, and Wil L. Kling. 2011. Impact of electrification of residential heating on loading of distribution networks. Paper presented at Power Tech Conference, Trondheim, Norway, 19–23 June 2011. doi:10.1109/PTC.2011.6019179.

Võsa, Karl-Villem, Andrea Ferrantelli, Dragomir Tzanev, Kamen Simeonov, Pablo Carnero, Carlos Espigares, Miriam Navarro Escudero, Pedro Vicente Quiles, Thibault Andrieu, Florian Battezzati, and et al. 2021. Building performance indicators and IEQ assessment procedure for the next generation of EPC-s. *E3S Web of Conferences* 246: 13003. doi:10.1051/e3sconf/202124613003.

Wang, Jidong, Zhiqing Sun, Yue Zhou, and Jiaqiang Dai. 2012. Optimal dispatching model of Smart Home Energy Management System. Paper presented at IEEE Innovative Smart Grid Technologies-Asia (ISGT Asia), Tianjin, China, 21–24 May 2012. doi:10.1109/ISGT-Asia.2012.6303266.

Weiss, Markus, Friedemann Mattern, Tobias Graml, Thorsten Staake, and Elgar Fleisch. 2009. Handy feedback: Connecting smart meters with mobile phones. Paper presented at MUM '09: 8th International Conference on Mobile and Ubiquitous Multimedia, Cambridge, UK, 22–25 November 2009. doi:10.1145/1658550.1658565.

Wijaya, Tri Kurniawan, Matteo Vasirani, and Karl Aberer. 2014. When Bias Matters: An Economic Assessment of Demand Response Baselines for Residential Customers. *IEEE Transactions on Smart Grid* 5: 1755–63. doi:10.1109/TSG.2014.2309053.

Witthaut, Dirk, and Marc Timme. 2012. Braess's paradox in oscillator networks, desynchronization and power outage. *New Journal of Physics* 14: 083036.

Xiao, Jin, Jae Yoon Chung, Jian Li, Raouf Boutaba, and James Won-Ki Hong. 2010. Near optimal demand-side energy management under real-time demand-response pricing. Paper presented at International Conference on Network and Service Management, Niagara Falls, ON, Canada, 25–29 October 2010. doi:10.1109/CNSM.2010.5691349.

Yan, Ye, Yi Qian, Hamid Sharif, and David Tipper. 2013. A Survey on Smart Grid Communication Infrastructures: Motivations, Requirements and Challenges. *IEEE Communications Surveys & Tutorials* 15: 5–20. doi:10.1109/SURV.2012.021312.00034.

Yang, Guang-Hua, and Victor O.K. Li. 2010. Energy Management System and Pervasive Service-Oriented Networks. Paper presented at IEEE International Conference on Smart Grid Communications (SmartGridComm), Gaithersburg, MD, USA, 4–6 October 2010. doi:10.1109/SMARTGRID.2010.5622001.

Yoon, Ji Hoon, Ross Baldick, and Atila Novoselac. 2014. Dynamic Demand Response Controller Based on Real-Time Retail Price for Residential Buildings. *IEEE Transactions on Smart Grid* 5: 121–129. doi:10.1109/TSG.2013.2264970.

Zhang, Di, Lazaros G. Papageorgioua, Nouri J. Samsatlib, and Nilay Shahb. 2011. Optimal Scheduling of Smart Homes Energy Consumption with Microgrid. *Energy* 1: 70–75.

Zhu, Ziming, Jie Tang, Sangarapillai Lambotharan, Woon Hau Chin, and Zhong Fan. 2012. An integer linear programming based optimization for home demand-side management in smart grid. Paper presented at Innovative Smart Grid Technologies (ISGT), Washington, DC, USA, 16–20 January 2012. doi:10.1109/ISGT.2012.6175785.

Part 3: Regional Transition

Social Innovation for Energy Transition: Activation of Community Entrepreneurship in Inner Areas of Southern Italy

Mariarosaria Lombardi, Maurizio Prosperi and Gerardo Fascia

1. Introduction

The current reliance on fossil fuels is unsustainable and harmful to the planet, being the main cause of climate change. It is well known that the use of renewable energy sources is one of the actions to pursue the energy transition towards zero-carbon fuels, capable of reducing the emission of greenhouse gases (GHGs), which are listed amongst the main causes of climate change.

In this context, inner areas, characterized by constant demographic decline and population aging, could play an important role in adopting measures of mitigation of and adaptation to climate change. Meanwhile, they could also benefit from this opportunity, through which they could valorize important and unique cultural assets and relevant environmental resources. The idea is to foster new forms of community entrepreneurship, based on collective renewable energy actions involving citizens in the energy system, as "renewable energy communities" or "citizen energy communities" (EU 2018). This basically means adopting a social innovation approach, capable of promoting a democracy process for ensuring environmental and economic benefits to the whole community.

This is particularly true for inner areas of Southern Italy, which have experienced widespread implementation of large-scale renewable energy plants without the engagement of the local community in the planning processes. This has led to limited acceptance of new investment projects in renewable energy by the citizens.

In light of these premises, the aim of this chapter is to propose an operational approach for developing community entrepreneurship in inner areas, where the financial resources obtained from the production, distribution and consumption of green energy are locally reinvested to valorize the cultural and natural resources, activating a comprehensive process of social and economic revitalization.

2. Political Pathway for Energy Transition at International, European and Italian Levels

Nowadays, access to energy represents one of the most central challenges and opportunities the world has to tackle for ensuring employment, mitigation of climate change and food production.

Thus, by the second half of this century, it is necessary to move towards a more affordable and clean global energy system (i.e., from fossil-based to zero-carbon systems) (SDG Tracker 2021). This implies the start of the energy transition process, which requires public support through adequate policy frameworks and financial instruments.

Hennicke et al. (2019) claimed that "the energy transition resembles an inter-generational contract in which the current generation pre-finances a gradual replacement of the entire fossil and nuclear energy system in the 21st century with energy efficiency, energy saving and renewable energies, and organizes the implementation processes in order to protect children, grandchildren, future generations and developing countries and its people from the risks of a non-renewable energy system" (ibid., p. 4). It is an enormous challenge, which requires the active involvement and commitment of various levels of governance, from the supranational ones (i.e., UN, EU) to the lower levels (i.e., national, regional and local), which have to include all individuals and the territorial communities. Some important political initiatives have been launched in this direction. Figure 1 shows the overall framework linking the global, European and Italian levels.

The 2030 Agenda, developed by the United Nations (UN) in 2015, represents a universal action plan. It defines 17 Sustainable Development Goals (SDGs) and 169 targets as strategies "to achieve a better and more sustainable future for all" (UN—United Nations 2021).

Among them, SDG 7, Affordable and Clean Energy, emphasizes the importance of changing the route of energy production and consumption for contrasting climate change. Specifically, this implies the following: achievement of universal access to modern energy; increase in the global percentage of renewable energy; doubling the improvement in energy efficiency; promotion of access, technology and investments in clean energy; and expanding and upgrading energy services for developing countries. These are the five targets defined by the UN taking into account that 13% of the global population does not have access to modern electricity, that 3 billion people depend on wood, coal, charcoal or animal waste for cooking and heating and that energy is the main factor responsible for climate change, accounting for around 60% of total GHG emissions (IEA et al. 2019).

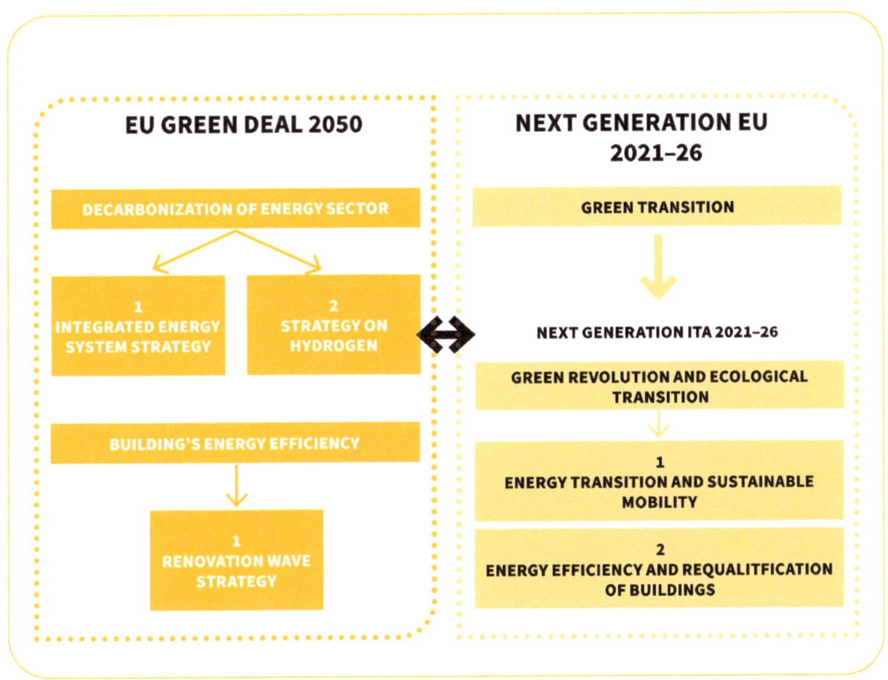

Figure 1. Main political initiatives for energy transition. Source: Graphic by authors, 2021.

In order to implement the UN 2030 Agenda, in 2019, the European Commission (EC) issued a plan called the Green Deal, a new growth strategy with the overarching aim of making Europe climate-neutral by 2050, through the deep decarbonization of all sectors of the economy (EC 2019). The approach consists of nine core policies that will bring tangible progress in the areas of the SDGs. Among them, "clean energy" specifically aims at decarbonizing the EU's energy system. Indeed, over 75% of the EU's GHG emissions come from the production and use of energy in the various economic sectors.

To reach this goal, some priorities have been identified, such as the increase in energy efficiency and use of renewable sources (RES). Regarding the latter, the strategy stresses the necessity of the transition from today's energy system to an integrated one largely based on RES. As specified in the impact assessment for the Climate Target Plan for a 55% GHG reduction, the 2030 share of renewables must reach 38–40% (EC 2020), compared with the 1990 levels. The first strategy aims at developing an integrated energy system based on multiple energy carriers, such as electricity, heat, cold, gas, solid and liquid fuels, infrastructure and end use sectors, such as buildings, transport or industry (EC 2020). The second strategy aims at a technology shift towards hydrogen, which refers to the production of hydrogen from RES and a subsequent deployment at a large scale through large-scale plants, above all large wind and solar plants (EC 2020).

Regarding the energy efficiency priority, particular attention is paid to buildings, taking into account that about 75% of the building stock is not energy-efficient, yet almost 85–95% of today's buildings will still be in use in 2050. Additionally, buildings are responsible for about 40% of the EU's total energy consumption, and for 36% of its GHG emissions from energy. Therefore, the renovation wave strategy has been launched for doubling the annual energy renovation rates in the next ten years both for public and private buildings (EC 2020).

Certainly, the implementation of the New Green Deal will be an important incentive to revitalize the European economy that has ended up in a deep recession following the pandemic crisis of COVID-19. This is why there is a profound link between the Next Generation EU-NGEU (known as the European Recovery Plan adopted in February 2021 following COVID-19) and the New Green Deal. In fact, it represents a temporary financial mechanism, for the period 2021–2026, to support reforms and investments promoted by member states, aimed at making European countries more sustainable, resilient and prepared for the challenges and opportunities of the ecological and digital transition (EU 2021). Indeed, among the six main pillars of the NGEU, one is dedicated to the green transition. In this regard, the president of the EC clarified that 37% of these funds will be allocated to green policies (i.e., expenditure related to climate change) in compliance with the systemic approach of the SDGs.

Specifically, the European Commission approved the Italian plan in June 2021 (Camera dei Deputati 2021). It identified six missions, among which the second one is related to the Green Revolution and Ecological Transition, aimed at improving the sustainability and resilience of the economic system, as well as ensuring a fair and inclusive environmental transition. It represents 31% of the total budget of the plan. This mission is structured in four components, in line with the European Green Deal, where two out of four are dedicated to the energy transition: (i) energy transition and sustainable mobility; and (ii) energy efficiency and building renovation.

Specifically, the first one is the component with the highest budget, equal to EUR 23.7 billion (40% of the total), and covers five lines of action: increase in the RES share in the system (biomethane, agrophotovoltaic, RES for energy communities, innovative RES plants, etc.); strengthening and digitalization of network infrastructure; promotion of the production, distribution and end uses of hydrogen;

developing more sustainable local transport; and developing international leadership at the industrial and research levels, as well as in the main transition supply chains.

Regarding the second component, with a budget of EUR 15.36 billion (29% of the total), there are three lines of action: energy efficiency of public buildings (schools); energy efficiency and seismicity of public and private residential constructions; and district heating systems (Governo Italiano 2021).

This new framework may represent a unique opportunity for the inner areas, as it may promote investments which are compatible with the potentials of rural regions and that local communities may activate. For instance, the renovation of public buildings or the realization of small-size energy systems may be the types of initiatives which can be easily carried out by the local community, by using the territorial resources and know-how and generating positive spillovers on the whole economic system.

3. Characteristics and Strategies of Inner Areas

The rapid economic growth from an economy based on agriculture towards the industrial sector, which occurred after World War II, has paved the way for a progressive migration of people from rural to urban areas. This rapid transformation has occurred due to better job opportunities, but also due to better opportunities to improve the quality of life, which could be captured by young and (relatively) most educated people.

This process occurred at different speeds across the EU territory and is still ongoing in rural areas, especially in those regions which are more isolated from urban areas and industrial settlements. In particular, local communities located in inner and mountainous areas are still dramatically shrinking, and there seems to be a lack of an effective strategy to contrast this phenomenon of depopulation and desertification with specific intervention policies. These areas are characterized by relevant distances from the main service centers (education, health and mobility). Specifically, in Southern Italy, they cover about 70% of the territory, underlining the importance to plan efficient policies and strategies.

The most important drivers of depopulation of rural areas have been widely investigated in the past (Zelinsky 1971). Among them, it is worth mentioning the transition from agricultural jobs (available in rural areas) to more appealing jobs of secondary and tertiary sectors (available in urban areas), which are socially more attractive, are well remunerated and offer some opportunities for career advancement. In addition, another important issue is related to the higher attractiveness of urban areas, due to a higher availability of public goods and services (i.e., education, transportation, access to information).

Despite the fact that the declining importance of rural areas with respect to urban areas is a typical phenomenon commonly occurring in all countries, it is worth noting that the complete abandonment of rural areas is not desirable, as it may cause several problems for the whole society.

The recognition of several functions of rural areas dates back to the 1990s, with the Buckwell report (Buckwell 1997), arguing the importance of a substantial Common Agricultural Policy (CAP) reform aimed at rural development incentives and environmental and cultural landscape payments. The basic principle was that, beside agricultural goods, rural areas would also produce stable semi-natural eco-system services, which are greatly valued by the public. This was the basic idea of the European Model of Agriculture (Swinbank 1999).

After two decades of policy interventions, addressed at supporting rural development with agricultural policies (i.e., CAP) and regional policies, there still remains a gap between rural and urban areas, probably due to the failure of the classic top-down approach, where developing projects do not require the active involvement of local communities. The problem arises when exogenous models (e.g., the establishment of an industrial settlement) are introduced in a socio- economic context which is not suitable for enhancing its correct and sustainable functioning and, furthermore, for generating positive externalities and spillovers, with positive rebound effects on the whole territorial system. The lack of active participation of local communities has caused high costs for the whole society. In fact, the territory will mainly provide the basic resources for the industrial operations (e.g., natural resources, labor), while the value added will be mainly transferred elsewhere (Hubbard and Gorton 2011). The impact on the livelihood of the local economy will be negligible, and there will be only limited chances for the emancipation of the rural communities.

On the contrary, the novel approach adopted since the 1990s by the EU, by means of the bottom-up approach, with the LEADER programs, has introduced the concept that local communities are key players in maintaining a full connection between needs, resources and economic and social activities and represent the "social fabric" paving the way for long-term sustainable development (Shucksmith 2000).

In fact, the local community detains the property rights of local resources and embeds tacit knowledge, that is, information, skills and abilities which are needed to valorize low-value and highly heterogeneous local resources (e.g., different types of biomass), according to development projects, which will preserve the territorial integrity.

Regarding the energy transition from fossil-based to zero-carbon systems, such as hydropower, wind power, solar radiation, geothermal and biomass, the involvement and active participation of the local community are crucial for several reasons. First, local resources are available at a low cost (e.g., labor, land). Second, the revenues generated with the new activities may activate the creation of new business opportunities and generate a multiplier effect and the revitalization of the local economy. Third, the extended redistributive effect of the project will facilitate the social acceptance of innovative initiatives, especially in the case of renewable energy, as rural communities do not urgently need the creation of additional sources of energy, while fearing the possible negative impacts on the environment and public health (Prosperi et al. 2019).

The well-known theoretical approach of total economic value is useful to grasp the relevance of the active involvement of the local community in the energy transition process, as it provides the basic understanding of different values which can be attributed to an economic good. This approach, widely used for the identification of different values of resources, goods and services (Adamowicz 1995; Perman et al. 2003), will help us to understand that the involvement of the local community may extend the value of renewable energy from the direct use value (i.e., revenue collected from the sale of energy) to many other categories of values. In short, the transition towards renewable energy systems, occurring with the engagement of the local community, will generate a multidimensional combination of impacts (Figure 2).

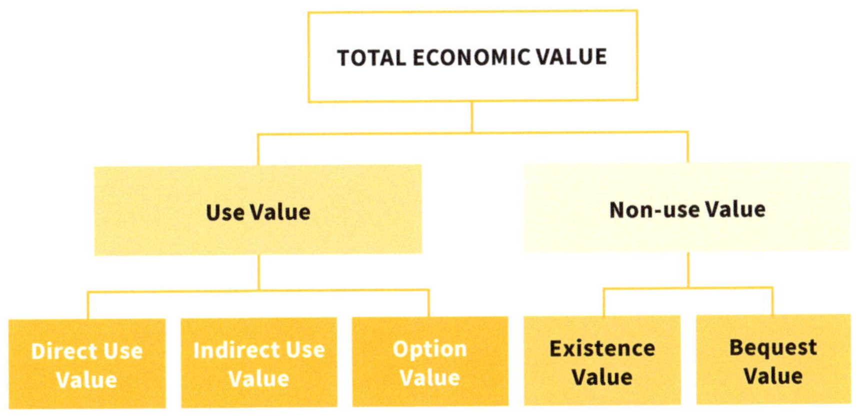

Figure 2. Conceptualization of total economic value. Source: Graphic by authors, 2021.

The total economic value of renewable energy can be conceived as the result of two categories of values: the use value, which is directly related to the generation of renewable energy, and the non-use value, which is more related to the wellbeing of local citizens.

The direct use value is quite self-evident, as it refers to the value of sales for the energy produced, to different types of consumers (i.e., households, public structures, industrial plants, commercial resellers).

The indirect use value may refer to the "greening effect" consequent to the reduction in carbon emissions and the impact that it may have on the reputation of the local community. Despite the fact that the local community may have a weak perception regarding to this type of value, it may be highly appreciated especially by the urban society, which is more sensitive towards global environmental problems.

The option value may refer to the strengthening of social ties and relationships (i.e., social capital), which provides the pre-condition for the development of further initiatives. Regarding the second group of values, they arise from the shared social dimensions (i.e., ethical and moral values) within the local community. The existence value refers to the higher level of wellbeing, which may be reached by a community when it proves it can undertake a global societal challenge (i.e., a rural community is able to achieve targets of renewable energy production and GHG reduction). The bequest value refers to the social preference towards innovative goods and services, which will (possibly) be beneficial to the future generations.

4. Community Entrepreneurship: The Italian Experience between Expectations and Disillusionment

The experience of community enterprise in Italy dates back to the 1960s, when its development spontaneously started from some bottom-up movements. A relevant example of an attempt to reinforce the local identity hindered by the massive depopulation of rural areas is the case of the small hamlet of Monticchiello (province of Siena, in Tuscany), where since 1967, theatrical performances have been held during the summer time to represent the changes affecting the local community over time (Andrews and Rosa 2005). The continuous confrontation among the community members paved the way to cooperative actions for facing common problems and new challenges.

Community enterprises have been developed in marginalized communities, characterized by large distances from lively urban settlements and profitable markets, and deprived of public services and infrastructure. The typical trends affecting these communities are aging, depopulation, economic stagnation, job insecurity, low income levels, incapacity to valorize available resources, deterioration of infrastructure and ecosystems and degradation and loss of natural resources caused by abandonment (e.g., farmland, forests, water streams and reservoirs). In addition, the investment of big energy companies has worsened the situation by causing serious impacts on the landscape and loss of farmland due to the installation of large-scale wind and photovoltaic plants.

Commonly, these problems are typical of inner rural and mountainous areas, but they also occur in some urban areas and suburbs in less developed regions, such as the case of Southern Italy (Confcooperative—Confederazione Cooperative Italiane 2018a).

In these situations, the key driver triggering the emergence of a community initiative is represented by the marginality and the vulnerability of the territory. Another relevant driver is represented by the existence of a system of mutual relationships among individuals, organizations, the environment and cultural ties within the same territory. Consequently, the so-called "virtual communities" are not taken into account, as they refer to occasional and weak relationships established within the context of the global community, as in the case of experiences referring to a shared and collaborative economy (MISE e Invitalia—Ministero dello Sviluppo Economico e Invitalia 2016). According to Mori (2015), the link with a specific territory is the basic requirement characterizing community initiatives.

Another requisite of community enterprise generation relies on the unmet needs, difficulties or problems faced by the local community, which may arise from the changes occurring over time. For instance, in recent decades, in order to pursue the enhancement of the efficiency of national public expenditure, several public services have been deactivated in rural areas and relocated in urban areas. Similarly, the depopulation of rural areas caused the closure of small and traditional firms and shops, with unavoidable consequences on the availability of job opportunities for local people, or, in some cases, the loss of symbolic social aggregation places (e.g., the closure of the last bakery shop or the largest manufacturing plant) (Confcooperative—Confederazione Cooperative Italiane 2018). Certainly, along with the closure of economic activities, the disappearance of other social institutions (e.g., civic association, political groups, sport clubs, parish) which play an important role for the aggregation of individuals and families may also occur.

The reaction against these difficulties requires the identification of strategies based on the valorization of idle local resources (material and immaterial common goods), that is, resources that are not used for any purpose, or resources that are inefficiently and unprofitably used. Furthermore, a system of strategic partnerships and social networks within the local community, and also with external agents, is the basic premise to activate an innovation process. A spontaneous process of revitalization of the local community cannot be expected, but, on the contrary, a promoting group of actors (i.e., a clique) may actively and deliberately pursue a transformation path, leading to the engagement of the community members in the definition of the vision and the mission of the community enterprise. The role of local public institutions is highly relevant, as they may endorse the initiative and they may also provide some asset to favor the establishment of the enterprise. In addition, local institutions may activate some public tendering regarding the provision of essential services, which will contribute to the business consolidation of the enterprise (Confcooperative—Confederazione Cooperative Italiane 2018).

The most relevant type of community enterprise diffused in Italy is the "cooperative of the community", whose diffusion has been moderately increased since the beginning of the 2000s. In fact, several Italian regional governments have enacted some specific regulations in order to promote its diffusion and strengthening. For instance, the Apulia region was among the first Italian regions enacting a specific regulation in this matter in (Bollettino Ufficiale della Regione Puglia n. 66 del 26/05/2014). However, at present, a national legislation is still missing, and, consequently, it is difficult to adequately monitor the diffusion process.

The cooperative of community seems a promising approach to address the development of marginal areas (Mastronardi et al. 2020). In fact, it can be conceived as a bottom-up initiative through which it is possible to boost social innovation, with the main purpose of satisfying local unmet needs, and to overcome the limitation of public interventions. In fact, the current public policy seems to be ineffective in addressing the problems of less favored areas. Similarly, it cannot be expected that the private sector, while pursuing market competitiveness and the maximization of profits, may always perform better than the public sector.

It appears that the expectations towards the cooperative of community are excessively optimistic. At present, there are still too limited successful cases confirming the adequacy of the model to face the development problems of regions lagging behind. Furthermore, there is still a literature gap regarding the lessons to be learned, in terms of possible solutions applied in different contexts (Bodini et al. 2016). In this regard, the adoption of evaluation tools would be desirable in order to perform economic assessment and social accountability of different initiatives, similar to what is occurring in the non-profit sector (i.e., the third sector) (Ministero del Lavoro e delle Politiche Sociali 2019). The evaluation exercises would be useful to verify the effects of the cooperation in the local context in which they operate.

The main features of the cooperative of community, which are included in the current legislation, are the following:

- It is an enterprise able to use idle and common resources, and to offer a steady and continuous provision of goods and services mainly to the local community;
- The membership is open, according to inclusive and democratic criteria;
- The cooperative is well rooted in the community, as its objective is the amelioration of the quality of life of the local community, conceived as the social group of the residents of a certain territory and people sharing values, culture and identity enshrined within a place, monuments, interests, resources and projects;
- The cooperative must guarantee that the provision of goods and services is accessible by the whole community (Bodini et al. 2016).

Despite the great expectations for the cooperative of the community, the Italian experience has not always been positive. The most frequent causes of failure are related to the incapacity of the management group in identifying an adequate business model through which to pursue the financial sustainability of the company, in order to guarantee the provision of goods and services to the local community. In fact, in several cases, the revenue of the cooperative mostly relies on public subsidies and contracts for public procurements, which are uncertain and discontinuous, as they derive from the political process. In addition, when the cooperative is too concentrated on public support, it lacks the capacity to adapt to the market conditions, and the emergence of economic inefficiencies. In other words, the cooperative, instead of acting as a firm, will gradually become similar to a public institution. Therefore, successful cases demonstrate that cooperatives have pursued financial sustainability through business diversification, including the market opportunities existing outside the local community (i.e., provision of goods and services to customers not belonging to the community) (MISE e Invitalia—Ministero dello Sviluppo Economico e Invitalia 2016).

Another strategy relies on the consolidation of a core business capable of ensuring a constant revenue stream through which to finance the general economic activity of the cooperative and provide either the ability to establish some investments (i.e., training of personnel, elaboration of new projects) or offset the temporal lag existing between the cost anticipation and the collection of payments (i.e., cash flow stabilization). For instance, almost all Italian cooperatives of community are involved in rural tourism activities, which are highly seasonal, or the organization of cultural activities and services, which are precarious and unprofitable but are still highly beneficial for the local community, in order to reinforce the cohesion among the population, and to create the conditions for the economic development of the territory. This is the reason why Italian cooperative communities rarely report successful stories. However, there are some instances of successful stories such as the case of the cooperative of Melpignano (in the Apulia region), which has been considered a best practice. In fact, it has been focused on the core business of energy production from renewable sources and represents an exemplary case for sustainability and profitability.

In this context, the energy transition towards renewable energy represents a very important opportunity which could be captured by rural communities, as they may be able to use local resources (e.g., agricultural and forest biomass, solar energy) with relatively small investments, but facing a constant and reliable demand, capable of generating a considerable amount of revenue flow. In this way, the energy transition will boost community entrepreneurship, will generate a positive impact on the local economy and will (indirectly) support the social wellbeing of the community.

A Cooperative of Community for Energy Transition: Bovino Municipality Case Study

The orientation of global policy makers to favor and support, through specific strategies, the energy transition towards more sustainable and accessible production systems may represent an important development opportunity for the communities of the inner areas of Southern Italy. Very soon, they will have to choose either to be protagonists of these changes, exploiting the opportunities offered by this paradigm shift in the energy sector, or to continue in passively suffering the effects. In the past, the populations of these territories have, indeed, been subjected to the consequences of energy policies at the national level without being able to participate in the decision-making processes which have defined and implemented these policies. Consequently, the realization of renewable energy plants (both photovoltaic and wind) by large industrial companies, as well as modifying the landscape aspect and compromising the naturalistic profile of these territories, was experienced by local populations as an exploitation of resources. Actually, the resulting benefits belonged to few people, and there was not an adequate refreshment system for mitigating the suffering. Thus, there was a rising natural distrust of local communities towards any attempt to address the energy issue, also in terms of local development. This is reminiscent of the importance of the local communities' social acceptance. This is a relevant determinant of the development of renewable energy systems because its absence can cause delays or even the abandonment of innovative projects. In other words, community engagement and the democracy of the energy policy processes at the local level have to be favored above all in this transition phase, learning by past mistakes (Prosperi et al. 2019).

In this section, the authors describe the experience of the Generative Communities project, funded by the local government of the Apulia region, in Southern Italy, under the public call Cooperative of Community 2018. This call aimed at supporting the establishment of new community-type entrepreneurial realities. The initiative, promoted by the CRESCO training department of Confcooperative Foggia[1], in partnership with the municipality of Bovino and two local non-profit organizations, concerned the creation of a participatory path in favor of the population of a small inner area community (Bovino) to achieve the basic requirements needed for the establishment of a local "cooperative of community" that would be, at the same time, an energy community for self-production and local distribution of renewable energy (see the EU Directive 2018/2001).

Bovino is a mountain municipality in the province of Foggia (Apulia region) with a population of 3256 inhabitants, included in the inner areas of the "Monti Dauni". Indeed, it is located 37 km northwest of the provincial capital Foggia (Figure 3).

Similar to most municipalities located in mountainous areas, Bovino has shown a constant decline in the number of residents in the last twenty years, overall equaling 20% (Istat 2020). Its main economic sectors are agriculture and services (commercial activities and professional firms) (IPRES 2016). All these factors denote the existence of territorial problems related to the marginalization of the area from the main lines of development with consequent phenomena of de-anthropization, economic decline and strong social hardship, as reported in Figure 4.

In this context, the model proposed by the authors, as reported in the Generative Communities project, aims at contributing to the improvement in the local socio-economic situation by reactivating the "local" economy, essentially leveraging latent territorial capital and the offer of some services to the resident population. The main development assets are based on the energy sector and cultural heritage.

The value proposition is to transform the endurance of the local community, meant as a passive adaptation to conditioning coming from outside the reference community context (selective market outcomes, consequences of administrative reorganization, etc.), into resilience, that is, a proactive attitude of catching opportunities, essentially by leveraging the territorial capital. The main impact,

---

[1]   Representative body of cooperatives in the province of Foggia.

expected at the local level, is mainly concerned with the empowerment and capacity building of the local community and the organizations involved.

The participatory approach consisted of the following steps:

a.   Public forum, open to the whole community, to favor the engagement of citizens;

b.   Focus groups with representatives of different stakeholders, to focus on the main strategic orientation of the cooperative of community;

c.   Expert committee, in order to analyze and perform a screening of proposals emerging from the focus groups, by considering the capacity and the resource endowment (project tailoring).

Figure 3. Location of the municipality of Bovino. Source: Graphic by authors, 2021.

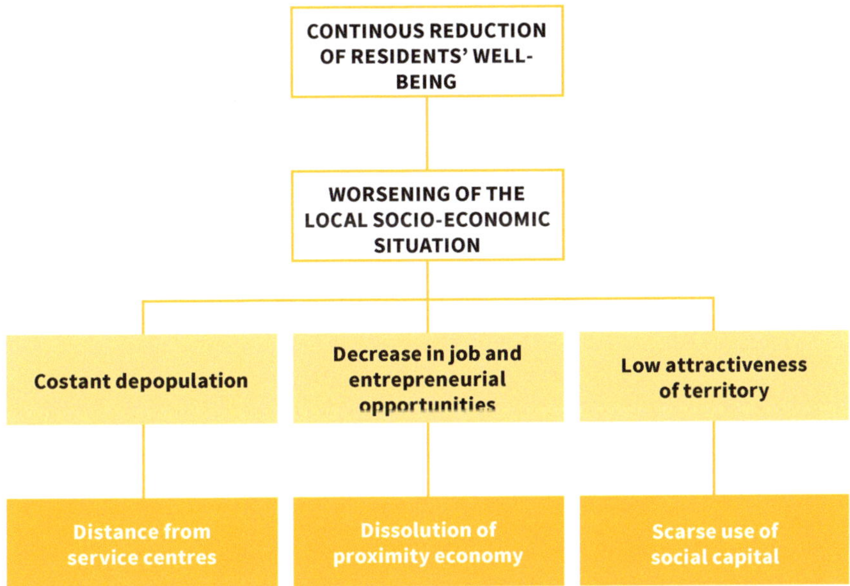

Figure 4. Typical issues of a rural marginalized area such as Bovino. Source: Graphic by authors, 2021.

The technical committee examined and selected the proposals coming from the participatory process for formulating some valid operational areas, which, coherently with the identified strategies, were suitable for the local context. This last phase represents the novelty of this process.

The project, which started in September 2019, has been structured in six phases, as reported in Table 1.

It is worth noting that the feasibility study, based on the energy balance, is a very important document for the whole sustainability of the cooperative. The drafting of the feasibility plan, relating to the investment in the energy sector, has been completed thanks to the information provided by the energy balance of the municipality and the suggestions derived from the thematic round tables.

From the experience acquired from this case study, it is possible to underline how this specific participatory approach has changed the general attitude of the community.

During the focus groups, some interesting results emerged from the participants' dialogues (30 informed individuals). In general, they expressed a strong distrust towards renewable energy plants managed by multinational companies. The main reason for this mistrust and opposition is due to the imbalance between the exploitation of local resources (mainly the landscape and farmland) and the economic compensation granted to the communities.

Table 1. Synthesis of the activities undertaken during the project.

Phase	Technique	People Involved
Participatory design	Forum, focus groups, technical committee (7 experts)	50 people attending the forum; 30 people participating in focus groups
Identification of a promoting group	Self-selection spontaneously occurred during the public events	15 people
Balance of competences of the cooperative founders	Evaluation of competences, skills and experiences, operated by the expert committee (2 members)	15 people
Training course	Lectures and study trip to other rural areas (100 hours)	13 people
Feasibility study based on the energy balance	Study performed by the expert committee (4 experts)	-
Dissemination of results	Open conference	50 people

Source: Table by authors, 2021.

On the contrary, at the end of the Generative Communities project, there was a radical change in attitude. In fact, during the final conference (through the open debate with 50 participants), it emerged that the acceptability of the renewable energy plants would significantly increase if there were more economic benefits for the community or if they are involved in the plants' management.

This change may be explained as follows: the ways in which citizens are effectively involved in the planning process; transparency and circulation of information; the development of the empowerment of organizations and individuals. In fact, with reference to the last aspect, the initiative has contributed to increasing the ability to influence and activate change through a process of participation, empowerment and awareness. Furthermore, it has facilitated a process of capacity building, that is, the construction of individual and collective skills, as well as the strengthening of the social cohesion of the community, which is also an intergenerational key.

Consequently, the proposed and implemented methodology can represent a useful tool to facilitate the energy transition in inner areas in terms of local development, overcoming the distrust of local communities, which, in this context, can assume the role of the driver of innovation and change.

5. Concluding Remarks

The energy transition is a technological change based on the shift from fossil to renewable sources, which may represent a unique opportunity for the development of marginalized rural areas and may generate multiple benefits for local communities. In fact, rural areas are endowed with abundant idle resources that are suitable for generating renewable energy, such as residual biomass from agriculture, agro-industry and forestry, and locations for the siting of geothermal, wind and solar energy plants. The operations needed for energy conversion may revitalize the local economy, by creating new job opportunities and by generating new sources of income.

The energy transition may be pursued according to two different approaches, that is, the typical top-down approach, where the investments are exogenous and the local community is marginally involved in the decision making, and the bottom-up approach, where the local community is engaged during all stages of the project development.

In this chapter, several arguments supporting the advantages arising from pursuing the bottom-up approach and, in particular, the formation of community enterprises were presented. First, the engagement of the local community may ease the social acceptance of new investments, leading to a reduction in transaction costs arising from opportunistic behavior, asymmetric information and idiosyncrasy. In fact, the local community, having a better understanding and knowledge of the local resources and having skills, know-how and capabilities to use them in more efficient and effective ways, may find low-cost and sustainable solutions. Second, the members of the local community may become, through community entrepreneurship, active actors of the energy generation, distribution and consumption processes, enabling them to revitalize their local economy and be able to consolidate a core business capable of ensuring a constant revenue stream, through which they can finance some precarious and seasonal activities and services which, though financially unprofitable, may be highly beneficial for the local community, in order to reinforce the cohesion among the population, and to create the conditions for the economic development of the territory.

The lesson learned from the study case of the municipality of Bovino is that the engagement of the local community in the energy transition process is not spontaneous, and that many efforts are needed in order to activate the public debate and to let citizens find their own solutions. Unfortunately, in the context of escalating social, environmental and economic challenges, business as usual, based on the top-down approach, and the introduction of exogenous industrial and business models are not suitable for pursuing a viable long-term development strategy. In addition, after several programming cycles occurred in the past, there is a risk of disillusionment and "community burn-out", leading to a diffused and generalized social opposition towards new development projects, which are worsening the already poor conditions of inner areas.

Author Contributions: Conceptualization, writing—original draft preparation, writing—review and editing, M.L., M.P. and G.F.; project administration, M.L. and G.F. All authors have read and agreed to the published version of the manuscript.

Funding: This research received no external funding.

Conflicts of Interest: The authors declare no conflict of interest.

References

Adamowicz, Vic. 1995. Alternative Valuation Techniques: A Comparison and Movement to a Synthesis. In *Environmental Valuation: New Persectives*. Edited by K. Willis and J. Corkindale. Oxford: CAB International.

Andrews, Richard, and Alberto Asor Rosa. 2005. *Teatro Povero di Monticchiello 1967–2004. Atto I*. Edited by R. M. Trentadue and U. Bindi. Montevarchi: Aska Editions, pp. 1–96.

Bodini, Riccardo, Carlo Borzaga, Pierangelo Mori, Gianluca Salvatori, Jacopo Sforzi, and Flaviano Zandonai. 2016. *Libro bianco—La Cooperazione di Comunità, Azioni e Politiche per Consolidare le Pratiche e Sbloccare il Potenziale di Imprenditoria Comunitaria*. Trento: Euricse Publisher, pp. 1–84.

Buckwell, Allan. 1997. *Towards a Common Agricultural and Rural policy for Europe*. Reports and Studies, No 5. Brussels: European Commission Directorate General for Economic and Financial Affairs DGII.

Camera dei Deputati. 2021. Il Piano Nazionale di Ripresa e Resilienza (PNRR). Available online: https://temi.camera.it/leg18/temi/piano-nazionale-di-ripresa-e-resilienza.html (accessed on 10 September 2021).

Confcooperative—Confederazione Cooperative Italiane. 2018a. La Cooperativa di Comunità: Un Circolo Virtuoso, Collana "Strumenti" n. 5. pp. 1–43. Available online: http://www.fontecchio.gov.it/wp-content/uploads/2018/02/N.5-La-cooperativa-di-comunita.pdf (accessed on 10 April 2021).

Confcooperative—Confederazione Cooperative Italiane. 2018. Processi Generativi di Sviluppo Territoriale: La Dimensione Comunitaria, Collana "Strumenti" n. 8. pp. 1–39. Available online: https://lavoro.chiesacattolica.it/wp-content/uploads/sites/27/2019/07/09/TENEGGI-Processi-di-sviluppo-territoriale.pdf (accessed on 10 April 2021).

Ministero del Lavoro e delle Politiche Sociali. 2019. Decreto del Ministero del Lavoro e delle Politiche Sociali del 23 Luglio, 2019. Linee Guida per la Realizzazione di Sistemi di Valutazione Dell'impatto Sociale delle Attività Svolte Dagli enti del Terzo Settore, Gazzetta Ufficiale Serie Generale n. 214 del 12/09/2019. Available online: https://www.gazzettaufficiale.it/eli/id/2019/09/12/19A05601/sg (accessed on 10 April 2021).

EC. 2019. *Communication from the Commission to the European Parliament, the European Council, the Council, the European Economic and Social Committee and the Committee of the Regions—The European Green Deal, Brussels, 11.12.2019 COM (2019) 640 Final*. Brussels: EC.

EC. 2020. *Report from the Commission to the European Parliament, the Council, the European Economic and Social Committee and the Committee of the Regions Renewable Energy Progress Report, Brussels, 14.10.2020 COM (2020) 952 Final*. Brussels: EC.

EC. 2020. *Communication from the Commission to the European Parliament, the Council, the European Economic and Social Committee and the Committee of the Regions Powering a Climate—Neutral Economy: An EU Strategy for Energy System Integration COM/2020/299 Final*. Brussels: EC.

EC. 2020. *Communication from the Commission to the European Parliament, the Council, the European Economic and Social Committee and the Committee of the Regions a Hydrogen Strategy for a Climate—Neutral Europe COM/2020/301 Final*. Brussels: EC.

EC. 2020. *Communication from the Commission to the European Parliament, the Council, the European Economic and Social Committee and the Committee of the Regions a Renovation Wave for Europe—Greening our Buildings, Creating Jobs, Improving Lives COM/2020/662 Final*. Brussels: EC.

EU. 2018. Directive 2018/2001 of the European Parliament and of the Council of 11 December 2018 on the Promotion of the Use of Energy from Renewable Sources. Available online: https://eur--lex.europa.eu/legal--content/EN/TXT/PDF/?uri=CELEX:32018L2001&from=fr (accessed on 5 May 2021).

EU. 2021. Regulation (EU) 2021/241 of the European Parliament and of the Council of 12 February 2021 Establishing the Recovery and Resilience Facility. *Official Journal of the European Union L* 57: 17–75.

Governo Italiano. 2021. Piano Nazionale di Ripresa e Resilienza—#NEXTGENERATIONITALIA. Available online: https://www.governo.it/sites/governo.it/files/PNRR_0.pdf (accessed on 19 May 2021).

Hennicke, Peter, Jana Rasch, Judith Schröder, and Daniel Lorberg. 2019. The Energy Transition in Europe: A Vision of Progress, Wuppertal Spezial, No. 54, Wuppertal Institut für Klima, Umwelt, Energie, Wuppertal. Available online: http://nbn-resolving.de/urn:nbn:de:bsz:wup4-opus-73368 (accessed on 4 May 2021).

Hubbard, Carmen, and Matthew Gorton. 2011. Placing agriculture within rural development: Evidence from EU case studies, Environ. Plann. C: Gov. *Policy* 29: 80–95.

IEA, IRENA, UNSD, WB, and WHO. 2019. Tracking SDG 7: The Energy Progress Report 2019. Washington DC. Available online: https://trackingsdg7.esmap.org/data/files/download-documents/2019-Tracking%20SDG7-Full%20Report.pdf (accessed on 5 May 2021).

IPRES. 2016. Imprese Attive e Localizzazioni D'impresa. Available online: https://www.ipres.it/it/news/item/49-imprese-e-addetti-in-puglia-2016 (accessed on 17 April 2021).

Istat. 2020. Popolazione Residente al 1° Gennaio: Puglia. Available online: http://dati.istat.it/Index.aspx?QueryId=18550 (accessed on 17 April 2021).

Bollettino Ufficiale della Regione Puglia n. 66 del 26/05/2014. Available online: http://burp.regione.puglia.it/documents/10192/4806320/LEGGE+REGIONALE+20+maggio+2014%2C%20n.+23+%28id+4806333%29/4ae8b341-640f-4e7b-be57-f48c694d87ff;jsessionid=C4D05BB4DF54C823E19A4943604050A1 (accessed on 15 April 2021).

Mastronardi, Luigi, Maria Giagnacovo, and Luca Romagnoli. 2020. Bridging regional gaps: Community-based cooperatives as a tool for Italian inner areas resilience. *Land Use Policy* 99: 104979. [CrossRef]

MISE e Invitalia—Ministero dello Sviluppo Economico e Invitalia. 2016. *La Cooperazione di Comunità per uno Sviluppo Locale Sostenibile. Studio di Fattibilità su "Lo Sviluppo delle Cooperative di Comunità"—Report Finale*. Roma: MISE e Invitalia, pp. 1–280. Available online: https://www.cooperativedicomunita.confcooperative.it/Portals/0/News/INVITALIA-LA%20COOPERAZIONE%20DI%20COMUNITA-Report%20finale.pdf (accessed on 20 May 2021).

Mori, Pier Angelo. 2015. *Community and Cooperation: The Evolution of Cooperatives towards New Models of Citizens' Democratic Participation in Public Services Provision*. Trento: Euricse, pp. 1–25.

Perman, Roger, Yue Ma, James McGilvray, and Michael Common. 2003. *Natural Resource and Environmental Economics*, 3rd ed. Boston: Pearson Addison Wesley, pp. 1–699.

Prosperi, Maurizio, Mariarosaria Lombardi, and Alessia Spada. 2019. Ex ante assessment of social acceptance of small-scale agro-energy system: A case study in southern Italy. *Energy Policy* 124: 346–54. [CrossRef]

SDG Tracker. 2021. Ensure Access to Affordable, Reliable, Sustainable and Modern Energy for All. Available online: https://sdg--tracker.org/energy (accessed on 10 May 2021).

Shucksmith, Mark. 2000. Endogenous development, social capital and social inclusion: Perspectives from LEADER in the UK. *Sociologia Ruralis* 40: 208–18. [CrossRef]

Swinbank, Allan. 1999. CAP reform and the WTO: Compatibility and developments. *European Review of Agricultural Economics* 26: 389–407. [CrossRef]

UN—United Nations. 2021. The Sustainable Development Agenda. Available online: https://www.un.org/sustainabledevelopment/development--agenda/ (accessed on 5 April 2021).

Zelinsky, Wilbur. 1971. The hypothesis of the mobility transition. *Geographical Review* 61: 219–49. [CrossRef]

Finnish Forest Industry and Its Role in Mitigating Global Environmental Changes

Ekaterina Sermyagina, Satu Lipiäinen and Katja Kuparinen

1. Introduction

Climate change is currently one of the greatest global threats. Mitigating global warming requires a significant reduction in greenhouse gas (GHG) emissions either by reducing emission sources or enhancing the sinks to remove these gases from the atmosphere. To succeed, long-term structural changes are needed for different sectors, thus creating more balanced and sustainable patterns of energy supply and demand. The forest industry is an energy-intensive sector that emits approximately 2% of industrial fossil carbon dioxide emissions worldwide. Considering the high share of biofuels already used within this sector, the forest industry may shortly become a significant user of bio-based carbon capture technologies. The possibility to implement these technologies can transform pulp and paper mills into negative CO_2 emitters. Moreover, the forest industry can also contribute to GHG emission mitigation outside the mill gates by producing biomass-derived heat, electricity and liquid biofuels, and by providing wood-based products, such as packaging materials, textiles and chemicals, which will substitute the fossil-based alternatives.

The global forest industry has managed to decrease its dependence on fossil fuels as a result of energy efficiency improvement, fuel switching and structural changes, and thus its fossil CO_2 emissions have decreased substantially in the 21st century (International Energy Agency 2020). However, the paper demand is expected to increase from the current 400 Mt/a to 750–900 Mt/a by 2050, and thus there is a huge need to develop towards more sustainable operation in order to avoid an increase in CO_2 emissions. Therefore, more understanding of the possibilities of the forest sector is essentially required. In Europe, the forest sector has ambitious targets to contribute to the mitigation of GHG emissions (CEPI 2011). The sector aims to emission reduction from the 1990 level of 60 $MtCO_2$/a to 12 $MtCO_2$/a in 2050 with an increasingly significant role of generated wood-based materials in substitution of fossil materials in different applications. This book chapter considers the forest industry's possibilities to contribute to the mitigation of environmental change using Finland as a target country. The Finnish Forest Industries have recently presented a roadmap towards low-carbon operation (Finnish Forest Industries 2020a). The roadmap takes into account CO_2 reduction from increased annual forest growth, increased production of bio-based materials and reduction in fossil fuel use in industrial processes, transportation and off-site production of energy. However, there is a lack of academic studies that discuss extensively the CO_2 mitigation possibilities of the forest sector.

Finland is an important producer of pulp, paper, and sawn wood. Besides these traditional goods, Finnish mills are generating a range of innovative wood-based products that will be discussed in this study. Even though the Finnish forest industry is an important energy producer on a national level, it has currently only 13% of fossil fuels in its fuel mix, and therefore fossil CO_2 emissions are already relatively low. The previous reports, however, claim that the Finnish forest industry can become fossil-free by 2035. Finland's government confirmed the National Energy and Climate Strategy for 2030, which should help to achieve a long-term goal of a carbon-free society and keep the stable course for reach 80—95% reduction in GHG emissions from the level of 1990 by 2050 (Ministry of Employment and the Economy 2014). Finland is among the world leaders in the utilization of renewable energy sources, especially bioenergy. The substantial forest resources, developed forest industry and the well-established forest infrastructure mean that wood-based bioenergy has a significant place within the renewable energy sector in Finland. According to Statistics Finland (2020a), the total consumption of energy in Finland amounted to 1362 petajoules (PJ) in 2019 with 38% covered by renewables. Figure 1 provides a historical trend on the total energy consumption from 1970 to 2019 (ibid.). The shift from fossil fuels towards renewable alternatives continues in different spheres, with a significantly increasing increased share of the latter ones in total consumption from the 2010s.

The energy-transition tendencies and general development in the Finnish pulp and paper sector have been analysed in a few recent papers. The efficiency of the Finnish pulp and paper industry has been evaluated and compared to the EU average level (Koreneff et al. 2019). The results of this study showed that while the production efficiency is on a high level, the dominancy of kraft pulp production leads to the notably higher energy intensity of the production process. Lipiäinen et al. (2022) have evaluated the main steps towards decarbonization in Finnish and Swedish pulp and paper industries, highlighting the essential steps performed there to reduce fossil CO_2 emissions and at the same time maintain competitiveness on a high level. While previous works limited their scope to the pulp and paper sector, the current study aims to help in better understanding of the general picture and provides the most updated knowledge on the situation with forest industry in Finland: structure, characteristics, fuel consumption and GHG emissions. In addition, we identify important research directions to facilitate the transition towards a sustainable future.

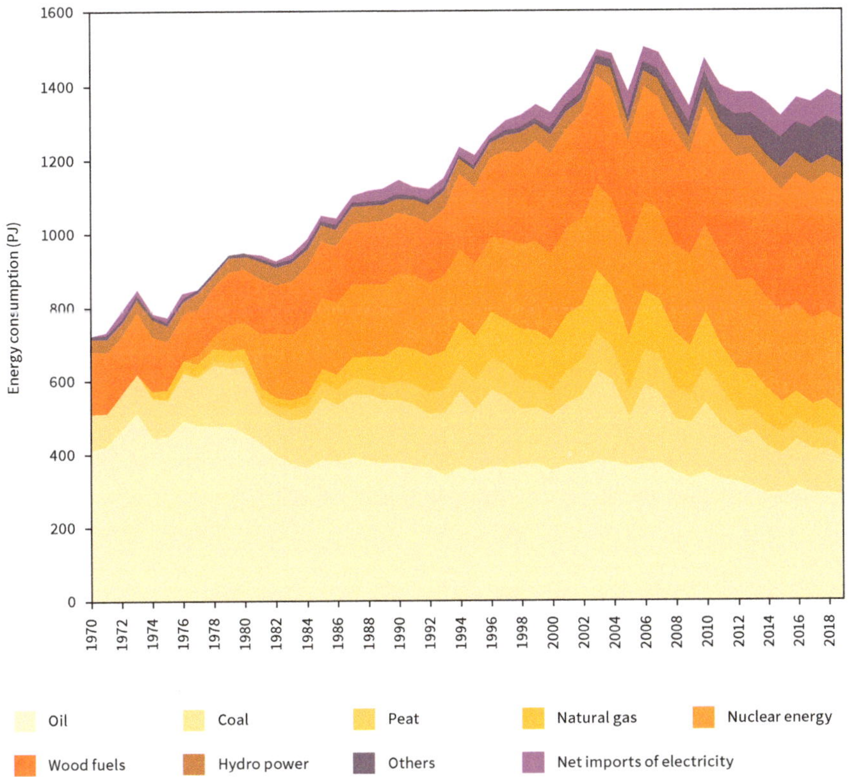

Figure 1. Total energy consumption in 1970–2019 in Finland. Source: Graphic by authors based on data from Statistics Finland (2020a).

2. Methods

2.1. Data Gathering

Data were gathered from several sources to evaluate both the current state and the development of the Finnish forest sector. The main sources for energy consumption, production and emissions were Finnish Forest Industries and Statistics Finland databases (Finnish Forest Industries 2020b; Statistics Finland 2020c). During the study, a large number of previous studies were reviewed and the Finnish forest sector's possibilities to participate in mitigation of climate change were evaluated.

Moreover, the emerging forest industry projects, such as the production of new bioproducts in Finland, have been identified and introduced.

2.2. Energy Efficiency Index Method

Energy efficiency development in the Finnish forest industry was studied using the energy efficiency index method. The method was applied in the previous study for assessing energy efficiency development in the pulp and paper industry (Lipiäinen et al. 2022). The present work follows the same method and uses the same reference-specific energy consumption values.

3. Finnish Forest Industry

The forest industry is a globally important industrial sector. Several wood-based products, such as paper, paperboard and sawn wood participate in people's everyday life. The forest industry can be divided into the chemical forest industry that produces mainly pulp and paper, and the mechanical one that focuses on producing sawn wood and wood-based panels. The global forest industry produced 188 Mt of wood pulp, 409 Mt of paper, 493 Mm^3 of sawn wood and 408 Mm^3 of wood-based panels in 2018 (Food and Agriculture Organization of the United Nations 2019).

In 2019, Finland produced 30%, 11% and 7% of European pulp, paper and board, and sawn wood, respectively (CEPI 2019; Food and Agriculture Organization of the United Nations 2019). 73% of produced pulp was kraft (sulphate) pulp. The rest consists of mechanical, chemi-mechanical, and semi-chemical pulp. Within paper production, the major shares had printing and writing papers (47%), and packaging materials (45%). A minor amount of newsprint and hygiene papers were produced. The Finnish forest industry has been recently going through a structural change (Kähkönen et al. 2019). The development of information technology has led to a decrease in the consumption of printing and writing papers (Johnston 2016), whereas the demand for packaging materials has increased globally (Hetemäki et al. 2013). The changes in the production volumes between 1990 and 2019 are presented in Table 1. Production of newsprint, printing and writing papers has declined and several mills have been closed, while substantial growth in volumes of chemical pulp, packaging materials and sawn wood has taken place. The role of new products, such as chemicals, biofuels, and energy, is increasing, and forest industry companies are looking for new product opportunities, which will be discussed later.

Table 1. Production volumes in the Finnish forest industry.

Grade	1990	2019	Change
Domestic wood usage [Mm^3]	44	61	+38%
Imported wood usage [Mm^3]	6	10	+66%
Chemical pulp [Mt]	5.16	8.72	+69%
Mechanical pulp [Mt]	3.73	3.28	−12%
Newsprint [Mt]	1.43	0.27	−81%
Printing and writing paper [Mt]	4.68	4.60	−2%
Packaging materials [Mt]	2.32	4.44	+91%
Household and sanitary papers [Mt]	0.16	0.20	+22%
Other paper and paperboard [Mt]	0.37	0.20	−46%
Sawn wood [Mm^3]	7.50	11.39	+52%
Wood-based panels [Mm^3]	1.34	1.29	−4%

Source: Data from Food and Agriculture Organization of the United Nations (2020) and Nature Resource Institute (LUKE) (2020).

The forest industry, especially the pulp and paper industry (PPI), is an energy-intensive sector. The PPI is the fourth largest industrial energy user and globally it consumed 7 EJ of energy in 2018 (International Energy Agency 2018). The share of biofuels was 48%, which is 18% higher than at the beginning of the 2000s. Energy demand per ton of produced paper has decreased substantially, and fuel switching together with increased efficiency has led to lower fossil CO_2 intensity. In Finland, the manufacturing of forest industry products is by far the largest industrial energy consumer, which accounts for more than half of total industrial energy use: 316.8 PJ in 2019 (Statistics Finland 2020b). The forest industry is also a significant energy producer: it can cover a major share of its heat demand and about half of its electricity demand. This highlights the importance of efficiency improvement and cost-efficient reduction in GHG emissions along with the promotion of clean and sustainable technologies for this energy-intensive sector. The characteristics of energy use and emissions in the Finnish forest industry are presented in Table 2.

Table 2. Energy and emissions in the Finnish forest industry in 2019.

Energy and Emissions	Finnish Forest Industry	Share in Total Domestic Values
Electricity consumption	19.3 TWh/a	22.4%
Electricity production	10.1 TWh/a [1]	15.3%
Energy consumption	316.8 PJ/a	23.3%
Dominant energy sources	Biomass 87%, NG 5% [2]	-
Fossil CO_2 emissions	2.7 Mt/a [2]	5.1%
NO_x	17.5 kt/a	10.9%
S	1.7 kt/a	10.9%

[1] 2016 electricity production used due to lack of data. [2] Include only on-site fuel use. NG = natural gas. Source: Data from Finnish Forest Industries (2020b) and Statistics Finland (2020c).

4. Climate Change Mitigation Opportunities in the Forest Industry

Climate change mitigation requires both reducing GHG emissions and enhancing carbon sinks. There is no single solution to solve the problem of climate change, but a wide range of solutions is needed, some of which are more mature than others. The forest industry has a significant role in promoting these in Finland. Sustainable forest management enables not only the efficient utilization of forest resources but also their increase thus enhancing the role of forests as carbon sinks. Many changes and challenges have already transformed the forest industry in Finland in recent years and there should be more developments to come to adapt to the rapidly changing global situation. Modern pulp mills are expanding the traditional concept of pulp mills by introducing the combination of multifunctional biorefineries and energy plants that utilize wood resources to produce not only pulp but also energy as well as new high-value products from side-streams and residues.

Previous studies have estimated the contribution of the forest sector to mitigation of climate change by direct or indirect CO_2 emission reduction (e.g., Finnish and Swedish roadmaps). Based on the results from the previous studies, the main possibilities for the forest industry are addressed in the present work:

- Direct CO_2 emissions reduction;
- Green energy production
- Bio-based materials.

The present study does not consider forest management but assumes that wood used by the forest industry is sustainably harvested.

4.1. CO$_2$ Emissions and the Reduction Possibilities

The Finnish forest industry emitted 2.7 Mt of fossil and about 22 Mt of biogenic CO_2 in 2019 (Finnish Forest Industries 2020b). The mechanical forest industry is a minor energy user and CO_2 emitter in comparison to the chemical forest industry. A large part of CO_2 produced in a pulp and paper mill comes from biomass combustion and can be considered carbon-neutral when the wood is from a sustainable origin. The fossil CO_2 emissions from the PPI are energy-related, which makes their reduction easier compared with for example the cement industry, since it does not require direct process modifications. The lime kilns also produce process-related CO_2 emissions, but they are largely biogenic. The fossil CO_2 emissions have decreased by 44% over the last 20 years (1999–2019) (Figure 2). Emissions per ton of product decreased by 37% from 194 to 122 kg$_{CO2}$/ton. The industrial strike in 2005 and the economic crisis around 2009 decreased absolute emissions strongly due to a significant drop in production volume. During those crises, emissions per ton of product did not change significantly. The reforms in energy taxation around 2003 and 2011 have most probably contributed to the drops in emissions. The European Union (EU) introduced Emission Trading System (ETS) in 2005. It may have affected the fossil CO_2 emissions, but it has been argued that ETS has had only a limited effect on the PPI due to excess emission allowances (Gulbrandsen and Stenqvist 2013). After a period of steady emission levels between 2012 and 2016, the emissions have been declining again. The emissions are expected to continue to decline in the future as the Finnish forest industry is aiming towards net-zero emissions. A recently published report argues that nearly zero emissions can be achieved in 2035 (Pöyry 2020).

The primary means to reduce CO_2 emissions of the PPI are switching to low-carbon fuels, electrification, and energy efficiency improvement. Carbon capture and storage (CCS) technologies that initially had been aimed mainly at the power sector, have significant potential in the PPI. In Finland, the role of the forest industry in mitigating environmental change and meeting the national carbon neutrality target by 2035 is not limited to emissions reduction in the mills (Lipiäinen and Vakkilainen 2021). The PPI supplies green electricity to the grid and produces other renewable energy carriers. Moreover, it has been evaluated that products of the Finnish forest industry can annually mitigate 16.6 Mt of CO_2 emissions (Finnish Forest Industries 2020b). However, forests are a significant carbon sink, and therefore they must be managed sustainably to achieve emission reductions.

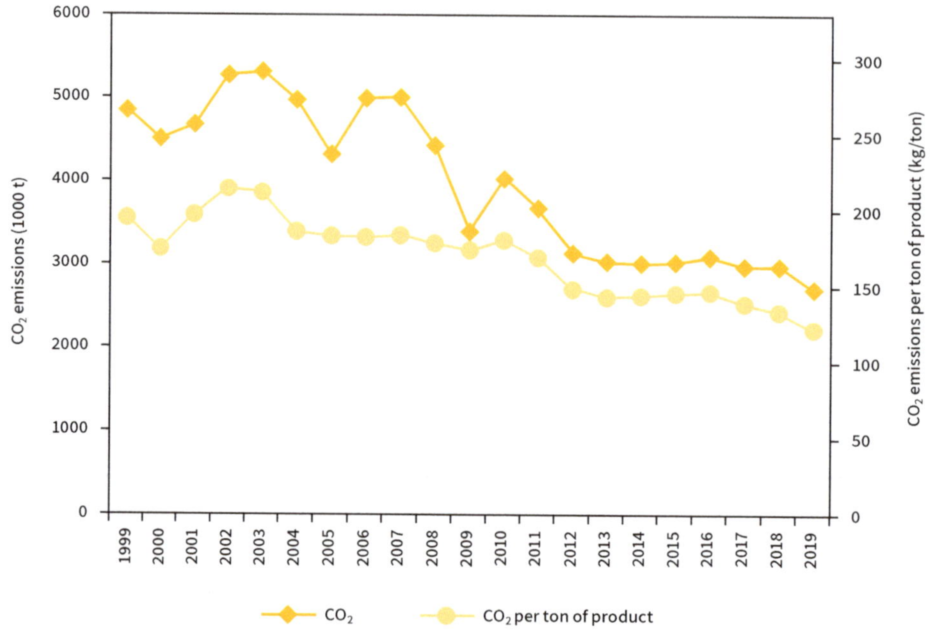

Figure 2. Trends in fossil CO_2 emissions in the Finnish pulp and paper industry. Source: Graphic by authors based on data from Finnish Forest Industries (2020b).

4.1.1. Fuel Switching

The largest part of fossil-based CO_2 emission savings is expected to come from replacing the current use of fossil fuels with renewables (Moya and Pavel 2018; Metsäteollisuus ry 2020). The forest industry typically meets a large share of its energy demand by its own production. A modern stand-alone pulp mill can even surpass its heat and electricity demand by combusting wood residues (IRENA et al. 2018). Black liquor, a side-stream from kraft pulping, is the most important biofuel in the PPI. Mills combust it in the recovery boilers, and many mills have an additional power boiler for combusting wood residues. The boilers supply steam for the processes and for turbines to generate electricity. In the modern stand-alone pulp mills, fossil fuel consumption can be limited to start-ups, shutdowns, and other exceptional situations, but many mills still combust fossil fuels in the lime kilns, mainly natural gas or oil (Kuparinen et al. 2019). Stand-alone paper mills and some integrated pulp and paper mills cannot meet their energy demand by combusting their own residues. These mills either combust fossil fuels or purchase energy. Many kraft pulp mills produce an excess of electricity that is typically sold but can be also used for hydrogen or e-fuels generation. However, the role of electrification is estimated to be minor in the emission reduction within the Finnish forest industry, mainly it is expected to substitute for natural gas use.

Biomass has been an important fuel in the Finnish forest industry for a long time, but the fuel mix includes also natural gas, oil, coal, and peat (Figure 3). In 1990, the share of biofuels was 64%, natural gas covered 15%, and oil, coal, and peat stand for approximately 7% each. In 2019, the share of biofuels had increased to 87%, and the shares of natural gas, peat, oil, and coal were decreased to 6%, 3%, 2% and 1%, respectively. Peat is a specific fuel in Finland that is typically co-fired with biomass in power boilers. The recent political decisions in Finland promote the replacement of coal and peat by biomass. Increased volumes of chemical pulp have increased the use of biofuels, but many mills have also actively looked for solutions to decrease the use of fossil fuels. Currently, biomass has been seen as the most potential alternative for fossil fuels in the Finnish forest industry, and other renewables have not played a large role. The possibility to use wind power for covering paper mills' energy demand has been recently realized with a long-term Wind Power Purchase agreement of Finnish company UPM with German wind park development company (wpd) (UPM Communication Papers 2020). This agreement will enable the decrease in CO_2 emissions by 200,000 tonnes annually starting from 2022 and help to achieve the company's ambitious 65% CO_2 emission reduction target by 2030. In addition, UPM is utilizing hydropower sources and upgrading the performance of the existing hydropower plants (UPM Energy 2021).

The lime kilns are the primary fossil fuel users in chemical pulp production. CO_2 is produced from both the combustion and the actual lime regeneration reaction during the calcining process. The CO_2 from the reaction originates mainly from wood and is thus biogenic. The lime kiln process requires stable combustion conditions and easily controllable hot-end temperature. Consequently, fuel characteristics and quality should be consistent (Isaksson 2007). In addition to fossil fuels, alternative fuels such as methanol, tall oil, strong odorous gases, tall oil pitch, hydrogen, and turpentine are often co-fired in the lime kilns. Technically, it is also possible to utilize existing side-streams to substitute for fossil fuels there (Kuparinen and Vakkilainen 2017). However, biomass fuels typically have lower adiabatic flame temperature and lower energy content than fossil oil or natural gas. Therefore, higher firing rates are required to maintain the kiln capacity in the case of biomass supply. Another problem is that the impurities that originated from the solid biomass tend to accumulate in such a closed cycle process. These non-process elements can cause, e.g., corrosion and ring formation in the kiln and decrease the lime quality. The use of alternative fuels can thus lead to increased use of make-up lime and should be evaluated thoughtfully.

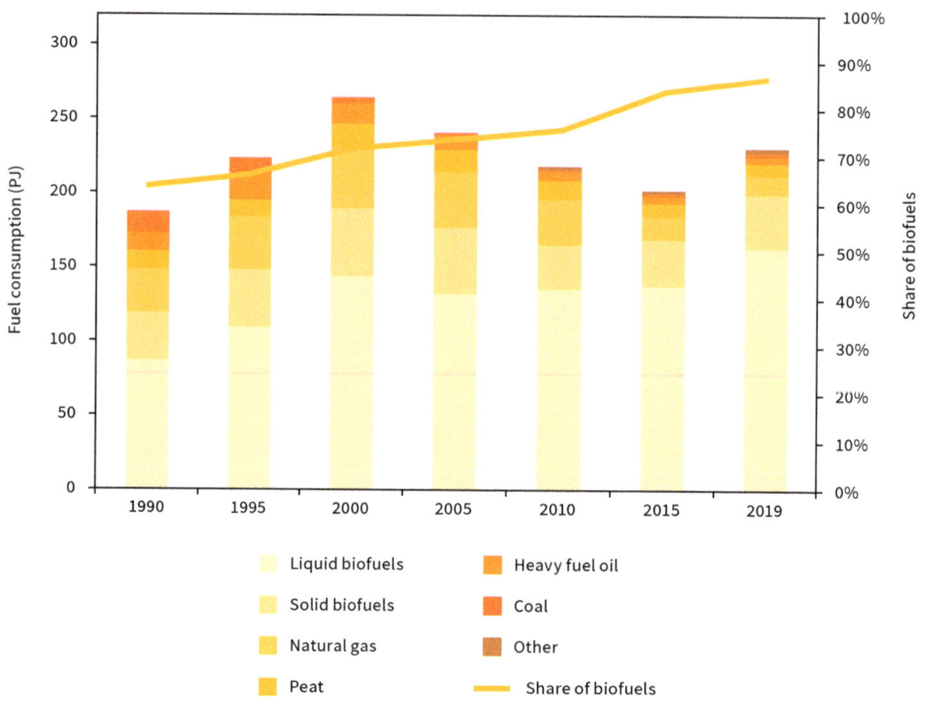

Figure 3. Development of fuels use in the Finnish forest industry. Source: Graphic by authors based on data from Finnish Forest Industries (2020b).

4.1.2. Energy Efficiency Improvement

Improvement of energy efficiency has played a notable role in the reduction in CO_2 emissions so far (European Commission 2018). Efficient reduction in heat losses, recovery of process heat and process optimization offer further possibilities for emissions reduction but require other concurrent actions to achieve the climate goals. The Best Available Technologies (BAT) reference document (European Commission 2015) presents state-of-the-art technologies. Energy efficiency is however not a straightforward concept. It is often measured by specific energy consumption (SEC), i.e., the energy consumed for the production of a unit of product. However, a change in SEC may result for example from increased utilization rate instead of improvements in energy efficiency. The forest industry is a heterogeneous sector with a diverse product portfolio and highly energy intensive production processes.

No uniform practice on collecting process information exists. The collection of reliable process information is necessary for efficiency improvement. While in general, the increased energy efficiency leads to a decrease in CO_2 emissions, an energy-efficient mill is not necessarily CO_2 efficient due to different products, process alternatives, and mill configurations. Stenqvist and Åhman (2016) noticed that the benchmark-based emission allowance allocation of the EU ETS does not result in the best performance in a heterogeneous sector like the PPI due to, e.g., lack of benchmark curves and biased reference values.

Energy efficiency within the forest industry can be improved by new technologies but also new modes of operation. Energy audits, motivated and competent employees, and process monitoring and control advance energy efficiency (Vakkilainen and Kivistö 2014). Finland has a long history of energy auditing, and both mandatory and voluntary energy auditing schemes are carried out to measure energy consumption and identify energy-saving opportunities (Ministry of Economic Affairs and Employment of Finland 2021). The Finnish know-how on energy audits has also been relied upon in other countries building their own audit schemes (Motiva Ltd. 2019). Enhanced process integration typically improves energy efficiency. The pulp and paper production processes result in secondary heat streams, whose further utilization would improve the total efficiency. The ongoing transformation from traditional pulp and paper mills to modern multi-product biorefineries offers a possibility to utilize these in the production of advanced bioproducts. Another viable option is improved drying techniques that help to reduce emissions from one of the most energy-intensive process stages.

The Finnish forest industry has been historically an energy-efficient operator (Fracaro et al. 2012). Finnish pulp and paper production is already rather efficient, despite the need for heating due to the cold climate. Compared to the EU average, the Finnish mills are large, efficient, and modern (Koreneff et al. 2019). Nevertheless, it is still possible to increase efficiency. The development of the forest industry's efficiency is presented in Figure 4. Between 2002 and 2019, primary energy efficiency and electricity efficiency improved 1.4% and 1.2% per year, respectively. The strike in 2005 and the economic crisis around 2009 decreased the efficiency because several mills were operating only part of their capacity. Several factors, such as closures of old mills, start-ups of new mills, technology development and increase in energy prices, have contributed to the efficiency improvement (Kähkönen et al. 2019).

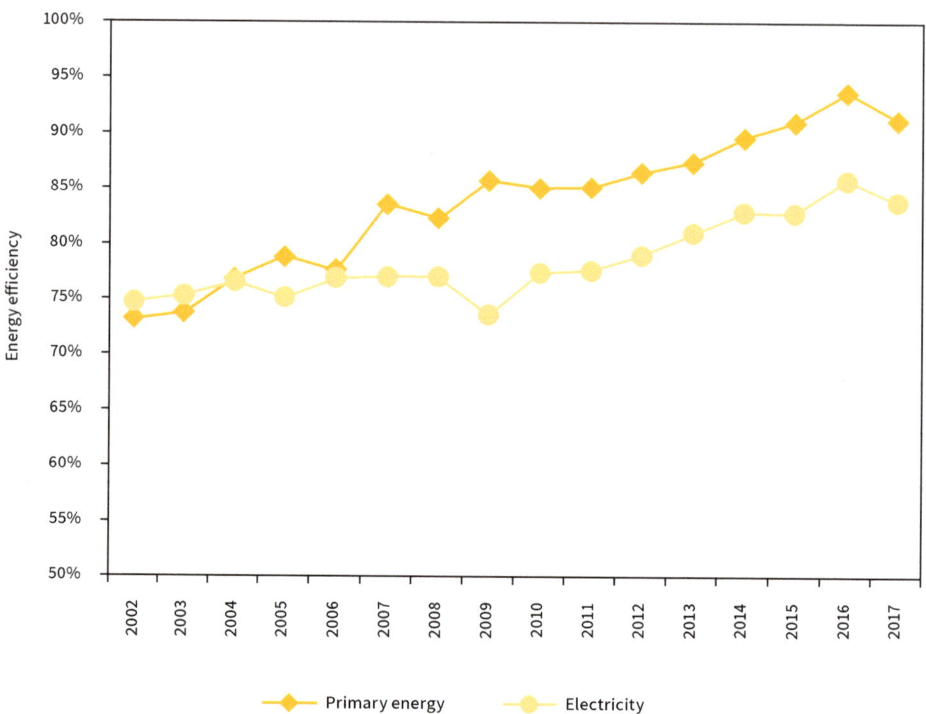

Figure 4. Trends in energy efficiency in the Finnish forest industry. Source: Graphic by authors based on data from Statistics Finland (2020c), Food and Agriculture Organization of the United Nations (2019) and Finnish Forest Industries (2020b).

4.1.3. Carbon Capture Technologies

When the target is net-zero or even negative emissions, carbon capture, utilization, and storage (CCUS) technologies have to be included in the palette. CCUS has a remarkable role in many decarbonizing scenarios, especially in the ones that include fossil fuels in the energy mix also in the future (European Commission 2018). Bioenergy with carbon capture and storage or utilization (BECCS/U) is one of the key negative emission technologies. The pulp and paper industry, being a significant bioenergy producer, has a unique possibility for the implementation of BECCS/U. The global technical capture potential from kraft pulp mills has been estimated at approximately 137 Mt_{CO2}/a (Kuparinen et al. 2019). One of the primary concerns regarding large-scale utilization of BECCS/U is increased land use since BECCS/U is often seen as promoting additional use of biomass for energy. Large existing pulp and paper units however offer the possibility to implement BECCS/U without additional biomass harvesting. The main weaknesses of BECCS/U currently are the lack of experience and political support. Public acceptance and uncertainties on the long-term behaviour of the stored carbon have also hindered their implementation in the EU (European Commission 2018). Besides, carbon capture processes require heat and electricity. Therefore, their integration into a mill affects the mill's energy balance and total CO_2 emissions.

The primary CO_2 sources in the PPI are the combustion processes, namely the recovery boilers, the power boilers, and the lime kilns. The minor CO_2 sources include, e.g., non-condensable gas destruction and biosludge treatment to produce biogas in some mills. The magnitude of the recovery process is often not fully appreciated. In 2016, the global sulphate pulp production was 137 Mt (FAO 2017). Consequently, more than 1300 Mt of weak black liquor was processed in recovery boilers globally and 206 Mt of black liquor dry solids were combusted to produce about 1.8 EJ of energy (Tran and Vakkilainen 2008). According to (International Energy Agency 2018), it makes black liquor the fifth most important fuel in the world after coal, oil, natural gas, and gasoline, and the most used biofuel globally. Therefore, the recovery boilers alone offer a notable possibility for BECCS/U.

The capture of biogenic CO_2 from the pulp and paper mill processes is a little-studied subject so far (Leeson et al. 2017). Recent publications (IEAGHG 2016; Kuparinen et al. 2019) have however indicated the technical feasibility of BECCS/U within the pulp mills. Several technologies including pre-combustion, post-combustion and oxy-combustion methods can be applied to pulp and paper mills. Many of these are in the development stage. The most studied capture method is the monoethanolamine (MEA) process (Onarheim et al. 2017; Leeson et al. 2017). The commercial MEA process is a post-combustion method and thus can be easily applied to existing mills. Based on earlier estimates, the cost of CO_2 avoided in pulp mills ranges between 20 and 92 EUR/t_{CO2} depending on the chosen processes and mill characteristics (Fuss et al. 2018; IEAGHG 2016).

CO_2 is typically seen as emission or waste; its value as a raw material for carbon-based products has not been recognized until recently. CO_2 conversion technologies include biotechnical and chemical or catalytic processes, a few of which have been commercialized so far (Lehtonen et al. 2019). The value of captured CO_2 as a raw material can be a push for cost-effective carbon capture. It has a wide range of potential utilization routes in industrial and chemical applications, where it currently comes mostly from fossil sources. CCU processes enable CO_2 recycling and therefore reduction in the CO_2 in the atmosphere. Net negative CO_2 emissions can be reached only if at least part of the captured CO_2 is stored or utilized in a process that permanently removes it from the atmosphere. Using it for fuel production, for instance, delays the release and enables the indirect reduction in the atmosphere, if the fuel is used to substitute for traditional fossil fuels.

The possibilities to utilize CO_2 in the PPI depend on mill-specific details. Currently, the chemical forest industry utilizes CO_2, e.g., for pH control and in brown stock washing. Instead of purchase, it can be captured from the mills (Ruostemaa 2018). Carbon capture from lime kiln flue gases and subsequent use as calcium carbonate paper filler (precipitated calcium carbonate, PCC) is a well-known and widely applied technology (Hirsch et al. 2013). In 2005, Teir et al. (2005) estimated that the potential to eliminate CO_2 emissions considering only the PCC used in the PPI in Finland would be 200 kt/a. Apart from this, softwood pulp mill typically produces tall oil as a by-product. Raw soap is separated from black liquor and converted to crude tall oil by acidulation, typically using sulfuric acid. Part of the acid, up to 50%, can be replaced by CO_2. Tall oil can be further converted to renewable fuels and used to substitute for fossil alternatives. Another relevant alternative is lignin separation from black liquor using sulfuric acid, which can be also replaced by CO_2: 150–250 kg_{CO2}/t_{lignin}.

The reduction in direct CO_2 emissions by fuel switching and energy efficiency improvement can avoid up to 2.5 $MtCO_2$ of emissions, which correspond to 5% of the Finnish total fossil CO_2 emissions. Moreover, capturing biogenic CO_2 has a significant potential to provide extended climate benefits.

4.2. Green Energy

Wood-derived energy made up 74% of total renewable energy in Finland in 2019 with the largest share covered by black liquor combustion (47.2 TWh) (Natural Resources Institute 2019). According to the Natural Resources Institute (2019), solid wood fuels used at power and heating plants accounted for 39.5 TWh, the small-scale combustion of wood comprised 16.8 TWh and other wood fuels covered 2.1 TWh in 2019. Wood energy resources for energy generation are typically used in highly efficient district heating (DH) systems and combined heat and power (CHP) plants (Alakangas et al. 2018). Several examples of biomass-fired CHP plants in Finland are given in Table 3. All presented plants rely mainly on woody biomass with a minor share of energy peat in consumption.

Table 3. Biomass-fired CHP plants in Finland.

Power Station	Location	Electricity	Heat	Fuel	Reference
Alholmens Kraft power plant	Jakobstad	265 MW	60 MW DH 100 MW process heat	forest residues peat	Alholmens Kraft (2020)
Vaasa power plant	Vaasa	230 MW	175 MW DH	wood and peat coal	Vaskiluodon Voima (2017)
Keljonlahti power plant	Jyväskylä	130 MW	260 MW DH	wood peat	Alva (2020)
Kaukaan Voima power plant	Lappeenranta	125 MW	110 MW DH 152 MW process heat	forest residues energy wood peat	Kaukaan Voima Oy (2019)
Seinäjoki power plant	Seinäjoki	120 MW	100 MW DH	forest chip swood residues recycled wood peat	Vaskiluodon Voima (2017)

Most of the bioenergy in Finland is produced in pulp and paper mills. The modern pulp and paper mills and sawmills operate with an integrated approach by using the residuals and by-products producing heat and power, biofuels and biomaterials (Kuparinen et al. 2019). Figure 5 presents some alternative technologies to produce biofuels or bioenergy by conversion of kraft pulp mill side streams. Many of these technologies are already used in Finnish mills and some, such as the production of synthetic hydrocarbons, are new possibilities.

Pulp and paper mills generate large amounts of sludges during wastewater treatment, which can be converted to renewable energy streams. Due to their high water content and poor dewaterability, pulp and paper mill sludges are extremely problematic streams, which are generally incinerated with low efficiency (Hagelqvist 2013). Anaerobic digestion is a noteworthy alternative to convert sludge into valuable commodities, i.e., biogas (mainly methane) and digestate (Bakraoui et al. 2019a, 2019b). Numerous studies have shown that most pulp and paper mill effluents can be to some extent anaerobically treated (Meyer and Edwards 2014; Bayr 2014). Hydrothermal carbonization (HTC) is another promising path to treat sludge, which has been actively studied at a laboratory scale (Saha et al. 2019; Areeprasert et al. 2015; Mäkelä et al. 2016). HTC converts sludge into hydrochar with upgraded properties that can be then combusted more effectively. World's first OxyPower HTC biofuel plant for sludge recycling is recently built by C-Green Technology in Heinola, Finland (C-Green Technology AB 2019). The facility will recycle 16,000 tons of biosludge annually at Stora Enso's fluting paper mill to reduce annual CO_2 emissions by 2500 tons.

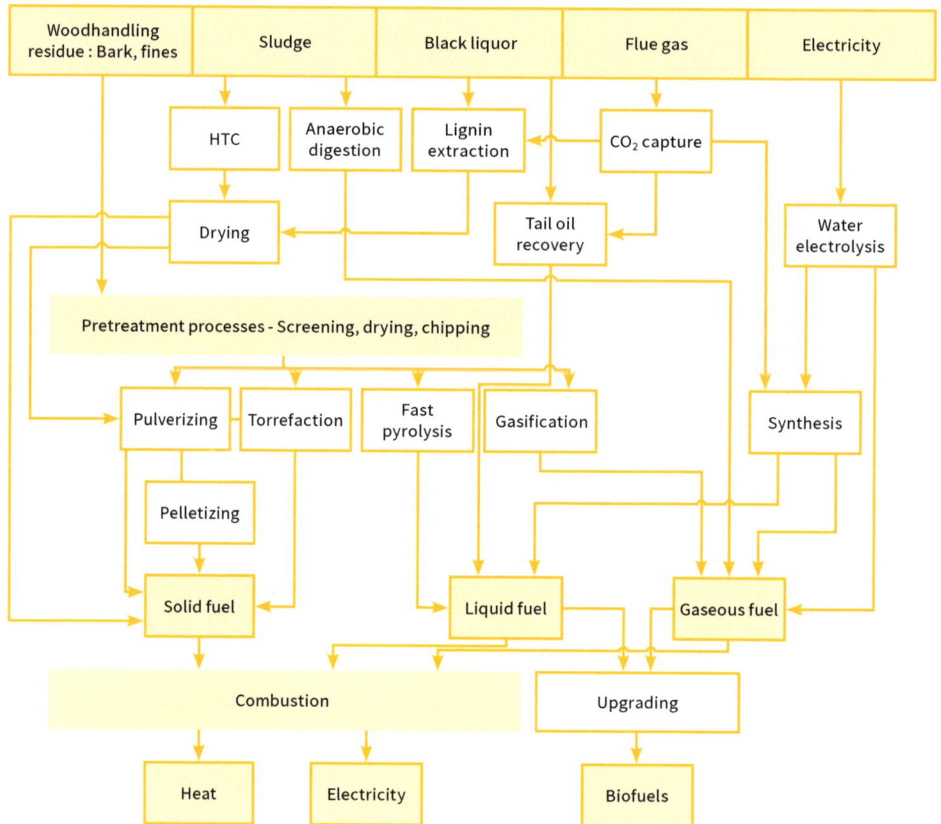

Figure 5. Alternative technologies for bioenergy and biofuel production by the usage of kraft pulp mill side streams. Source: Graphic by authors.

The most widely proposed utilization for captured CO_2 is the production of synthetic hydrocarbons from H_2 and CO_2 (Lehtonen et al. 2019). These hydrocarbons represent possible substitutes for fossil fuels in energy generation. Currently, H_2 is mostly produced from fossil fuels, usually, natural gas, using a steam reforming process, but renewable H_2 can be produced via electrolysis. Synthetic hydrocarbons produced from H_2 and CO_2 can be considered carbon neutral if H_2 is produced using renewable electricity and CO_2 originates from biomass or direct air capture. E-fuels and advanced biofuels have the significant benefit of being suitable for conventional engines and the traditional distribution infrastructure. Pulp mills offer an attractive option for integration of these processes due to their own production of renewable electricity, availability of biogenic CO_2, and abundant secondary heat streams.

Some examples of emerging technologies that may change the PPI in the future:

- Black liquor gasification is a promising technology to increase self-generation capabilities in the PPI. The process aims to replace the recovery boiler and produce energy, chemicals and fuels (Naqvi et al. 2010; Consonni et al. 2009). Despite of some technical difficulties, this technology can possibly provide significant investment returns along with energy and environmental benefits (Consonni et al. 2009; Bajpai 2016).
- Hydrogen production. The conversion of mill's streams to hydrogen-rich gases is actively studied recently. Supercritical water gasification (SWG) of black liquor is an innovative method to produce H_2-rich gases (Cao et al. 2020; Casademont et al. 2020; Özdenkçi et al. 2019, 2020). SWG is a potentially cost-effective way to improve pulp mill profitability (Özdenkçi et al. 2019). The applicability of SWG towards sludges was also intensively studied. This is a promising way to produce high-quality fuels like methane, H_2 and heavy oils (Zhang et al. 2010; Rönnlund et al. 2011).

Increased production of green electricity can contribute to grid decarbonization, and thus reduce CO_2 emissions from electricity generation. The new biofuels generated by the forest industry can play an important role in the decarbonization of other sectors. Some sectors are considered challenging or even impossible to electrify, and consequently, renewable fuels present a viable solution for their emission reduction. Decreasing the fossil fuel dependency in the heavy road, marine and air transport, for instance, most probably requires large amounts of bio-based and synthetic fuels.

4.3. Bio-Based Materials

Wood-derived materials are actively used in various applications, such as the production of paper, packaging, cosmetics, construction materials, composite and textile products. New innovative products are being constantly developed alongside the traditional ones in response to global challenges. Using wood to substitute intensive materials and fossil fuels can provide significant climate benefits (Leskinen et al. 2018). In addition, bio-based materials have a major role in climate change mitigation through temporary carbon storage (Jørgensen et al. 2015). Nowadays, besides the range of the standard products, such as pulp, tall oil, bark, turpentine, electricity and process steam, the forest industry mills can offer additional bioproducts such as textile fibres, biocomposites, fertilizers, biofuels and various cellulose- and lignin-derivatives. Finnish companies are actively expanding their product portfolio to new value chains thus allowing the Finnish forest industry to be successfully transformed into a bioproduct industry (Ministry of Economic Affairs and Employment 2017). One good example is Metsä Fibre bioproduct mill in Äänekoski, Finland (Metsä Group 2018). This unique mill is based on traditional kraft pulping technology and produces typical pulp mill products as well as a range of other commodities, such as product gas from bark gasification, sulphuric acid from odorous gases, biogas, biopellets from sludge digestion and biocomposites. The mill is highly self-sufficient in terms of electricity, producing 2.4 times more electricity than needed for mill operation.

The development of novel biomaterials from wood and produced with a reduced carbon footprint contributes significantly to the sustainable approach and brings higher flexibility to the forest industry. The most promising wood-based products for emerging markets are discussed in the following:

- Biochemicals and biofuels: Wood-based chemicals are considered as one of the main possibilities to compensate for the decline in revenues of PPI from reduced demand for graphic papers (Ignatius 2019). The main route is producing acids and alcohols by fermenting monomeric sugars from sawdust and chips (Hurmekoski et al. 2018). Biochemicals can be used for biofuels production, the need of which is expected to increase towards 2030 (Hurmekoski et al. 2018). The Finnish Parliament has approved a law that sets a gradually increasing 30% biofuels target for 2030. The tall oil-based technology route to generate renewable diesel seems effective and economically competitive (Heuser et al. 2013; UPM 2021b). The world's first UPM biorefinery in Lappeenranta, Finland uses the hy)drotreatment technology to produce UPM BioVerno diesel and naphtha from crude tall oil. The BioVerno diesel shows superior fuel properties in comparison with regular diesel and first-generation ester-type diesel fuel (Heuser et al. 2013).

- Biocomposites: A combination of biomass-derived fibres (mostly based on natural cellulosic fibers) with either virgin or recycled polymers offers a valuable alternative to oil-based plastics. Lignin-reinforced bioplastics have recently gained attention worldwide (Yang et al. 2019; Thakur et al. 2014). Finnish company Woodio has recently developed the world's first 100% waterproof wood composite of wood chips and resin-based adhesives that has a wide range of possible applications (Woodio 2021).

- Bio-based textiles: Dissolving wood pulp is a sustainable replacement for cotton and synthetic fibres in the textile and clothing industries, and its global production is increasing steadily (Kallio 2021). New textile wood-based fibre production technologies to manufacturing such materials as viscose and lyocell are actively investigated and applied. Innovation company Metsä Spring is launching a demo phase project in Äänekoski, Finland to produce the textile fibre Kuura®, which is produced by a novel direct-dissolution method (Metsä Group 2021). The carbamate, BioCelSol and Ioncell-F are sustainable and safe cellulose dissolution technologies recently developed in Finland (VTT 2017).

- Lignin-based materials: The majority of lignin is consumed as a fuel on-site, and only about 5% of lignin is currently utilized for the production of value-added products (Dessbesell et al. 2017). At the same time, lignin has potential for numerous applications: as a precursor for carbon fibers (Souto et al. 2018; Mainka et al. 2015), resins and adhesives (Cheng et al. 2011) and within different other applications (Kienberger 2019). Stora Enso's Sunila pulp mill, Finland is the largest integrated kraft lignin extraction plant in the world (50,000 tonnes of extracted lignin annually) (Stora Enso 2020). The new pilot plant to use lignin for manufacturing a graphite replacement for energy storage applications is currently under construction there.

- Nanocellulose: Cellulose nanomaterials possess a range of promising properties that enable their utilization in diverse applications, including packaging, filtering, biomedical applications, energy and electronics, construction and so forth (Lin and Dufresne 2014; Dhali et al. 2021). Stora Enso runs the world's largest micro-fibrillated cellulose (MFC) production facility at Imatra, Finland (Stora Enso 2019). MFC is used to produce an MFC-enhanced liquid packaging board New Natura™, which has extra strength and low weight. Another Finnish company, UPM, is implementing a novel method to produce wood-based cellulose nanofibril hydrogel GrowDex® for 3D cell culturing and other biomedical applications (UPM 2016). UPM is also producing a nanocellulose-based wound dressing FibDex® that provides an optimal environment for wound healing (UPM 2021a).

- Hemicellulose products: Hemicelluloses are generally burned along with lignin in the kraft pulp mills; however, the cost-effective extraction method would enable their more efficient utilization. Water solubility, biodegradability and amorphous structure make hemicellulose a promising precursor for high-value applications. It can be used as an environmentally friendly and inexpensive emulsifier to stabilize food, cosmetic and pharmaceutical products (Carvalheiro et al. 2008). Also, it can be hydrolyzed to produce biofuels and chemicals

(Hurmekoski et al. 2018). CH-Bioforce, a Finnish start-up company, has developed a method of biomass fractionation, which can be used to extract effectively biomass components (Bioforce 2020).

In addition to the aforementioned products, other concepts are being actively developed, such as advanced materials (porous cellulosic materials, coatings, films, foams), fertilizers and earthwork materials (Fabbri et al. 2018). Novel products bring additional flexibility to the mills, however, the production of new commodities may be significantly constrained by the availability of by-product flows within the mills (Hurmekoski et al. 2018).

The forest industry has significant potential to participate in the substitution of fossil-based materials in various sectors, and thus it plays a major role in creating and developing bioeconomy. The products of the Finnish forest industry already provide substantial climate benefits, but it is expected that the benefits will expand even more in the future (Finnish Forest Industries 2020a).

5. Conclusions

Climate change increases the significance of forest energy. The circular economy goals outlined globally, and within the EU, include improvements in material and energy efficiency, a realization of industrial symbiosis potential and a significant increase in the use of residues and wastes as valuable raw materials. Combining different technologies and using the potential of wood resources to the highest extent can boost energy and economic efficiency by providing fuel flexibility along with a wide range of products generated with a reduced carbon footprint.

Sustainability and life cycle thinking play a major role in the development of a circular economy in the forest sector in Finland. Finnish forest industry companies are constantly improving their energy efficiency and decreasing their dependency on fossil sources. The structural changes affecting the forest industry sector bring simultaneously new opportunities through novel outputs. The use of the best available techniques, tighter emissions regulations and emission-related costs are enabling a more effective transition of the forest industry towards effective biorefineries. Modern Finnish pulp and paper mills and sawmills operate with an integrated approach by utilizing the process residuals for producing renewable heat, power, and bioproducts. Among the most promising wood-derived products are biofuels, textile fibres, biocomposites, fertilizers, various cellulose- and lignin-derivatives. Carbon capture technologies have a remarkable potential within the PPI. While producing a significant amount of bioenergy, pulp and paper mills integrated with CO_2 capture technologies can become major sources of negative

CO_2 emissions. At the same time, as long as the political environment for bioenergy carbon capture is uncertain, the future potential is extremely challenging to evaluate.

The climate benefit from forest industry products in Finland has been estimated to be currently 16.6 $MtCO_2$, and it is expected to increase in the future. Many of the new bioproducts are in an early stage of development and thus further studies are needed to bring them to the markets. Replacement of fossil fuels used and efficiency improvement in the Finnish pulp and paper mills can lead to roughly 2.5 $MtCO_2$ emission reduction, which corresponds to approximately 5% of the domestic CO_2 emissions. An increase in demand for biomass sources might be the major challenge in the replacement of fossil fuels. While the role of forests as a carbon sink was out of the scope of this study, it is worth noting that the improved forest management can substantially enhance carbon removal from the atmosphere. As the aforementioned examples show, the possibilities of the forest industry to contribute to the mitigation of environmental changes evaluated on the Finnish example are certainly impressive. However, further studies are needed to enhance understanding of climate benefits of different solutions and to release the potential of the forest sector by overcoming technical, economic and political barriers.

Author Contributions: Conceptualization, S.L.; Validation, K.K., S.L. and E.S.; Investigation, K.K., S.L. and E.S.; Writing—Original Draft Preparation, K.K., S.L. and E.S.; Writing —Review and Editing, K.K., S.L. and E.S.; Visualization, K.K. and S.L.; Supervision and Project Administration, E.S. All authors have read and agreed to the published version of the manuscript

Funding: This research was funded by the Academy of Finland project "Role of Forest Industry Transformation in Energy Efficiency Improvement and Reducing CO_2 Emissions", grant number 315019.

Conflicts of Interest: The authors declare no conflict of interest.

References

Alakangas, Eija, Tomi J. Lindroos, Tiina Koljonen, and Jeffrey Skeer. 2018. *Bioenergy from Finnish Forests: Sustainable, Efficient and Modern Use of Wood.* Abu Dhabi: International Renewable Energy Agency. Available online: www.irena.org (accessed on 12 April 2021).

Alholmens Kraft. 2020. Alholmens Kraft Is the Largest Bio-Fuelled Power Plant in the World. Available online: https://www.alholmenskraft.com/en/company/bio-fuelled_power_plan (accessed on 10 March 2021).

Alva. 2020. Energiantuotanto. Available online: https://www.alva.fi/alva/yhtio/energiantuotanto/ (accessed on 27 March 2021).

Areeprasert, Chinnathan, Antonio Coppola, Massimo Urciuolo, Riccardo Chirone, Kunio Yoshikawa, and Fabrizio Scala. 2015. The Effect of Hydrothermal Treatment on Attrition during the Fluidized Bed Combustion of Paper Sludge. *Fuel Processing Technology* 140: 57–66. [CrossRef]

Bajpai, Pratima. 2016. Emerging Technologies. In *Pulp and Paper Industry*. Edited by Pratima Bajpai. Amsterdam: Elsevier, pp. 189–251. [CrossRef]

Bakraoui, Mohammed, Fadoua Karouach, Badr Ouhammou, Mohammed Aggour, Azzouz Essamri, and Hassan El Bari. 2019a. Kinetics Study of the Methane Production from Experimental Recycled Pulp and Paper Sludge by CSTR Technology. *Journal of Material Cycles and Waste Management* 21: 1426–36. [CrossRef]

Bakraoui, Mohammed, Mohammed Hazzi, Fadoua Karouach, Badr Ouhammou, and Hassan El Bari. 2019b. Experimental Biogas Production from Recycled Pulp and Paper Wastewater by Biofilm Technology. *Biotechnology Letters* 41: 1299–307. [CrossRef] [PubMed]

Bayr, Suvi. 2014. Biogas Production from Meat and Pulp and Paper Industry By-Products. Ph.D. thesis, University of Jyväskylä, Jyväskylä. Available online: https://jyx.jyu.fi/dspace/bitstream/handle/123456789/43316/978-951-39-5680-6_vaitos23052014.pdf?sequence=1 (accessed on 15 February 2021).

Bioforce. 2020. Biomass Consists of Cellulose, Hemicellulose and Lignin. Available online: https://www.ch-bioforce.com/ (accessed on 15 February 2021).

Cao, Changqing, Yupeng Xie, Liuhao Mao, Wenwen Wei, Jinwen Shi, and Hui Jin. 2020. Hydrogen Production from Supercritical Water Gasification of Soda Black Liquor with Various Metal Oxides. *Renewable Energy* 157: 24–32. [CrossRef]

Carvalheiro, Florbela, Luís C. Duarte, and Francisco M. Gírio. 2008. Hemicellulose Biorefineries: A Review on Biomass Pretreatments. *Journal of Scientific and Industrial Research* 67: 849–64.

Casademont, Pau, Jezabel Sánchez-Oneto, Ana Paula Jambers Scandelai, Lúcio Cardozo-Filho, and Juan Ramón Portela. 2020. Hydrogen Production by Supercritical Water Gasification of Black Liquor: Use of High Temperatures and Short Residence Times in a Continuous Reactor. *Journal of Supercritical Fluids* 159: 104772. [CrossRef]

CEPI. 2011. The Forest Fibre Industry. Available online: https://www.cepi.org/the-forest-fibre-industry-2050-roadmap-to-a-low-carbon-bio-economy/ (accessed on 2 February 2022).

CEPI. 2019. European Pulp & Paper Industry. Available online: http://www.cepi.org/topics/statistics (accessed on 29 April 2021).

C-Green Technology AB. 2019. C-Green Builds World's First OxyPower HTC Biofuel Plant for Sludge Recycling in Finland. Available online: https://static1.squarespace.com/static/5df0c803d3862346753f82d3/t/5e2895ff443f762e37439b76/1579718144028/20190927_C-Green_Heinola1_EN.pdf (accessed on 13 April 2021).

Cheng, Shuna, Ian D'Cruz, Zhongshun Yuan, Mingcun Wang, Mark Anderson, Mathew Leitch, and Chunbao Charles Xu. 2011. Use of Biocrude Derived from Woody Biomass to Substitute Phenol at a High-Substitution Level for the Production of Biobased Phenolic Resol Resins. *Journal of Applied Polymer Science* 121: 2743–51. [CrossRef]

Consonni, Stefano, Ryan E. Katofsky, and Eric D. Larson. 2009. A Gasification-Based Biorefinery for the Pulp and Paper Industry. *Chemical Engineering Research and Design* 87: 1293–317. [CrossRef]

Dessbesell, Luana, Chunbao Xu, Reino Pulkki, Mathew Leitch, and Nubla Mahmood. 2017. Forest Biomass Supply Chain Optimization for a Biorefinery Aiming to Produce High-Value Bio-Based Materials and Chemicals from Lignin and Forestry Residues: A Review of Literature. *Canadian Journal of Forest Research* 47: 277–88. [CrossRef]

Dhali, Kingshuk, Mehran Ghasemlou, Fugen Daver, Peter Cass, and Benu Adhikari. 2021. A Review of Nanocellulose as a New Material towards Environmental Sustainability. *Science of the Total Environment* 775: 145871. [CrossRef]

European Commission. 2015. *Integrated Pollution Prevention and Control, Best Available Techniques (BAT) Reference Document for the Production of Pulp, Paper and Board.* Luxembourg: Publications Office of the European Union, vol. EUR 27235. [CrossRef]

European Commission. 2018. A Clean Planet for All—A European Long-Term Strategic Vision for a Prosperous, Modern, Competitive and Climate Neutral Economy. *COM (2018)* 773: 114. Available online: https://ec.europa.eu/clima/sites/clima/files/docs/pages/com_2018_733_en.pdf%0Ahttps://ec.europa.eu/clima/sites/clima/files/docs/pages/com_2018_733_en.pdf?utm_campaign=AktuellHållbarhet-Direkten_181129_Username&utm_medium=email&utm_source=Eloqua&elqTrackId (accessed on 1 February 2021).

Fabbri, Paola, Davide Viaggi, Fabrizio Cavani, Lorenzo Bertin, Melania Michetti, Erika Carnevale, Juliana Velasquez Ochoa, Gonzalo Martinez, Micaela Esposti, and Piret Fischer. 2018. *Top Emerging Bio-Based Products, Their Properties and Industrial Applications.* Berlin: Ecologic Institut. Available online: www.ecologic.eu (accessed on 22 March 2021).

FAO. 2017. *FAOSTAT—Forestry Production and Trade.* Roma: FAO.

Finnish Forest Industries. 2020a. *A Green and Vibrant Economy: Finnish Forest Industry Climate Roadmap.* Helsinki: Finnish Forest Industries. (In Finnish)

Finnish Forest Industries. 2020b. Statistics. Available online: https://tem.fi/documents/1410877/2132212/IRENA_FinnishForest.pdf/8159091a-8fad-49b4-bfae-2266d2c32b84/IRENA_FinnishForest.pdf?t=1521204974000 (accessed on 15 February 2021).

Food and Agriculture Organization of the United Nations. 2019. *Forestry Production and Trade.* Roma: FAO.

Fracaro, Guilherme, Esa Vakkilainen, Marcelo Hamaguchi, and Samuel Nelson Melegari de Souza. 2012. Energy Efficiency in the Brazilian Pulp and Paper Industry. *Energies* 5: 3550. [CrossRef]

Fuss, Sabine, William F. Lamb, Max W. Callaghan, Jérôme Hilaire, Felix Creutzig, Thorben Amann, Tim Beringer, Wagner de Oliveira Garcia, Jens Hartmann, Tarun Khanna, and et al. 2018. Negative Emissions—Part 2: Costs, Potentials and Side Effects. *Environmental Research Letters* 13: 63002. [CrossRef]

Gulbrandsen, Lars H., and Christian Stenqvist. 2013. The Limited Effect of EU Emissions Trading on Corporate Climate Strategies: Comparison of a Swedish and a Norwegian Pulp and Paper Company. *Energy Policy* 56: 516–25. [CrossRef]

Hagelqvist, Alina. 2013. *Sludge from Pulp and Paper Mills for Biogas Production Treatment and Sludge Management*. Karlstad: Karlstad University.

Hetemäki, Lauri, Riikka Hänninen, and Alexander Moiseyev. 2013. Markets and Market Forces for Pulp and Paper Products. In *The Global Forest Sector: Changes, Practices, and Prospects*. Edited by Eric Hansen, Rajat Panwar and Richard Vlosky. Boca Raton: CRC Press, Taylor and Francis Group, pp. 57–92. [CrossRef]

Heuser, Benedikt, Ville Vauhkonen, Sari Mannonen, Hans Rohs, and Andreas Kolbeck. 2013. Crude Tall Oil-Based Renewable Diesel as a Blending Component in Passenger Car Diesel Engines. *SAE International Journal of Fuels and Lubricants* 6: 817–25. [CrossRef]

Hirsch, Georg, Antje Kersten, Hans-Joachim Putz, Brigitte Bobek, Udo Hamm, and Samuel Schabel. 2013. $CaCO_3$ in the Paper Industry—Blessing or Curse? Paper presented at Tappi PEERS Conference 2013, Green Bay, WI, USA, 15–18 September 2013; p. 44.

Hurmekoski, Elias, Ragnar Jonsson, Jaana Korhonen, Janne Jänis, Marko Mäkinen, Pekka Leskinen, and Lauri Hetemäki. 2018. Diversification of the Forest Industries: Role of New Wood-Based Products. *Canadian Journal of Forest Research* 48: 1417–32. [CrossRef]

IEAGHG. 2016. *Techno-Economic Evaluation of Retrofitting CCS in a Market Pulp Mill and an Integrated Pulp and Board Mill*. Cheltenham: IEA Greenhouse Gas R&D Programme, vol. 2016/10.

Ignatius, Heikki. 2019. *Wood-Based Biochemical Innovation Systems: Challenges and Opportunities*. Helsinki: University of Helsinki.

International Energy Agency. 2018. *Statistics*. Paris: International Energy Agency.

International Energy Agency. 2020. *Pulp and Paper*. Paris: International Energy Agency.

IRENA, VTT, and MEAE. 2018. *Bioenergy from Finnish Forests*. Abu Dhabi: International Renewable Energy Agency.

Isaksson, Juhani. 2007. *Meesauunikaasutin (Lime Kiln Gasifier)*. Recovery Boiler Seminar 18 October 2007. Helsinki: Finnish Recovery Boiler Committee.

Johnston, Craig M.T. 2016. Global Paper Market Forecasts to 2030 under Future Internet Demand Scenarios. *Journal of Forest Economics* 25: 14–28. [CrossRef]

Jørgensen, Susanne V., Michael Z. Hauschild, and Per H. Nielsen. 2015. The Potential Contribution to Climate Change Mitigation from Temporary Carbon Storage in Biomaterials. *International Journal of Life Cycle Assessment* 20: 451–62. [CrossRef]

Kallio, Maarit. 2021. Wood-Based Textile Fibre Market as Part of the Global Forest-Based Bioeconomy. *Forest Policy and Economics* 123: 102364. [CrossRef]

Kaukaan Voima Oy. 2019. Kaukaan Voima, Lappeenranta—Pohjolan Voima. Available online: https://www.pohjolanvoima.fi/voimalaitokset/kaukaan-voima-lappeenranta/ (accessed on 12 April 2021).

Kienberger, Marlene. 2019. Potential Applications of Lignin. In *Economics of Bioresources*. New York: Springer International Publishing, pp. 183–93. [CrossRef]

Koreneff, Göran, Jarno Suojanen, and Pirita Huotari. 2019. *Energy Efficiency of Finnish Pulp and Paper Sector—Indicators and Estimates*. Espoo: VTT.

Kuparinen, Katja, and Esa Vakkilainen. 2017. Green Pulp Mill: Renewable Alternatives to Fossil Fuels in Lime Kiln Operations. *BioResources* 12: 4031–48. [CrossRef]

Kuparinen, Katja, Esa Vakkilainen, and Tero Tynjälä. 2019. Biomass-Based Carbon Capture and Utilization in Kraft Pulp Mills. *Mitigation and Adaptation Strategies for Global Change* 24: 1213–30. [CrossRef]

Kähkönen, Satu, Esa Vakkilainen, and Timo Laukkanen. 2019. Impact of Structural Changes on Energy Efficiency of Finnish Pulp and Paper Industry. *Energies* 12: 3689. [CrossRef]

Leeson, Duncan, Niall Mac Dowell, Nilay Shah, Camille Petit, and Paul S. Fennell. 2017. A Techno-Economic Analysis and Systematic Review of Carbon Capture and Storage (CCS) Applied to the Iron and Steel, Cement, Oil Refining and Pulp and Paper Industries, as Well as Other High Purity Sources. *International Journal of Greenhouse Gas Control* 61: 71–84. [CrossRef]

Lehtonen, Juha, Sami Alakurtti, Antti Arasto, Ilkka Hannula, Ali Harlin, Tiina Koljonen, Raija Lantto, Michael Lienemann, Kristin Onarheim, Juha-Pekka Pitkänen, and et al. 2019. *The Carbon Reuse Economy—Transforming CO₂ from a Pollutant into a Resource*. Espoo: VTT Technical Research Centre of Finland Ltd. [CrossRef]

Leskinen, Pekka, Giuseppe Cardellini, Sara González-García, Elias Hurmekoski, Roger Sathre, Jyri Seppälä, Carolyn Smyth, Tobias Stern, and Pieter Johannes Verkerk. 2018. *Substitution Effects of Wood-Based Products in Climate Change Mitigation. From Science to Policy 7.* Joensuu: European Forest Institute.

Lin, Ning, and Alain Dufresne. 2014. Nanocellulose in Biomedicine: Current Status and Future Prospect. *European Polymer Journal* 59: 302–25. [CrossRef]

Lipiäinen, Satu, and Esa Vakkilainen. 2021. Role of the Finnish Forest Industry in Mitigating Global Change: Energy Use and Greenhouse Gas Emissions towards 2035. *Mitigation and Adaptation Strategies for Global Change* 26: 1–19. [CrossRef]

Lipiäinen, Satu, Katja Kuparinen, Ekaterina Sermyagina, and Esa Vakkilainen. 2022. Pulp and Paper Industry in Energy Transition: Towards Energy-Efficient and Low Carbon Operation in Finland and Sweden. *Sustainable Production and Consumption* 29: 421–31. [CrossRef]

Mainka, Hendrik, Olaf Täger, Enrico Körner, Liane Hilfert, Sabine Busse, Frank T. Edelmann, and Axel S. Herrmann. 2015. Lignin—An Alternative Precursor for Sustainable and Cost-Effective Automotive Carbon Fiber. *Journal of Materials Research and Technology* 4: 283–96. [CrossRef]

Metsä Group. 2018. Next-Generation Bioproduct Mill in Äänekoski. Available online: www.metsagroup.com (accessed on 12 February 2021).

Metsä Group. 2021. Metsä Group's New Textile Fibre Is Kuura. European Commision. Available online: https://www.metsagroup.com (accessed on 12 February 2021).

Metsäteollisuus ry. 2020. *Vihreä Ja Vireä Talous: Metsäteollisuuden Ilmastotiekartta 2035 (Green and Dynamic Economy: The Climate Road Map of Forest Industry 2035)*. Helsinki: Metsäteollisuus ry. (In Finnish)

Meyer, Torsten, and Elizabeth A. Edwards. 2014. Anaerobic Digestion of Pulp and Paper Mill Wastewater and Sludge. *Water Research* 65: 321–49. [CrossRef]

Ministry of Economic Affairs and Employment of Finland. 2021. *Energy Efficiency Agreements 2017–2025*. Helsinki: Ministry of Economic Affairs and Employment.

Ministry of Economic Affairs and Employment. 2017. *Wood-Based Bioeconomy Solving Global Challenges*. Helsinki: Ministry of Economic Affairs and Employment.

Ministry of Employment and the Economy. 2014. *Energy and Climate Roadmap 2050. Report of the Parliamentary Committee on Energy and Climate Issues*. Helsinki: Ministry of Employment and the Economy.

Motiva Ltd. 2019. Energy Audits. Available online: https://www.motiva.fi/en/solutions/policy_instruments/energy_audits (accessed on 2 February 2022).

Moya, Jose A., and Claudiu C. Pavel. 2018. *Energy Efficiency and GHG Emissions: Prospective Scenarios for the Pulp and Paper Industry*. Luxembourg: Publications Office of the European Union, vol. EUR 29280. [CrossRef]

Mäkelä, Mikko, Verónica Benavente, and Andrés Fullana. 2016. Hydrothermal Carbonization of Industrial Mixed Sludge from a Pulp and Paper Mill. *Bioresource Technology* 200: 444–50. [CrossRef]

Naqvi, Muhammad, Jinyue Yan, and Erik Dahlquist. 2010. Black Liquor Gasification Integrated in Pulp and Paper Mills: A Critical Review. *Bioresource Technology* 101: 8001–15. [CrossRef]

Natural Resources Institute. 2019. Total Energy Consumption by Energy Source by Unit. Available online: http://statdb.luke.fi/PXWeb/pxweb/en/LUKE/LUKE__04Metsa__08Muut__Energia/09.01_Energian_kokonaiskulutus.px/?rxid=dc711a9e-de6d-454b-82c2-74ff79a3a5e0 (accessed on 22 April 2021).

Nature Resource Institute (LUKE). 2020. *Metsäteollisuuden Puunkäyttö Väheni 2019 (Wood Usage Decreased in Forest Industry in 2019)*. Helsinki: Nature Resource Institute (LUKE). (In Finnish)

Onarheim, Kristin, Stanley Santos, Petteri Kangas, and Ville Hankalin. 2017. Performance and Cost of CCS in the Pulp and Paper Industry Part 2: Economic Feasibility of Amine-Based Post-Combustion CO_2 Capture. *International Journal of Greenhouse Gas Control* 66: 60–75. [CrossRef]

Özdenkçi, Karhan, Cataldo De Blasio, Golam Sarwar, Kristian Melin, Jukka Koskinen, and Ville Alopaeus. 2019. Techno-Economic Feasibility of Supercritical Water Gasification of Black Liquor. *Energy* 189: 116284. [CrossRef]

Özdenkçi, Karhan, Mauro Prestipino, Margareta Björklund-Sänkiaho, Antonio Galvagno, and Cataldo De Blasio. 2020. Alternative Energy Valorization Routes of Black Liquor by Stepwise Supercritical Water Gasification: Effect of Process Parameters on Hydrogen Yield and Energy Efficiency. *Renewable and Sustainable Energy Reviews* 134: 110146. [CrossRef]

Pöyry. 2020. *Tiekartta Metsäteollisuudelle Vähähiilistyvässä Yhteiskunnassa—Osa: Päästöt. (Road Map for the Forest Industry in a Low-Carbon Society—Part: Emissions)*. Available online: https://www.metsateollisuus.fi/uutishuone/ilmastoteot-metsateollisuuden-keskiossa (accessed on 2 June 2021). (In Finnish).

Ruostemaa, Seppo. 2018. Talteenotetun Hiilidioksidin Käyttö Kemiallisen Metsäteollisuuden Tuotantolaitoksessa (Utilization of Locally Captured Carbon Dioxide in a Chemical Forest Biorefinery). Master's Thesis, Aalto University, Espoo. (In Finnish).

Rönnlund, Ida, Lillemor Myréen, Kurt Lundqvist, Jarl Ahlbeck, and Tapio Westerlund. 2011. Waste to Energy by Industrially Integrated Supercritical Water Gasification and Effects of Alkali Salts in Residual By-Products from the Pulp and Paper Industry. *Energy* 36: 2151–63. [CrossRef]

Saha, Nepu, Akbar Saba, Pretom Saha, Kyle McGaughy, Diana Franqui-Villanueva, William Orts, William Hart-Cooper, and M. Reza. 2019. Hydrothermal Carbonization of Various Paper Mill Sludges: An Observation of Solid Fuel Properties. *Energies* 12: 858. [CrossRef]

Souto, Felipe, Veronica Calado, and Nei Pereira. 2018. *Lignin-Based Carbon Fiber: A Current Overview. Materials Research Express*. Bristol: Institute of Physics Publishing. [CrossRef]

Statistics Finland. 2020a. Statistics Finland—Energy Supply and Consumption. Available online: http://www.stat.fi/til/ehk/2019/ehk_2019_2020-12-21_tie_001_en.html (accessed on 14 March 2021).

Statistics Finland. 2020b. Statistics Finland—Energy Use in Manufacturing. Available online: http://www.stat.fi/til/tene/2019/tene_2019_2020-11-12_tie_001_en.html (accessed on 14 March 2021).

Statistics Finland. 2020c. *Statistis Finland's PxWeb Databases*. Helsinki: Statistics Finland.

Stenqvist, Christian, and Max Åhman. 2016. Free Allocation in the 3rd EU ETS Period: Assessing Two Manufacturing Sectors. *Climate Policy* 16: 125–44. [CrossRef]

Stora Enso. 2019. New Natura Enhanced with MFC. Available online: https://www.storaenso.com/en/newsroom/news/2019/6/new-natura-enhanced-with-mfc (accessed on 22 April 2021).

Stora Enso. 2020. Trees Can Become Batteries for Electric Cars—A Unique Pilot Plant Is under Construction at Our Sunila Mill in Finland. Available online: https://www.storaenso.com/en/newsroom/news/2020/6/trees-can-become-batteries-for-electric-cars----unique-pilot-plant-in-sunila-mill (accessed on 22 April 2021).

Teir, Sebastian, Sanni Eloneva, and Ron Zevenhoven. 2005. Production of Precipitated Calcium Carbonate from Calcium Silicates and Carbon Dioxide. *Energy Conversion and Management* 46: 2954–79. [CrossRef]

Thakur, Vijay Kumar, Manju Kumari Thakur, Prasanth Raghavan, and Michael R. Kessler. 2014. Progress in Green Polymer Composites from Lignin for Multifunctional Applications: A Review. *ACS Sustainable Chemistry and Engineering* 2: 1072–92. [CrossRef]

Tran, Honghi, and Esa Vakkilainen. 2008. The Kraft Chemical Recovery Process. In *Tappi Kraft Pulping Short Course*. Available online: https://www.tappi.org/content/events/08kros/manuscripts/1-1.pdf (accessed on 2 February 2022).

UPM Communication Papers. 2020. UPM Takes next Step towards 65% CO2 Emission Reduction Goal by 2030. Press Release. February 12. Available online: https://www.upmpaper.com/knowledge-inspiration/news/2020/02/upm-takes-next-step-towards-65-co2-emission-reduction-goal-by-2030/ (accessed on 23 April 2021).

UPM Energy. 2021. Renewable Hydropower Generates Flexible Power. Press Release. May 6. Available online: https://www.upm.com/articles/energy/21/renewable-hydropower-generates-flexible-power/ (accessed on 23 April 2021).

UPM. 2016. Growing Cells in Nanocellulose. Available online: https://www.upm.com/news-and-stories/articles/2016/08/growing-cells-in-nanocellulose/ (accessed on 22 April 2021).

UPM. 2021a. Advanced Nanocellulose Dressing for Natural Wound Healing. Accessed April 12. Available online: https://www.upm.com/about-us/this-is-biofore/bioforecase/fibdex-nanocellulose-dressing-for-wound-healing/ (accessed on 24 April 2021).

UPM. 2021b. UPM BioVerno-Diesel. Available online: https://www.upmbiofuels.com/fi/liikennepolttoaineet/upm-bioverno-diesel-polttoaine/ (accessed on 23 April 2021).

Vakkilainen, Esa, and Aija Kivistö. 2014. *Forest Industry Energy Consumption—Trends and Effects of Modern Mills*. Report. Lappeenranta: LUT University.

Vaskiluodon Voima. 2017. Vaskiluodon Voima. Available online: https://www.vv.fi/wp-content/uploads/sites/8/2017/03/Vaskiluodon_Voima_ENG_FINAL-1.pdf (accessed on 2 February 2022).

Woodio. 2021. Shop Woodio—A Sustainable Choice—Woodio. Available online: https://woodio.fi/ (accessed on 12 April 2021).

VTT. 2017. Environmentally Friendly Textile Fibres. Available online: https://www.chemistryviews.org/details/news/10513472/Environmentally_Friendly_Textile_Fibres.html (accessed on 16 March 2021).

Yang, Jianlei, Yern Chee Ching, and Cheng Hock Chuah. 2019. Applications of Lignocellulosic Fibers and Lignin in Bioplastics: A Review. *Polymers* 11: 751. [CrossRef]

Zhang, Linghong, Chunbao Xu, and Pascale Champagne. 2010. Energy Recovery from Secondary Pulp/Paper-Mill Sludge and Sewage Sludge with Supercritical Water Treatment. *Bioresource Technology* 101: 2713–21. [CrossRef]

Public Transit Challenges in Sparsely Populated Countries: Case Study of the United States

Warren S. Vaz

1. Introduction

Transportation has long been recognized both as a critical pillar of developed societies and a major contributor to pollution. Transportation is critical for access to food and other resources, employment, communication, and, thus, providing access to transportation is key to eliminating discrimination and socioeconomic barriers that limit several marginalized populations (Dostál and Adamec 2011). According to a 2020 report by the U.S. Environmental Protection Agency, transportation globally accounts for about 28% of CO_2 emissions (US EPA 2020). Thus, transportation today provides salient benefits to society, but also exacts a cost in terms of health and impact on the environment. With the global population continuing to grow and several large countries like Brazil, China, India, Indonesia, and most of Africa continuing to develop, the demand for transportation is only projected to increase.

To meet this growth in the most sustainable way possible, the answer must be clean transportation. The poster child is the electric vehicle (EV) powered by clean energy (e.g., solar, wind). An alternative is the hydrogen fuel cell vehicle powered by green hydrogen, which is hydrogen generated by electrolysis using clean electricity. These technologies also tend to be more efficient, from an energy standpoint, than conventional vehicles. For example, consider the 2020 versions of the Tesla Model 3 and the Toyota Camry, two passenger sedans of comparable size (~1550 kg). The fuel efficiency of the electric Tesla is about 5.7 km/kWh. For the Toyota, it is about 1.3 km/kWh, when the energy content of gasoline is factored in. Another way to consider efficiency is considering the efficiency of moving cargo or people. For example, the same Toyota can transport at most five passengers, but typically closer to 1–2. Thus, its efficiency would be about 0.4–0.8 kWh/km/passenger. For a typical mass transit bus that can transport 40 passengers, this efficiency is about 0.13 kWh/km/passenger, assuming the bus is full. Thus, the conclusion here is that mass transit is a critical piece in the transition towards a clean and sustainable future. Ideally, this mass transit would be fueled by clean energy sources, but even using conventional sources would result in a reduction in emissions (Yuan et al. 2019).

The focus in this chapter is on passengers, but a similar argument can be posited for goods or cargo. This chapter focuses on public transit challenges in sparsely populated countries. The case study of the United States is used to demonstrate how and why the historical factors that shape a country's policies are critical to planning any future improvements. Accordingly, some suggestions for the future of public transit policy are presented, together with a selection of recent projects and upcoming projects taking shape.

2. Sparsely Populated Countries

While mass transit may be an effective way to handle the increasing need for transportation in the most efficient manner, this would require a significant investment in infrastructure. Individual vehicles only require a road network. Transit buses also require bus stops, depots, drivers, transit schedules, coordinators, etc., not to mention the huge initial cost of the actual buses. Rail is more efficient, but even more capital intensive. Governments or private industry are willing to invest in these projects if there is an economic case. Additionally, this typically is a function of the population: the larger the target market, the larger the expected revenue. However, if the population density is too low, then there are additional challenges. Short, efficient trips become impossible and an expansive infrastructure leads to a prohibitive upfront investment. Compounding these factors are areas that house historically poor populations. While these populations have the greatest dependency on cheap transportation and would benefit the most, such areas are the least likely to see significant public investment as they are typically underrepresented in government and policymaking. On the other end of the spectrum are affluent areas. There is a strong correlation between vehicle ownership and per capita gross domestic product (GDP) (IRF 2013). Similarly, one would expect a strong correlation between per capita emissions and per capita GDP. However, other factors might be worth considering.

Figure 1 shows five indicators for 28 countries, which account for about two-thirds of the total global population as well as about two-thirds of the total global GDP (PPP or purchasing power parity). For each of these, the per capita emissions are plotted in relation to the population, population density (per square kilometer), GDP (PPP), GDP per capita, and percentage of urbanization. A trendline for the plot is also shown. Indeed, it can be concluded, based on the R^2-values obtained, that the GDP per capita has the strongest correlation to emissions. The next most important factor is urbanization, or the fraction of the population that lives in cities compared to rural areas. Another interesting trend is in the population density plot. It can be observed that countries with very low population densities tend to produce far more emissions than those with high population densities. Taken together, the data point to the conclusion that the biggest polluters are countries with a high per capita GDP and urbanization, but low population density. Such countries also have high levels of vehicle ownership. All these factors contribute to high emissions per capita. In addition, it is argued here that such countries face significant challenges to adopting or expanding mass transit. To demonstrate this, the case of North America is examined, particularly focusing on the United States. The countries are highly developed and very rich in natural resources. They have a lot in common, including being sparsely populated and relatively isolated, both internally and externally.

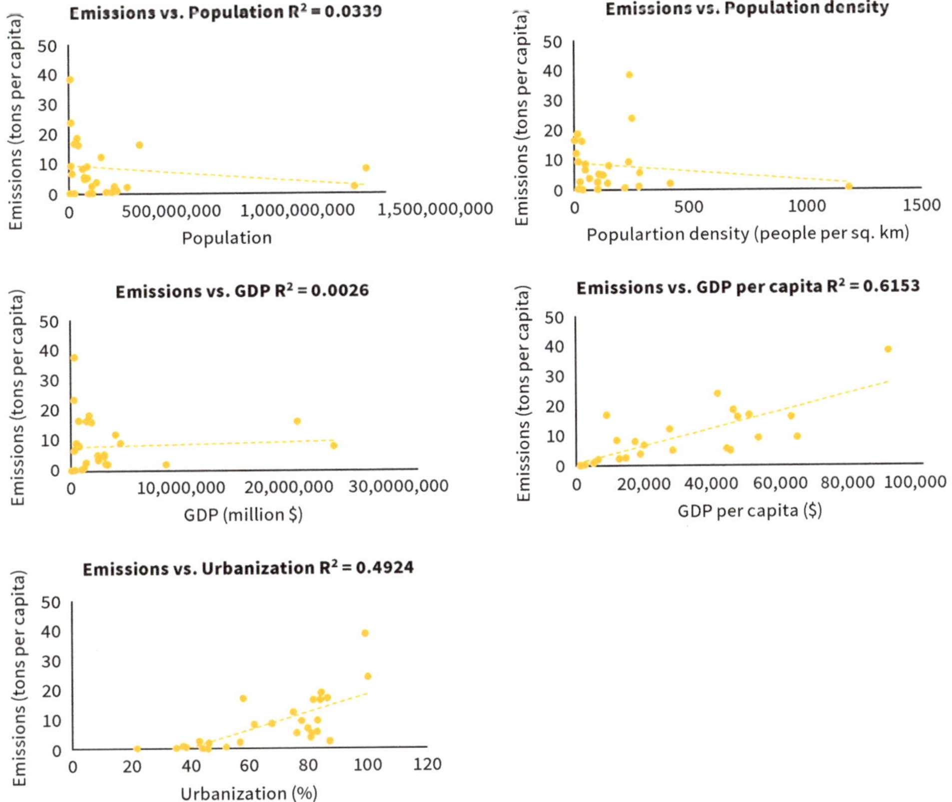

Figure 1. Influence of population, population density, GDP (PPP), GDP per capita, and urbanization on per capita emissions. Source: Graphic by author, data from (WDI 2021; Ritchie 2019; IMF 2020; UN 2019).

North America has three countries: Canada, Mexico, and the United States. These countries have a similar history and share many common characteristics. All three are relatively young, ranging from 1776 (United States) to 1867 (Canada). These countries were inhabited by various Native American civilizations, then settled by Europeans, who also brought several enslaved peoples to the continent. By the middle of the 19th century, slavery had been abolished in all three countries and each one experienced waves of immigration from various parts of the globe continuing to this day. This has only added to, and in many ways catalyzed, the natural population and industrial growth that is virtually unprecedented in history.

Table 1 shows statistics for various economic and transportation categories for Canada, the United States, and Mexico. There are some commonalities, like the degree of urbanization. However, Canada and the United States have more in common with each other than Mexico. Canada and the United States are considerably wealthier than Mexico. Part of the reason is because both countries are huge: second and fourth, by area. The United States is also third in the world by population, but its population density is still about half that of Mexico. Even if only the contiguous 48 states are considered, the population density only changes from 33.6 to 40 per sq. km. With wealth comes a higher standard of living and energy consumption, resulting in greater emissions. The United States and Canada emit about four times more than Mexico. The per capita vehicle ownership is similar. Even though the degree of urbanization is about the same, a closer look at the distribution is instructive. The top 100 combined statistical areas (CSA) account for 81% and 66% for Canada and the United States, respectively, compared to just 45% for Mexico. Summing the total for all cities above 100,000 residents, they account for 74% and 85%, respectively, compared to just 48% for Mexico. This leads to the conclusion that populations in Canada and the United States are much more concentrated than in Mexico. Part of this can be due to the presence of huge suburbs around a major city.

While Canada and the United States have a lot in common, this chapter will focus on the United States for several reasons. The United States has almost nine times the population, over 12 times the GDP, and is seen as a global leader and influencer in the world and has been that way since World War II. Furthermore, the United States has a history of invention and innovation in several critical sectors, including transportation: railways, automobiles, including mass production, road infrastructure, traffic laws, etc., aircraft, rockets, and more. Countries have looked to it for benchmarks and manufacturing standards and lessons learned and technologies discovered have often translated to international markets. Thus, if the United States were to, for example, stop manufacturing internal combustion vehicles and fully switch to electric vehicles, that would put significant pressure on other countries to follow suit considering how huge a market space is controlled by the United States. Finally, the United States is also one of the largest consumer markets in the world in several sectors such as automobiles, electricity, petroleum, etc. Changes here would have a much greater impact than virtually any other country.

Table 1. Various economic and transportation statistics for North American countries.

Country	Canada	United States	Mexico
Area (million sq. km)	9.984	9.833	1.973
Population (millions)	38.01	328.24	126.01
Population density (per sq. km)	3.92	33.6	61.0
Population of top 100 CSAs (fraction of total)	0.81	0.66	0.45
Population of >100K cities (fraction of total)	0.74	0.85	0.48
GDP (nominal, trillion US dollars)	1.6	20.807	1.322
GDP (per capita, US dollars)	42,080	63,051	10,405
Emissions (per capita, metric tons of CO_2)	16.3	16.1	3.8
Urbanization (%)	81.6	82.7	80.7
Vehicle ownership (per 1000 people)	685	838	297

Source: Table by author, data from (IMF 2020; UN 2019; Statistics Canada 2016; US Census Bureau 2019; INEGI 2010; IOMVM 2013).

3. History of the United States: An Overview

The United States declared its independence from colonial rule in 1776. In its early years, it was not universally recognized as a nation for several years. Several established nations still viewed it as a potential colonial acquisition. Most importantly, the entire territory of the first 13 colonies was east of the Appalachian Mountains, north of Florida—about 11% of its extent today. Gradually, several territories were acquired, organized, and formalized as states. This included territories east of the Mississippi, then the entire Mississippi River basic (Louisiana Purchase), the Oregon Territory, Florida and the Republic of Texas, and Spanish

Territories all the way to the Pacific Ocean. Much later came the additions of Alaska and Hawaii. There were numerous smaller changes to the territory of the United States, but these were predominately outside the contiguous or lower 48 states—in the Caribbean and Pacific.

From the earliest colonial days, settlers engaged in importing slaves from Africa. From the 17th century up to the American Civil War that ended in 1865, millions of slaves worked predominantly in the Southern states on large cash crop plantations (cotton, tobacco, sugar, etc.). By contrast, the economy of the Northern states was predominantly predicated on agriculture and industry. Innovations in manufacturing and transportation technology—such as the steam engine—propelled the rise of factories and factory workers with several Northern states abolishing slavery in their territories. Southern fears over the abolishment of slavery in the Union led to the Civil War, which culminated in the permanent abolishment of slavery in the United States in 1863 and the defeat of the Confederacy of Southern States in 1865.

After the war, the conditions in the South degenerated for virtually the entire population. Black Americans were now free and full citizens of the Union. However, their White counterparts, some of whom were their former slaveowners, rejected all attempts of the federal government to integrate free slaves into society, even requiring military invention in certain cases. Simultaneously, unable to profit from free slave labor, several White landowners, including women, had to work the land themselves—a hardship not previously endured. Those in power engaged in various tactics to restrict the rights of Blacks, keep a stranglehold on political power, restrict their access to education, land ownership, etc. They set up various mechanisms that ensured a system very similar to the slave-owner system before the war, known as sharecropping. The Jim Crow Era in the South ensured segregation on the basis of race.

Meanwhile, the new territories in the West continued to grow and develop at the expense of the native population, who were continually pushed further west. Initially, the Indian Territory was west of the 13 colonies, then moved to the Missouri Territory, and finally reduced to the area of the modern State of Oklahoma. On the back of the railroad and steam locomotive, American industry grew prodigiously. Total track length progressed from about 60,000 miles (96,000 km) to about 160,000 miles (100,000 km) in the 1890s with the completion of the Transcontinental Railroad. This figure peaked during World War 1 (254,000 miles) and is about 140,000 miles today (Stover 1999). Some of the most iconic and enduring American companies were founded during the latter half of the 19th century and the early 20th century: Standard Oil (later Marathon, ExxonMobil, BP, and Chevron), General Electric, AT&T, Emerson

Electric, Carnegie Steel (later U.S. Steel), Ford, General Motors, etc. The vast majority of these were headquartered and had most of their operations in Northern States.

After World War I, there was a period of booming economic growth in the country. The transportation section grew with major innovations in aviation, automobile technology, rail (e.g., diesel locomotive), and shipping. Then, came the Great Depression followed by the New Deal, a series of major economic and infrastructure reforms aimed to lift the country out of the Depression. World War II continued to drive the growth of the massive American industrial complex accompanied by a surge in population, the "baby boom". The Interstate System that was signed into law during the war made road travel easier than ever. Simultaneously, little progress towards equity was happening in the American South. Multiple incidents of racial violence, voter suppression, the rise of the Ku Klux Klan, lynching, segregation, etc. made it virtually impossible for Blacks to gain political power or good jobs or education or even improve their lives in any tangible way. This prompted a mass exodus of millions of Black families from the South to the North and West. Coupled with continued immigration, several cities like Philadelphia, Detroit, Chicago, Cleveland, Baltimore, and New York City saw a huge increase in their population and a change in their demographics. This lasted roughly from 1916 to 1970. On the other hand, after World War II, thousands of veterans left the cold harsh winters of the North and migrated to the Sun Belt states. Cities like Los Angeles, San Diego, Las Vegas, Dallas, Houston, and Phoenix grew significantly.

3.1. Urban Sprawl

During this period of urbanization, a pattern began to emerge: the gradual move of large populations from city centers to small satellite towns and communities within close proximity, commonly called the suburban development or suburbs. Spurred by legislation like the Federal Home Loan Bank Act of 1932 and the National Housing Act of 1934 along with contributions from the Veterans Administration, families were able to buy homes in newly developed areas outside cities. No longer did they need to live in cramped, densely packed apartments in crowded, polluted, and noisy city centers. The United States transitioned from a primary economy (natural resources) to a secondary one (manufacturing). To travel to these well-paying jobs in the cities, workers bought increasingly affordable cars, one of the most important products of this new economy. The total number of miles of paved road continued to increase, especially between cities.

Thus, what happed was that the traditional city centers absorbed the majority of the Black and immigrant population, while most of the middle-class White residents flocked to the suburbs. Under the National Housing Act, the Federal Housing Authority (FHA) created several guidelines and minimum standards for new housing developments. Unfortunately, it also strongly promoted racial segregation in its guidelines. The design of suburbs focused on car-friendly, wide streets and left out pedestrians. Instead of grid plans, curved and dead-end or cul-de-sac designs were adopted. These were designed to slow cars down, limit traffic through streets, but also ensure that cars were necessary to get around. The FHA rated any designs submitted within these guidelines as "good plans" and the rest as "bad plans", making it riskier for developers to pursue. Suburbs grew, their design made walking and public transit inefficient, and existing public transit systems in the city centers fell out of favor and became neglected. Their use was associated with being of the poor class. The people who depended on these systems the most had little power to influence their improvement.

Even if Black families wanted to move out of the city centers, 'redlining' made it impossible for them to secure affordable housing. This was the practice by government agencies of systematically denying goods or services to particular groups by a number of tactics ranging from selectively raising interest rates or prices to placing strict criteria on specific goods and services. The Fair Housing Act of 1968 put a gradual end to this practice, but the intervening decades had done their damage. Whole generations of minority populations, such as Blacks, were unable to afford homes or had to buy homes in less desirable or segregated neighborhoods. For most middle- and low-income families, home ownership is the surest way to build wealth. For those unable to do so, renting is the only option and does not result in any increase in equity. Similarly, minorities were also denied access to education and political power with suppression of voting. It was only after the Civil Rights Acts of the 1960s that some of these practices slowly began to decrease. Still, the country is recovering from these effects to this day. Some studies show that segregation in some cases, such as the workplace, is worsening (Hall et al. 2019).

Figure 2 shows the urban sprawl for two major cities: Dallas, Texas located in the southern United States and Detroit, Michigan located in the northern United States. The figures show not only growth of population within the city center itself (Dallas County for Dallas and Wayne County for Detroit), but the surrounding counties as well. Some of these experienced triple-digit growth rates. The average home in the United States grew in size from 1500 sq. ft. in the 1970s to over 2,000 sq. ft. nowadays (Friedman and Krawitz 2001). Interestingly, the average lot size has continued to decrease as pressure for suburban housing continues to increase.

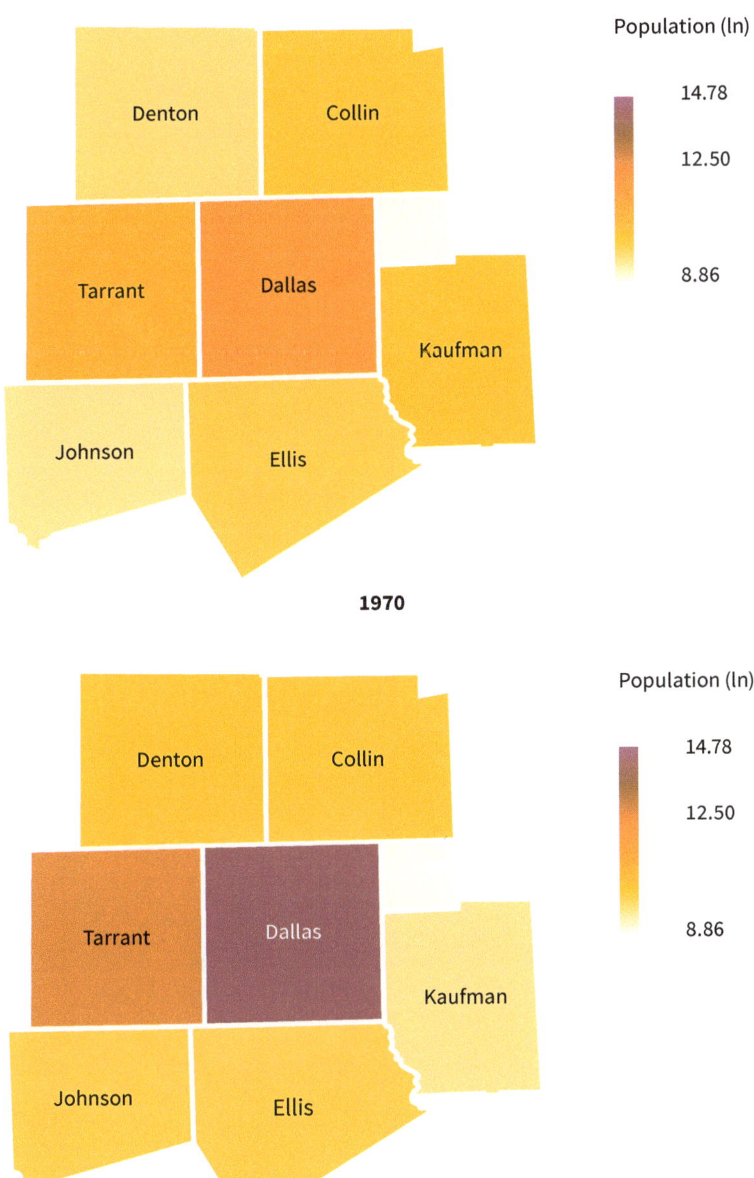

Figure 2. *Cont.*

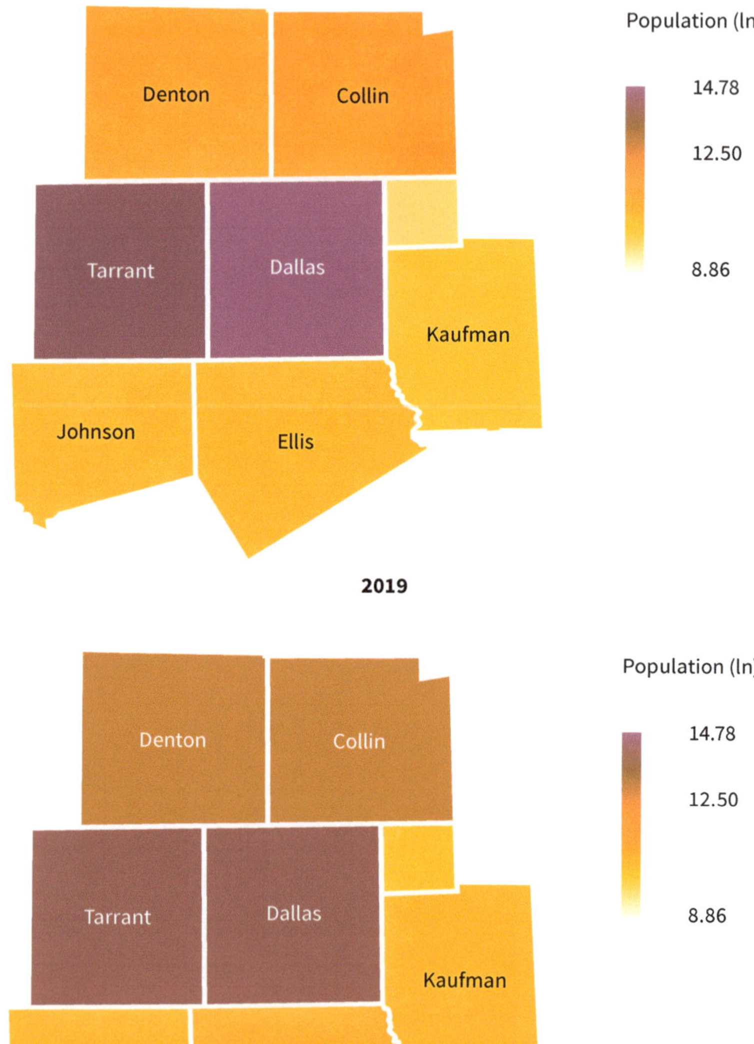

Figure 2. *Cont.*

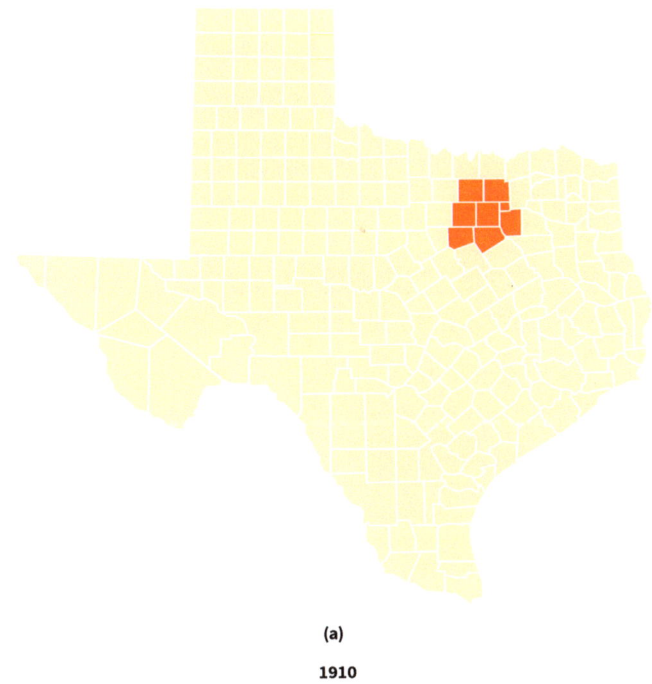

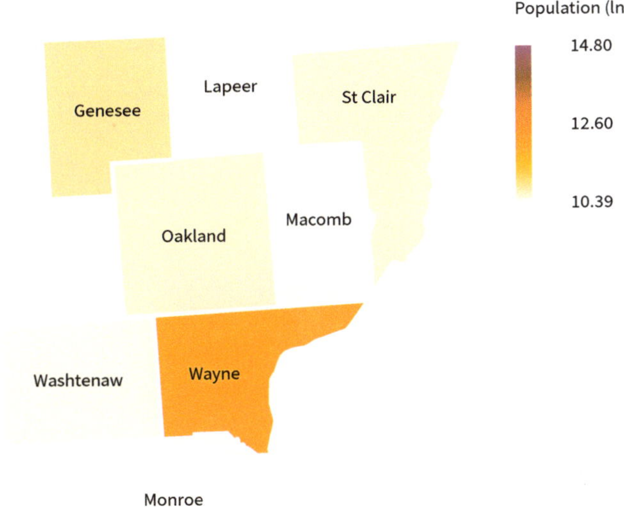

Figure 2. *Cont.*

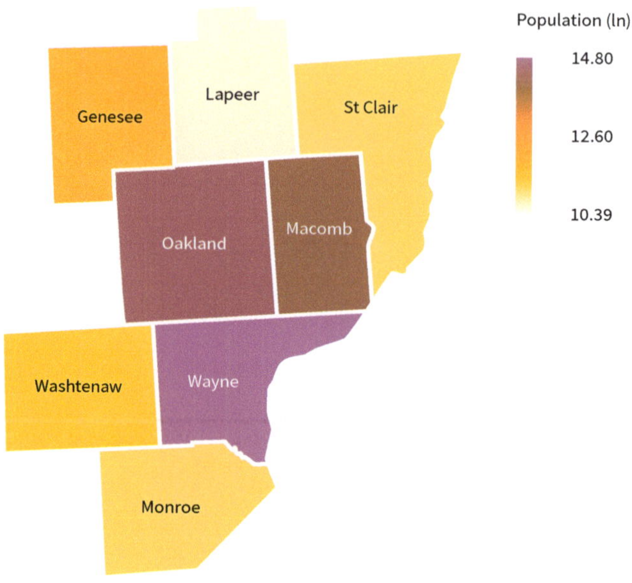

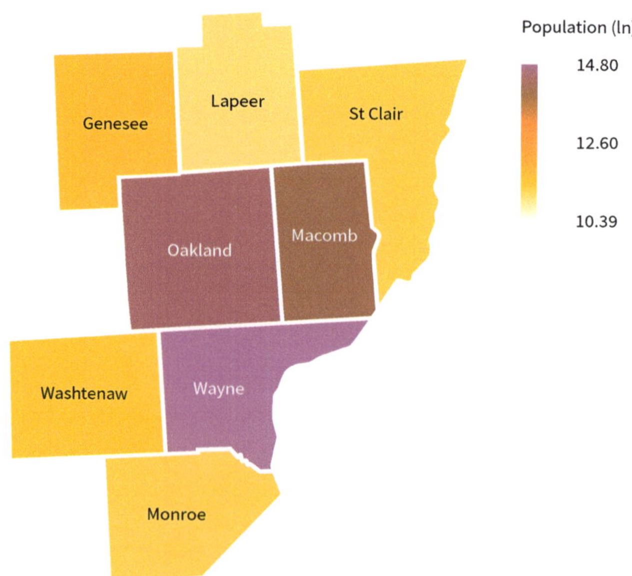

Figure 2. *Cont.*

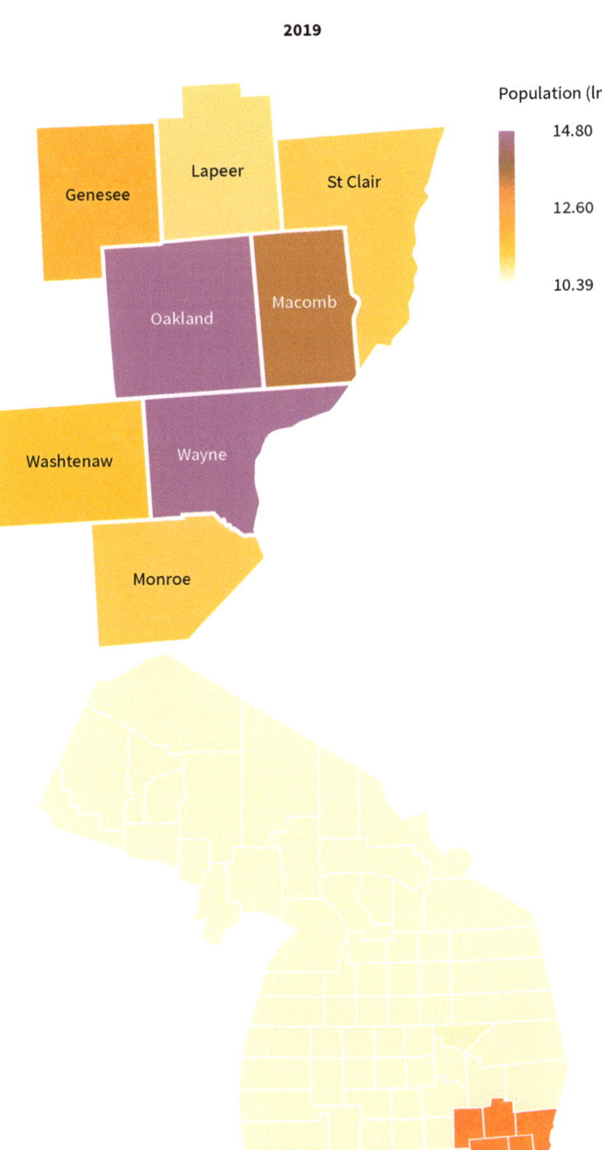

Figure 2. Growth of suburbs in the 20th century in (**a**) Dallas, Texas and (**b**) Detroit, Michigan. Source: Graphic by author, data from (US Census Bureau 2020).

3.2. Public Transit Challenges

Therefore, the United States arrived at the situation in which it is today: declining urban centers surrounded by sprawling suburban development. The coronavirus pandemic of 2020 has only continued to accelerate a general move out of major cities as companies allow employees to work from home. This has shaped transportation policy and consumer trends. According to the Bureau of Transportation Statistics, 76% of employees who commute to work do so in a passenger vehicle alone, whereas only 5% use public transit (US DOT 2020). Similarly, total public transit ridership has continued to trend down in recent years, reaching a peak in 2014. Part of this has been attributed to the rise in ride-hailing and vehicle-sharing apps. Average commute time has gone from under 22 minutes in 1980 to a peak of 27 minutes in 2018. Passenger vehicles continue to become bigger and more expensive. For example, the number of midsize sedans sold in 2020 was about 50% of what it was in 2012. Sport utility vehicles (SUV) made up 47.4% of all sales in 2019 and that is expected to rise to 78%. Expectedly, the average transaction price of a new vehicle continues to trend upwards, reaching about USD 39 thousand in 2020, up from USD 35.5 thousand in 2016. In 1960, about 80% of households had one car or less. In 2019, it was only about 40%. As for local travel, defined as trips of 50 miles or less, there were 3140 average person-trips per household in 2017. Of those, 329 were walking, 2592 were by passenger vehicle, and only 80 by public transit. As for the totality of domestic travel, between 2010 and 2018, air travel, passenger vehicle travel, and public transit increased—though the increase in public transit was only about 2%. Inter-city rail saw a slight decrease. When looking at passenger-miles traveled in 2018, inter-city rail and public transit combined were 62 times lower than passenger vehicles.

The above statistics provide sufficient evidence on the preference of personal passenger vehicles over more efficient public transport. Personal vehicles are less environmentally friendly and more expensive than using public transit. However, they offer more convenience and flexibility. Given the lack of demand, there is a very weak economic incentive to improve existing public transit infrastructure. The people who rely on it the most typically have the least political power or financial capital to do so. This leads to a vicious cycle of further neglect and depredation. Transportation has been identified as one of the critical factors in helping people to escape poverty (Department for Transport 1997). The transformative Interstate System was a huge boom to the country's economy. However, in several cases, poor communities were displaced and bypassed by the new highways. Transportation is also the second biggest expense for the average family. Thus, minorities and immigrants continue to have reduced access to transportation options. This only worsens the inequity within society.

The United States has led the world in innovation in several key transportation-related technologies. However, implementation of the next generation of clean mass transit projects is severely lacking. There are plenty of demonstrations and projects to prove the efficacy of technologies that are viewed skeptically by the public in this country (Bamwesigye and Hlavackova 2019; Behrendt 2019; Fialová et al. 2021; Freudendal-Pedersen et al. 2019; Łukaszkiewicz et al. 2021). Of course, with transportation, proper land use is another critical issue in United States. There are several recent studies and analyses on smart and sustainable land use (Al-Thani et al. 2018; Hammad et al. 2019; Tobey et al. 2019) that can inspire policymakers in the United States as well.

Another important thing to keep in mind is that, like Canada and Mexico, the United States has a federal government with strong state governments. Critical sectors like education and transportation are funded and regulated by all levels of government. Most importantly, there is no federal transit authority that operates one bus or rail network for the entire country. This includes Amtrak, which provides inter-city rail service in the United States. It receives funding through a combination of state and federal subsidies, but is a for-profit organization. This makes it very difficult to improve existing transit services or invest in the development of new projects, whether by government or private industry. Approvals, cooperation, and shared resources need to come from multiple agencies and governing bodies, and this often becomes the prohibiting factor. For example, any bus service across multiple states would have to abide by emissions regulations, safety regulations, disability services, etc. in each state. Further, minimum pay, employee benefits, levels of funding from each state authority involved, etc. would all serve to complicate the project.

Arguably the biggest challenge to implementing widescale public transit in the United States is geography or population distribution across its geography, which can be understood with Figure 3. It shows the locations of all the CSAs with a population greater than 200,000 in both Canada and the United States. The marker size is scaled, the smallest being about 200,000 and the largest being about 8.4 million. Almost the entire population of Canada lives within 100 miles of the Canada–United States border. For this reason, a lot of companies and services operate across the border. Examples are sports leagues, companies, particularly in the Seattle–Vancouver, Detroit–Windsor, and Toronto–Buffalo areas, and service by Amtrak railway. The Quebec City–Windsor corridor, which includes Ottawa, Montreal, and Toronto, has a population of about 18 million, nearly half of Canada. The Great Lakes region has about 55.5 million people and the Northeast corridor (Boston to Richmond, Virginia) has about 52 million people. Other notable megalopolises include the California (north and south) region (38 million) and the Texas Triangle–Gulf Coast region (33 million). All these regions are clearly visible in Figure 3. Given this distribution and considering that these megalopolises are hundreds and thousands of miles apart, it is not easy to create a nationwide public transit system like the Eurozone. Americans prefer the flexibility of air travel and road trips to travel between major cities and everywhere in between. Unfortunately, both these forms are the worst emitters of greenhouse gases.

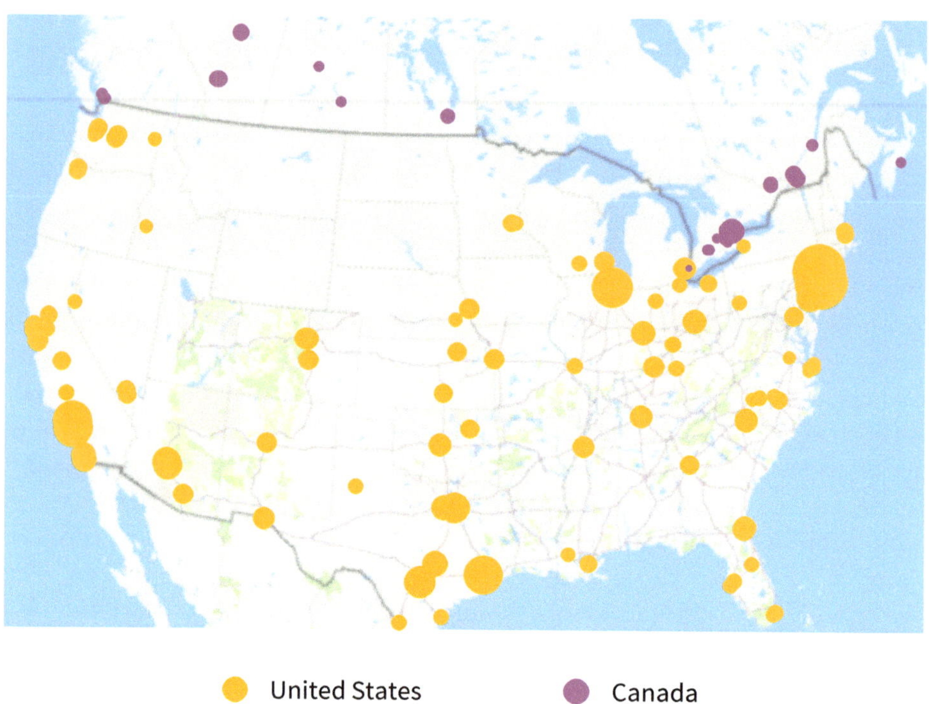

Figure 3. CSAs with population over 200,000 in Canada and the United States—marker size scaled from 200,000 to 8,400,000. Source: Graphic by author, data from (Statistics Canada 2016; US Census Bureau 2019).

4. Potential Solutions

4.1. Sustainable Transportation: A Multiobjective Approach

Despite the challenges described, climate change appears to be the extreme coercion driving promising new projects. This section briefly introduces a multi-objective approach to selecting sustainable transportation options. The ideal public transit system would satisfy multiple objectives:

- Accommodate multiple passengers;
- Efficient, use the least amount of fuel to cover the most distance;
- Safe;
- Affordable;
- Level of service;
- Low maintenance and operating costs;
- Low initial capital requirement;
- Technologically feasible;
- Sustainable, environmentally friendly.

Of course, there are other considerations; these are just examples. Note that some of these are 'cooperating', meaning improving one also improves another. Using a more efficient powertrain contributes to being sustainable. Other objectives are 'conflicting', meaning improving one worsens another. Having a high level of service typically means a more expensive system due to having a larger fleet or perhaps longer travel times due to frequent stops. In classical multi-objective optimization, it is desirable to optimize conflicting objectives to produce a series of optimal solutions. This technique can be extended to the choice of public transit type in order to demonstrate how such a technique might be adopted in such cases.

Table 2 shows some key transit metrics for the most common types: buses and trains. Electric buses are also shown along with conventional diesel buses. As for trains, the two most widely used categories are shown: rapid transit rail and commuter or inter-city rail. The operating cost refers to how expensive it is to operate the vehicle. The capital refers to the upfront cost to purchase one additional vehicle for an existing transit system. Speed and trip time are average or typical values. The efficiency is the amount of energy consumed per passenger-mile. The cost refers to the price the customer pays per passenger-mile. Note that, given the complexity in determining the values in Table 1, a lot of assumptions are made by the reporting agency. Ranges are provided wherever appropriate. Thus, these should be taken as representative values only for comparison. However, by comparing the numbers directly, an informed decision can be made.

Table 2. Various economic and transportation statistics for North American countries.

Mode Objective	Bus	Electric Bus	Rapid Transit Rail	Commuter/ Inter-City Rail
Operating cost (USD/vehicle revenue hour)	166.51	41	312.09	562.96
Capital (million USD)	1.1	1.7	30.4	30.4
Speed (mph)	9	9	16	40
Trip time (min)	22–45	22–45	47	47–120
Efficiency (kWh/passenger-mile)	3319	1107	3228	1688
Cost (USD/unlinked passenger-mile)	1.31	1.31	0.92	0.51

Source: Table by author, data from (US DOT 2010, 2018, 2020).

When multiple conflicting objectives are in play, improving one often worsens another. Thus, it may be difficult for a customer or a transit authority to choose one mode over another. A graphical representation of the decision-making criteria is shown in Figure 4. Only three modes of transportation are shown. More can be added as needed. Only two objectives are plotted: average trip cost to the passenger and the average emissions per passenger-mile. It is desirable to minimize both these objectives, so values of zero and zero would be ideal. However, this is not practical. However, whichever mode can approach this point would be the closest to ideal. Accordingly, it is clear that trains are the best mode, having the lowest cost and emissions. Personal passenger cars are the worst. Note that electric buses could prove to be the best if powered by clean, renewable energy. For Figure 4, it was assumed that the electricity was the average grid mix for the United States. Even so, the choice between electric buses and conventional buses is clear: the former have lower emissions and lower average trip costs. This methodology can be extended to an arbitrary number of objectives and "solutions" (modes of transportation).

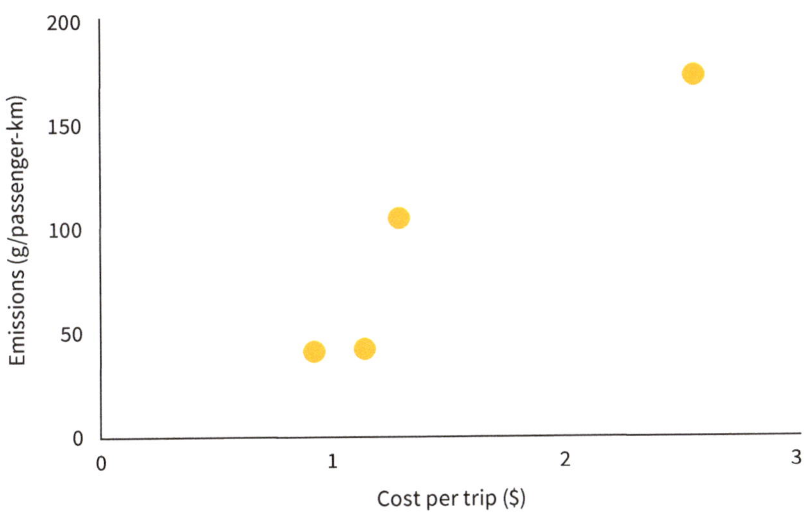

Figure 4. Emissions versus trip costs for common modes of transportation. Source: Graphic by author, data from Table 2.

4.2. Transportation Policy

In the United States, only about 17% of the total cost of public transit projects (capital only) are funded by the Federal government (Mallett 2021). The rest comes from state and local governments and, in some cases, private investors. The Federal Public Transportation Program has gradually increased its funding from 2011 ($10 billion) to 2018 ($13 billion). For 2020 and 2021, the COVID-19 relief package provided an additional $39 billion in total. As climate change and its effects, both immediate and long term, come into sharper focus, federal and state governments are increasing funds allocated to public transit projects. These are geared to reduce carbon emissions. They also increase access to transportation across the different segments of population, increasing equity. This, too, has moved into the spotlight during the social unrest in the United States during the COVID-19 pandemic.

The next section presents some of the largest upcoming public projects in the United States. However, none of these are interstate projects and there are few inter-city projects in the works. Consider that, between 2010 and 2019, the United States added 1203 miles (1936 km) of transit compared to 21,950 miles (35,318 km) of new highway and arterial roadway (US DOT 2019). This is emblematic of transportation trends in the United States. Given the trends in oil prices, consumer vehicle purchases, and the complicated nature of funding and politics in the country, passenger vehicles and planes will continue to dominate long-distance and even local travel.

For cross-country or even interstate transit, the country's geography and population distribution will continue to pose a significant challenge. This is also true for Canada. In the short term, the best bet would be for the United States to improve domestic transit infrastructure and electrify its fleet. The reason China has about 99% of the world's electric buses and over 500,000 charging stations is because electric vehicles are part of the transportation policy at a national level. The United States has not prioritized electrification in the same way. The current administration has emphasized electrification of the federal fleet, a good way to lead by example. This means conversion or replacement of some 645,000 vehicles, comparable to the annual emissions of countries like Haiti or DR Congo. Once the federal government takes the lead, state and local governments would follow. Consumers are already adopting electric vehicles, with several automakers planning for all-electric fleets as early as 2035.

However, this is only one piece of the puzzle. The other piece is public transit. Cities and suburbs in the United States are not pedestrian-friendly. Transit maps are oriented to facilitate travel from the suburbs to the city centers. There is little interconnectedness between suburbs and neighborhoods. This makes it very difficult for people to use the transit system outside of work or trips to the city. This means that owning at least one personal vehicle is virtually a requirement for living in and around major cities. In rural communities, the vast majority of the land area of the United States, personal vehicles are the only way to travel. Future transit projects need to address this disparity or shortcoming to reduce the dependence on personal vehicles. This is reflected in places like New York City. In the borough of Manhattan, 22% of households have at least one vehicle compared to the nationwide rate of 838 per 1000 residents.

The final facet of sustainable transportation for sparse countries is land use. Infill development is the rededication of vacant parcels or plots of land within large urban spaces. City planners need to focus on transit-oriented development. City blocks and spaces need to be designed around the concept of the "15-minute-city". This is the idea where the majority of daily needs are located within a 15-minute walk. This includes mass transit. Rather than removing housing from work, new developments should integrate housing with businesses and office spaces. Furthermore, developers need to ensure that new development has sufficient variety to serve the needs of young, single professionals, small families, and also the retired community.

The following are some of the largest public transit projects that are being undertaken or are soon to be undertaken.

1. Heavy rail and subway—this project was completed in the San Francisco Bay area in 2020, the largest public infrastructure project undertaken in Santa Clara County. Its project cost was USD 2.4 billion and added 10 miles to the existing coverage. It is set to link San Jose and Santa Clara, two neighboring cities. This project required the cooperation of multiple transit authorities, such as the Bay Area Rapid Transit and the Santa Clara Valley Transit Authority. Issues pertaining to platform configuration and additional training for operators using ramped tunnels had to be overcome during the project.

2. Light rail—the cities of Boston, Los Angeles, San Diego, and Seattle have light rail projects totaling USD 8.5 billion opening in 2021. These projects account for over 28 miles of light rail. COVID has impacted all these projects, as can be expected. Even outside the pandemic, some of these projects faced technical challenges, cost overruns, issues with contractors and unions, etc. All of these have been delayed in coming online. This is the nature of such mass transit multi-billion-dollar projects.

3. Electric bus—Indianapolis' Bus Rapid Transit system has a planned extension, the Purple Line, under construction and slated to open in 2022 or 2023. This would double the existing capacity of the Red Line. The 15-mile extension is expected to cost USD 155 million. The entire transit system is electric, specifically using the BYD K11 electric bus. The fate of the project is a bit uncertain—including future extensions. This is because of legislation introduced that challenges the financial obligations of the county and the operator, IndyGo (Indianapolis Public Transportation Corporation).

4. Hyperloop—this refers to the proposed mode of passenger transportation where a high-speed train travels in a sealed tube at vacuum or very low pressure. This low air resistance allows the train to travel at very high speeds very efficiently, making it competitive with air travel over distances under 1,500 km in terms of travel time. White papers have proposed a project along the Los Angeles–San Francisco corridor with an estimated cost of USD 6 billion. Virgin Hyperloop conducted its first human passenger trial at a speed of 172 km/h at its test site in Las Vegas. SpaceX built a one-km track in Hawthorne, California. This technology has the potential to be a true interstate transit system. Various countries are also investigating this new transportation technology. In the United States, significant political and economic challenges would need to be

overcome to make this idea a reality. In California, for instance, the California High-Speed Rail Authority is already working on its first project, slated to open in 2029, costing upwards of USD 64 billion for Phase I. It would be very difficult to justify a second similar project. It must be noted that in the United States, while there is technically high-speed rail, only about 34 miles of track actually allow trains to reach up to 150 mph (241 km/h) for only 34 mi (55 km) of the 457 mi (735 km) track.

5. Closing Remarks

The future of clean transportation and public transit in the United States is far from decided. While the United States has been a leader and innovator in several areas, it severely lags in this one. It need not be stated that this is because of its unique history and particular circumstances. This chapter examines the complex history and multitude of factors quite briefly and with a focus on trying to understand the state of affairs from an engineering and technological standpoint. Public transit and urban sprawl are far from the only public interest items shaped by the country's history. Additionally, it would be overly simplistic to assert that what was presented herein were the only factors. Indeed, the influence of geopolitics, the progress of technology, political and economic cycles, public opinion, etc. all have left their mark on the United States.

Thus, it is not prudent to generalize the lessons learned here to other regions with similar characteristics. It is not sufficient to just consider the statistics, but also the history and societal factors. Consider neighboring Canada: despite the vast similarities, it has a considerably different climate. Its history was also shaped differently, having been largely influenced by its relationship with the United Kingdom. Recent developments, such as the effect of heat waves and wildfires and the discovery of thousands of Indigenous children buried in unmarked graves, will, no doubt, have an effect going forward. Perhaps this is the key takeaway from the foregoing chapter: when applying policy to populations, one must do so in consultation with the population while being respectful of their history and culture.

Funding: This work received no external funding.

Acknowledgments: The author is grateful to the University of Wisconsin-Oshkosh for providing resources to support this work.

Conflicts of Interest: The author declares no conflict of interest.

References

Al-Thani, Soud K., Cynthia P. Skelhorn, Alexandre Amato, Muammer Koc, and Sami G. Al-Ghamdi. 2018. Smart Technology Impact on Neighborhood Form for a Sustainable Doha. *Sustainability* 10: 4764. [CrossRef]

Bamwesigye, Dastan, and Petra Hlavackova. 2019. Analysis of Sustainable Transport for Smart Cities. *Sustainability* 11: 2140. [CrossRef]

Behrendt, Frauke. 2019. Cycling the Smart and Sustainable City: Analyzing EC Policy Documents on Internet of Things, Mobility and Transport, and Smart Cities. *Sustainability* 11: 763. [CrossRef]

Department for Transport. 1997. A New Deal for Transport: Better for Everyone. Available online: https://www.gov.uk/government/organisations/department-for-transport (accessed on 11 March 2021)

Dostál, Ivo, and Vladimir Adamec. 2011. Transport and its Role in the Society. *Transactions on Transport Science* 4: 43–56. [CrossRef]

Fialová, Jitka, Dastan Bamwesigye, Jan Łukaszkiewicz, and Beata Fortuna-Antoszkiewicz. 2021. Smart Cities Landscape and Urban Planning for Sustainability in Brno City. *Land* 10: 870. [CrossRef]

Freudendal-Pedersen, Malene, Sven Kesselring, and Eriketti Servou. 2019. What is Smart for the Future City? Mobilities and Automation. *Sustainability* 11: 221. [CrossRef]

Friedman, Avi, and David Krawitz. 2001. *Peeking Through the Keyhole: The Evolution of North American Homes*. Montreal: McGill-Queen's University Press.

Hall, Matthew, John Iceland, and Youngmin Yi. 2019. Racial Separation at Home and Work: Segregation in Residential and Workplace Settings. *Population Research and Policy Review* 38: 694. [CrossRef]

Hammad, Ahmed W., Ali Akbarnezhad, Assed Haddad, and Elaine Garrido Vazquez. 2019. Sustainable Zoning, Land-Use Allocation and Facility Location Optimisation in Smart Cities. *Energies* 12: 1318. [CrossRef]

International Monetary Fund (IMF). 2020. *World Economic Outlook Database*; Report for Selected Countries and Subjects: October 2020. Washington, DC: World Economic and Financial Surveys. Available online: https://www.imf.org/en/Publications/WEO/weo-database/2020/October/ (accessed on 22 January 2021).

INEGI. 2010. *Censo de Población y Vivienda 2010—Consulta*. Available online: https://www.inegi.org.mx/sistemas/scitel/Default?ev=5 (accessed on 14 February 2021). (In Spanish)

International Organization of Motor Vehicle Manufacturers (IOMVM). 2013. World Vehicles in Use All Vehicles (PDF). Available online: https://www.oica.net/wp-content/uploads/total-inuse-2013.pdf (accessed on 14 February 2021).

International Road Federation (IRF). 2013. *IRF World Road Statistics*. Geneva: International Road Federation Press.

Łukaszkiewicz, Jan, Beata Fortuna-Antoszkiewicz, Łukasz Oleszczuk, and Jitka Fialová. 2021. The Potential of Tram Networks in the Revitalization of the Warsaw Landscape. *Land* 10: 375. [CrossRef]

Mallett, William J. 2021. *Federal Public Transportation Program: In Brief*; Report R42706; Washington, DC: Congressional Research Service.

Ritchie, Hannah. 2019. Where in the World Do People Emit the Most CO_2? *Our World in Data*. Available online: https://ourworldindata.org/per-capita-co2 (accessed on 22 January 2021).

Statistics Canada. 2016. Population Estimates, July 1, by Census Metropolitan Area and Census Agglomeration, 2016 Boundaries. Available online: https://www150.statcan.gc.ca/t1/tbl1/en/tv.action?pid=1710013501 (accessed on 30 January 2021).

Stover, John F. 1999. *The Routledge Historical Atlas of The American Railroads*. New York: Routledge.

Tobey, Michael B., Robert B. Binder, Soowon Chang, Takahiro Yoshida, Yoshiki Yamagata, and Perry P. J. Yang. 2019. Urban Systems Design: A Conceptual Framework for Planning Smart Communities. *Smart Cities* 2: 522–37. [CrossRef]

United Nations (UN). 2019. 2019 Revision of World Population Prospects. Department of Economic and Social Affairs Population Dynamics. Available online: https://population.un.org/wpp/ (accessed on 22 January 2021).

United States (US) Census Bureau. 2019. Metropolitan and Micropolitan Statistical Areas Population Totals and Components of Change: 2010–2019. *Population Division*. Available online: https://www.census.gov/data/tables/time-series/demo/popest/2010s-total-metro-and-micro-statistical-areas.html (accessed on 14 February 2021).

United States (US) Census Bureau. 2020. *Population and Housing Unit Estimates Tables*; Archived Data; Sutland: United States (US) Census Bureau.

United States Department of Transportation (Hodges, Tina) (US DOT). 2010. *Public Transportation's Role in Responding to Climate Change*; Washington, DC: Federal Transit Administration.

United States Department of Transportation (US DOT). 2018. *2018 National Transit Summaries and Trends*; Washington, DC: Federal Transit Administration, Office of Budget and Policy.

United States Department of Transportation (US DOT). 2019. *Highway Statistics 2019*; Washington, DC: Federal Highway Administration, Office of Highway Policy and Information.

United States Department of Transportation (US DOT). 2020. *Transportation Statistics Annual Report 2020*; Washington, DC: Bureau of Transportation Statistics. [CrossRef]

United States Environmental Protection Agency (US EPA). 2020. *Inventory of U.S. Greenhouse Gas Emissions and Sinks 1990–2018*; Environmental Protection Agency 430-R-20-002. Washington, DC: US EPA.

World Development Indicators (WDI). 2021. *CO$_2$ Emissions (Metric Tons per Capita)*. Washington, DC: Data Bank: The World Bank Group, Available online: https://databank. worldbank.org/reports.aspx?source=2&series=EN.ATM.CO2E.PC&country=# (accessed on 22 January 2021).

Yuan, Rui-Qiang, Xin Tao, and Xiang-Long Yang. 2019. CO$_2$ emission of urban passenger transportation in China from 2000 to 2014. *Advances in Climate Change Research* 10: 59–67. [CrossRef]

A Systematic Analysis of Bioenergy Potentials for Fuels and Electricity in Turkey: A Bottom-Up Modeling

Danial Esmaeili Aliabadi, Daniela Thrän, Alberto Bezama and Bihter Avşar

1. Introduction

1.1. Global Warming: A Thread Ahead

For decades, scientists have been warning about the negative consequences of climate change on society. In late 2015, as the public demand from authorities grew, a considerable majority of countries decided to boost their actions to restrict a global-mean temperature rise to 1.5 °C above pre-industrial levels. This ambitious target in the Paris Agreement was designed to be achieved through countries' contributions (i.e., nationally determined contributions). Unlike the Kyoto Protocol, which expired in 2012, the Paris Agreement differentiates countries' responsibilities by distinguishing "developed" countries from "developing" countries. Although this approach enables developing countries to improve their future contributions, relying on self-imposed contributions may result in countries declining to make ambitious targets (Pauw et al. 2019). That is why some countries have requested to be recognized as a member of non-Annex I countries (UNFCCC 2018), while others find targets not bold enough. For instance, the European Union (EU) has further propelled the actions of members in the "Green Deal" by committing to slash emissions by half from 1990 levels, by 2030 (European Commission 2019). The United States also rejoined the Paris climate agreement after leaving it for a short interval in the previous administration (Pedaliu 2020).

Many solutions have been proposed to slow down, and eventually stop, global warming, among which are decarbonizing societies using renewable resources and demand response management. Moving toward a bio-based economy is one of the proposed solutions that promise a cleaner production of energy for various sectors such as industries and transportation. This is particularly important since most efforts in the past were solely focused on the power sector. As stated by experts and policymakers around the world, the underlined targets in the Paris Agreement are difficult to reach if we rely on renewable electricity alone (FSR 2019). This stems from the fact that emissions from energy-intensive industries (such as iron and steel, and cement), as well as the transport sector, comprise a considerable share of the total greenhouse gas (GHG) emissions every year. These hard-to-abate sectors require storable energy sources with high energy density (Friedmann et al. 2019). For overcoming this challenge, currently, there are several technological concepts under development (e.g., all-electric commercial jets).

1.2. Gas-Power Network

According to the World Energy Outlook, which is published by the International Energy Agency (IEA), natural gas is expected to play a significant role in the global energy landscape in the future by replacing coal (IEA 2020b). Using gas-fired peaker power plants, natural gas can also hedge against the intermittency of renewable sources such as solar and wind (United Nations 2020). In fact, the EU believes that any cost-optimal solutions to achieve a near-zero carbon energy system by the mid-century should consider a "dual" gas-power network (Bowden 2019). Yet, combusting fossil natural gas is destructive for the environment, even though it pollutes less than coal and crude oil; therefore, in long-term solutions, natural gas should be replaced with other renewable alternatives.

Methane (CH_4) production from biological origins (bio-CH_4) and (green) hydrogen (H_2) can be suitable substitutes for natural gas. Due to their similar compositions, bio-CH_4 is easier to implement than H_2, since it does not require huge investments in infrastructures. Indeed, existing natural gas facilities for storage and transportation can also be used for bio-CH_4 (Matschoss et al. 2020). Thus, through investment grants and tax incentives, many countries invest heavily in the decarbonization of the gas sector using bio-CH_4 (Brémond et al. 2020). While bio-CH_4 can partially fulfill the energy demand, we should also prevent fugitive CH_4 emissions in various sectors, as the detrimental impact of CH_4 on the climate is much higher than carbon dioxide (CO_2). As a matter of fact, the sixth assessment report of the Intergovernmental Panel on Climate Change (IPCC) emphasizes that methane emissions are responsible for 0.5 °C of warming to date (IPCC 2021, Figure SPM.2).

1.3. Turkey's Current Status

Turkey is a developing country and a member of the Organization for Economic Co-operation and Development (OECD), which is growing to be an influential player in West Asia and southeastern Europe. Current Turkey's energy consumption per capita falls short when compared with other OECD peer countries, although this is expected to change (Difiglio et al. 2020). Turkey is investing heavily in the energy sector to support its rapidly growing economy. Unfortunately, Turkey's domestic reserves are not adequate to fulfill its demands; thus, the nation has to rely on oil and natural gas imports from neighboring countries such as Russia, Iraq, Azerbaijan, and Iran (Esmaeili Aliabadi 2020). Owing to imported fossil fuels, Turkey's energy trade deficit is soaring to a massive amount, with natural gas being the second important cause after crude oil and petroleum products (Erkoyun et al. 2020).

In order to ameliorate the trade deficit and achieve supply security, Turkey is diversifying its energy generation portfolio by investing in domestic renewable resources under law No. 6094, including bioenergy. The Turkish government introduced technology-specific feed-in tariffs (FITs), the so-called renewable energy resources support mechanism (YEKDEM, by its Turkish acronym). According to YEKDEM, the government is obligated to purchase the generated power for a decade with fixed prices from the renewable facilities that are commissioned prior to July 2021 (EMRA 2005; IEA 2021). As appeared in Table 1, this support mechanism also provides incentives to local energy technologies.

Table 1. Technology-specific feed-in tariffs according to the renewable energy resources support mechanism (YEKDEM).

Technology	Base Incentive (US ¢/kWh)	Local Equipment (US ¢/kWh)	Total (US ¢/kWh)
Concentrated solar power	13.3	9.2	22.5
Solar photovoltaics	13.3	6.7	20
Biomass and waste	13.3	5.6	18.9
Geothermal	10.5	2.7	13.2
Wind	7.3	3.7	11
Hydro	7.3	2.3	9.6

Source: Table by authors, data from EMRA Law No. 5346 (EMRA 2005).

The YEKDEM scheme caused a boom in the deployment of clean energy technologies such as wind and solar; however, as depicted in Figure 1, bioenergy did not receive the same attention (Esmaeili Aliabadi 2019). As of January 2021, the total installed capacities of solar power plants are more than six times of biogas, biomass, and waste heat power plants combined.

Despite the low utilization of biogas as a fuel source for electric power production, which is not on par with other renewable resources, Turkey's biogas production potential is estimated to be over 221 PJ per year in 2016 (Daniel-Gromke et al. 2016). The biogas production efficiency is a function of both biological matter properties (e.g., lignocellulose content) and the production parameters (e.g., technology). In Turkey, firewood, as the classic biomass fuel, attracts more attention for energy production, as it is convenient to process and have a high production rate; however, other biomass sources such as agricultural residues including hazelnut shell, tea waste, and wheat straw have been utilized to meet the energy demand. Maltsoglou et al. (2016) provide regional assessments for the availability and potential of agricultural residues for heat and power production in Turkey. Preparing an up-to-date regional plan for bioenergy is vital since there is no silver bullet that can solve the energy issue in every province.

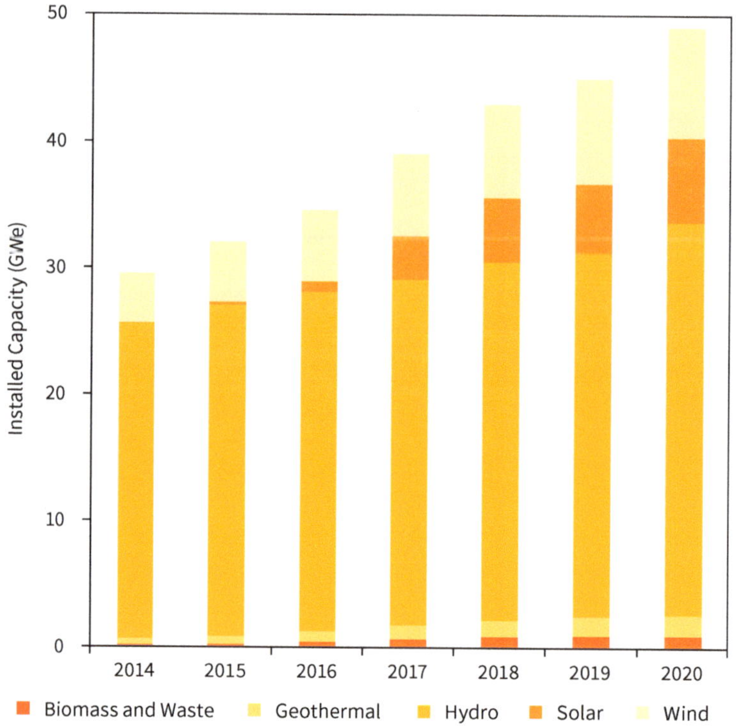

Figure 1. Breakdown of installed capacity by renewable technologies from 2014 to 2020. Source: Graphic by authors.

In the transport sector, biodiesel and bioethanol production is again supported by law. From 2018, Turkey's energy market regulatory authority made it compulsory to blend a minimum of 0.5% biodiesel in diesel oil (Tiryakioglu 2017). In 2014, Turkey increased the percentage of bioethanol blended with gasoline to 3% from the initial value of 2% in 2013. Thus far, bioethanol is produced in Turkey almost entirely from sugar beet, corn, and barley (AGWeek 2011; Ozdingis and Kocar 2018). Producing bioethanol fuel from limited energy crops can be disrupted frequently in the long term, as it competes with the food and medical supply chains. For example, due to the high demand for disinfectants amid the pandemic, bioethanol production for gasoline was suspended in 2020 (Erkul 2020).

Supporting biofuels is not limited to Turkey. In Germany, similar policies are in place since 2015 through a GHG-based quota, which requests fuel suppliers to mix biofuels with conventional fuels such that the resulting mixture achieves a specific (i.e., 6% in 2020) GHG mitigation (Meisel et al. 2020).

While Turkey is facing unique challenges, it shares a common ground with other developing countries; hence, successful practices can be adopted by the rest of the world. To this end, we calculated Turkey's sectoral GHG emissions until 2040, using a bottom-up technology-oriented energy model. Analyzing the collected information, we propose strategies to curtail GHG emissions by promoting bioenergy production and consumption.

2. Materials and Methods

2.1. Energy Systems Modeling

In order to model intertwined energy systems, there are two fundamental modeling perspectives: the top-down macroeconomic approach and the bottom-up engineering approach. Both of these approaches have their own advantages and disadvantages: for instance, top-down models assume an unalterable world, whereas bottom-up models account for technological breakthroughs over time. Bottom-up models serve as the practical method to estimate energy trends in mid- and long term (Esmaeili Aliabadi et al. 2021).

In order to evaluate the total GHG emissions at the country level, the whole energy system should be considered, with all components and their interactions. To this end, a bottom-up technology-rich optimization model based on TIMES[1] (Loulou et al. 2005) has been developed, in which parallel technologies compete to satisfy end-use demand with minimum costs and within the frame of financial, environmental, and technological constraints. As illustrated in Figure 2, the developed model creates a complex network of processes, in which commodities are being acquired (imported or extracted), transformed (through conversion processes),

---

[1] TIMES is the acronym for The Integrated MARKAL-EFOM (MARKet ALlocation-Energy Flow Optimization Model) System. The TIMES source code can be acquired from https://github.com/ etsap-TIMES/TIMES_model (accessed on 21 September 2021).

and transferred to be used by others. The results determine the optimal technology mix for each demand service until 2040, using 1200 processes, 181 commodities, and over 50 thousand data values.

In the proposed model, years are divided into eight time slices distinguishing day and night in each season. Hourly available datasets, such as electricity demand, wind speed, precipitation, and solar irradiation in each region, are transformed to be in accordance with the specified setting.

While the developed model consists of a single region (i.e., Turkey), it respects the regional characteristics of its elements. For instance, the availability factor of wind farms considers the local wind speed and employed technologies.

Considering the technical properties, we assigned emission factors (EFs) to processes for various gases (CO_2, CH_4, and N_2O). To assess the potency of GHGs in trapping heat, experts employ a relative measure called global warming potential, by which the greenhouse effects of GHGs are compared with those of CO_2 as the reference gas. We used coefficients mentioned in Gillenwater et al. (2002, Table 2) to calculate the total annual CO_2-equivalent emissions.

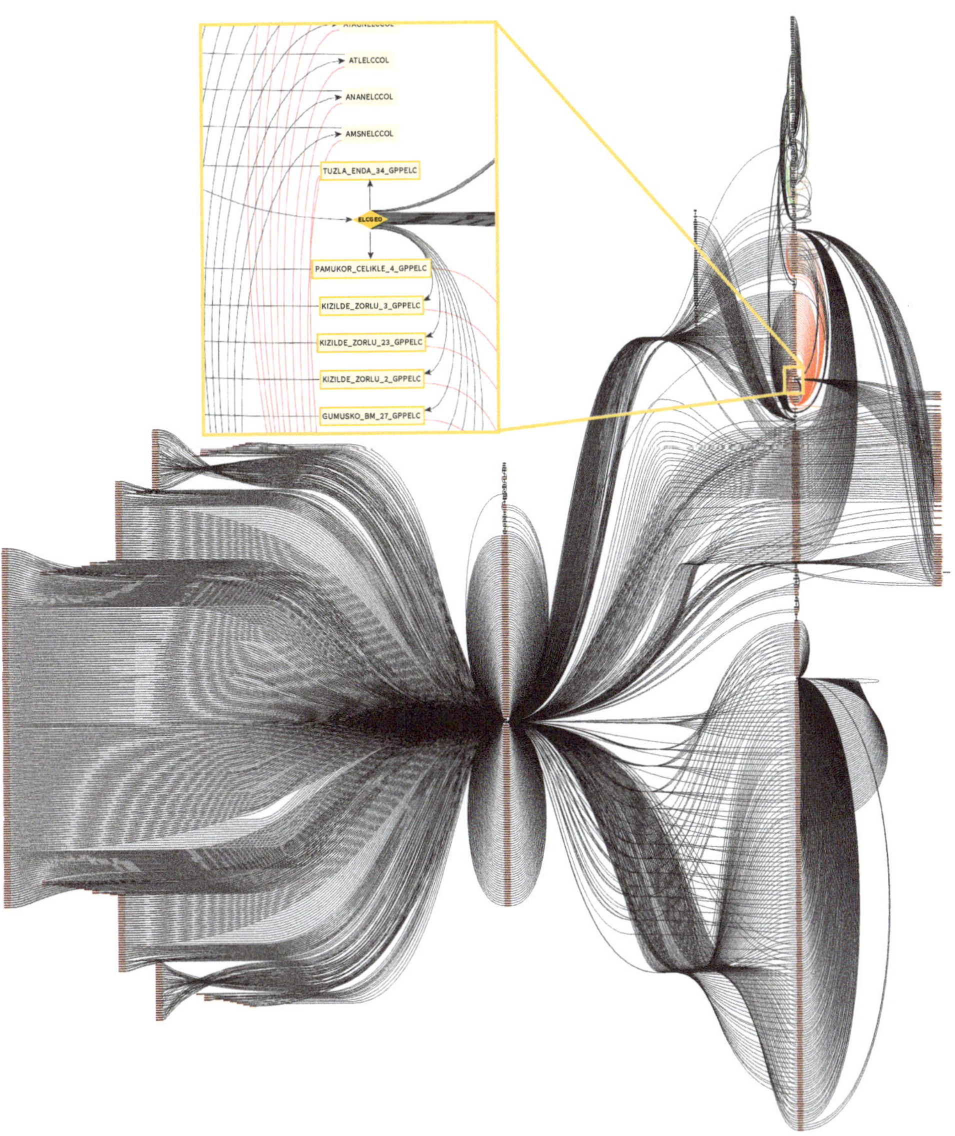

Figure 2. Magnifying a section of the reference energy system. In the magnified section of the network, one can see the exploitable geothermal energy is redirected to individual geothermal power plants, which illustrates the level of details in processes. Interested readers are invited to zoom into the digital version of this figure for the details. Source: Graphic by authors.

2.2. Scenarios

For this study, we devised two scenarios: (1) the current policy (CP) scenario, in which established regulations and available technologies were taken into account, and (2) the pro-bioenergy (Pro-Bio) scenario, in which bioenergy was used intensely for producing liquid fuels and electricity.

In the pro-bioenergy scenario, we assumed that YEKDEM (or the replacing mechanism) continues supporting bioenergy (IEA 2021) and that an aggressive approach is adopted to produce biofuels in the transport sector. We adopted technology descriptions of the BioENergy OPTimization (BENOPT) model (Millinger 2020; Millinger et al. 2021) for technologies that convert crops and biomass residues to liquid biofuels, heat, and electricity. Specifically speaking, under the Pro-Bio scenario, methane emissions from dairy and non-dairy cattle and fugitive emissions from natural gas reserves were redirected to power and heat sectors. Furthermore, waste cooking oil was purchased and transformed into biodiesel using two technologies:

- Hydro treating vegetable or waste cooking oil;
- Producing fatty-acid methyl ester via transesterification of cooking oil.

Due to an active tourist sector in Turkey, centralized management of waste oil is possible to a great extent via unions[2]. Turkey's government can further support this policy using tax incentives. Nonetheless, producing biodiesel from waste cooking oil is limited to consumption in the country. To relax this constraint, oil (i.e., lipid) production from microalgae is assumed to become economically viable in the future.

To achieve commercial-scale production of algae-based biodiesel, scientists and engineers should overcome many techno-economic challenges such as enhancing lipid production and designing an efficient dewatering method. In order to enhance lipid production, biologists are genetically engineering microalgal strains using state-of-the-art technologies such as CRISPR-cas9, which can decrease the cost of biodiesel downstream (Lü et al. 2011; Ng et al. 2017; Radakovits et al. 2010). Furthermore, open pond systems for microalgae cultivation are prone to parasites and invasive species; therefore, scientists need to investigate gene editing strategies to increase microalgae strain's tolerance against a/biotic stressors. The current problem with oil production from microalgae is dewatering, which is energy intensive.

---

[2]   In Turkey, the alternative energy and biodiesel manufacturers union (ALBIYOBIR by its Turkish acronym) is a major association that collects waste cooking oil for biodiesel production.

Researchers are investigating energy efficient dewatering methods that can be combined with the lipid extraction step (Ghasemi Naghdi et al. 2016).

By enhancing the technology and increasing yield, we can make microalgae an attractive alternative to produce oil on a commercial scale. Under the Pro-Bio scenario, we assumed a higher cost for producing energy from microalgae than waste cooking oil that drops rapidly over time due to the learning effect. All in all, we introduced five new technologies that produce heat and electricity from biogas, biomethane, and vegetable oil. Furthermore, we added four new technologies to produce bioethanol and biodiesel from lignocellulose, sugar beet, microalgae, and waste cooking oil.

Finally, no upper bound was set for the GHG emissions under the Pro-Bio and CP scenarios. Table 2 summarizes the differences between the two scenarios.

Table 2. Differentiating components under each scenario.

Component	Type	Current Policy Scenario	Pro-Bio Scenario
Fugitive gases	Technology	Release into atmosphere	Utilized for heat production
Technologies	Technology	Available technologies	Adopted from BENOPT
YEKDEM	Policy	Ends at 2021	Continues until 2040
Blending bioethanol with gasoline	Policy	increasing the blending ratio to about 8% by 2040	increasing the blending ratio to about 20% by 2040
Blending biodiesel with diesel	Policy	increasing the blending ratio to 0.7% by 2040	increasing the blending ratio to 20% by 2040

Source: Table by authors.

It is noteworthy to mention that the results of the presented model in this manuscript should not be compared with Difiglio et al. (2020), as the presented model was modified thoroughly to include only publicly available information and datasets.

3. Results

3.1. Current Policy Scenario

The total GHG emissions (CO_2-equivalent) from each sector under the CP scenario are illustrated in Figure 3a. Due to the massive deployment of solar photovoltaics and wind turbines, the GHG emission intensity of the power sector is expected to decrease from ~490 g CO_2-eq./kWh, in 2020, to 252 g CO_2-eq./kWh, in 2040. This declining trend results in dropping GHG emissions from the electric power sector despite higher electricity demand.

As one can see from Figure 3b, the total GHG emissions of Turkey (including non-energy sectors such as agriculture and chemical industries) are steadily rising. The color bars in Figure 3b are adopted from the Climate Action Tracker website[3]. The colors represent different ranges in the selected years. The emission abatement strategies within these ranges are likely to cause global warming mentioned in two ends of that particular color if all countries follow similar approaches.

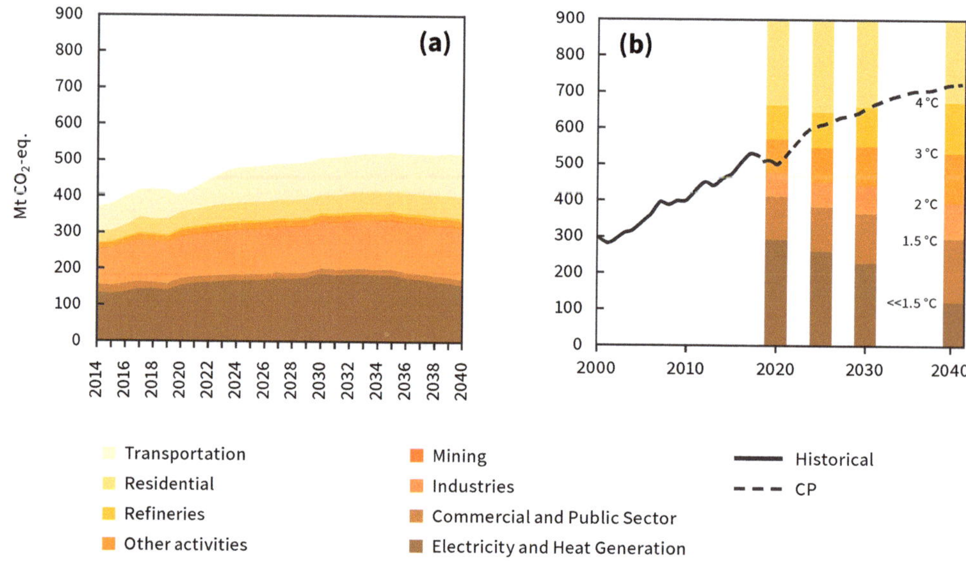

Figure 3. (**a**) Sectoral GHG emissions (expressed in Mt CO$_2$-eq.) from 2014 until 2040; (**b**) comparing the total GHG emissions with the Paris climate agreement. The solid line shows the historical emissions, and the dashed line displays the projected emissions. Source: Graphic by authors.

According to our calculations, the existing GHG abatement strategies in Turkey are highly insufficient. Although the current trend is above an acceptable level, it is below Turkey's intended nationally determined contribution, which was 929 Mt CO$_2$-eq. in 2030 (Republic of Turkey 2015).

---

[3]   Please see https://climateactiontracker.org/countries/turkey/ (accessed on 20 September 2021).

Further investigation in Figure 4a shows that methane, which can be used as a fuel source, comprises a considerable amount of emissions. By exploiting emitted CH_4, we can prevent polluting the atmosphere with a gas multiple times stronger than CO_2 in trapping heat, and simultaneously, combust less of imported (and relatively expensive) natural gas. Among various CH_4 emission sources in Figure 4b, the agriculture and mining sectors contribute between 69% and 91% of the total CH_4 emissions from 2014 to 2040. It is estimated that Turkey holds 679 billion cubic meters of shale gas reserves in the Thrace region and southeastern Anatolia[4]. Extracting these resources as planned, with similar EFs as the current technology, can emit a considerable amount of CH_4 into the atmosphere. Among other sources of bio-CH_4 emissions in the agriculture sector, dairy and non-dairy cattle can be counted as big-ticket items.

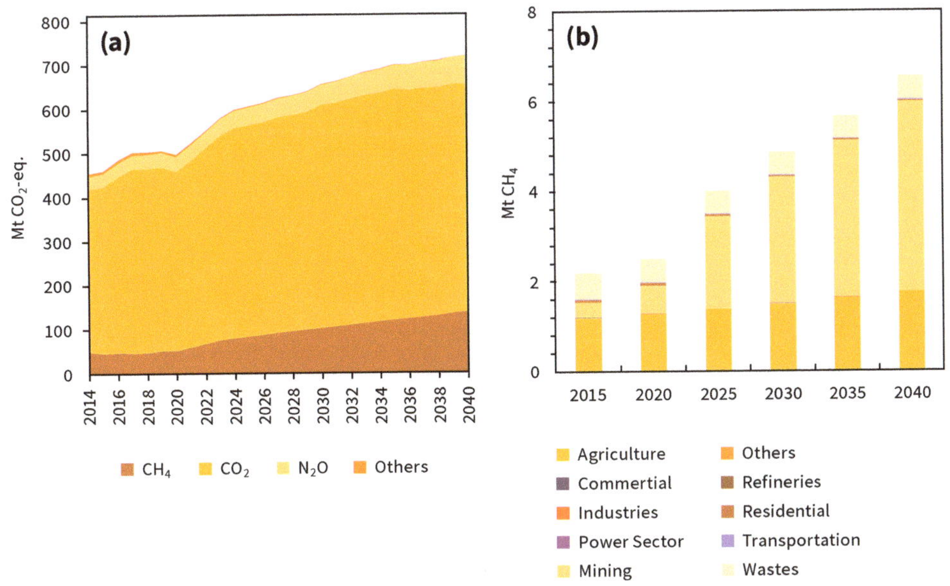

Figure 4. (a) Total GHG emissions by type (expressed in Mt CO_2-eq.) from 2014 until 2040; (b) methane emissions from various sectors. Source: Graphic by authors.

---

4   Please see https://politicstoday.org/what-is-turkeys-shale-gas-potential/ (accessed on 20 September 2021).

In the transportation sector, vehicle ownership growth leads to higher fuel consumption. Subsequently, there are intentions to increase the blending ratio of bioethanol and biodiesel. Increasing the blending ratio of bioethanol to about 8% and biodiesel to 0.7% by 2040 can double these biofuels consumptions (15.7 PJ)[5]. To increase the blending ratios, while not competing with food crops, it is required to adopt better production technologies such as second-generation bioethanol.

In rural areas, biomass is still being exploited for residential space heating, water heating, and cooking. However, we expect biomass consumption to decline, driven by electrification and gasification in the residential sector.

3.2. Pro-Bio Scenario

According to our calculations in Figure 5a, the total GHG emissions of Turkey can reach 570 Mt in 2040. Under the Pro-Bio scenario, Turkey can produce approximately 250 PJ biodiesel and bioethanol per annum by 2040, using waste cooking oil, microalgae, lignocellulosic biomass, and food crops (see Figure 5c).

In the power sector, the installed capacity of biomass-based power plants can raise to 11 GW by 2040, of which combined heat and power using waste cooking oil would play a significant role (see Figure 5b). Subsequently, the GHG emission intensity in the power sector would drop to 226 g CO_2-eq./kWh in 2040.

Finally, as illustrated in Figure 5d, the biomass consumption in the building sector in 2040 is negligible ($\sim$48 PJ), which paves the way for better management of forests. As reported by the IEA (2020a) and Schimschar et al. (2016), biomass consumption share in buildings for space heating is decreasing, owing to better standards, and electrification and gasification of the residential sector, especially in rural areas.

---

[5]   Producing biodiesel over 0.7% ratio is difficult to achieve when waste cooking oil is solely considered.

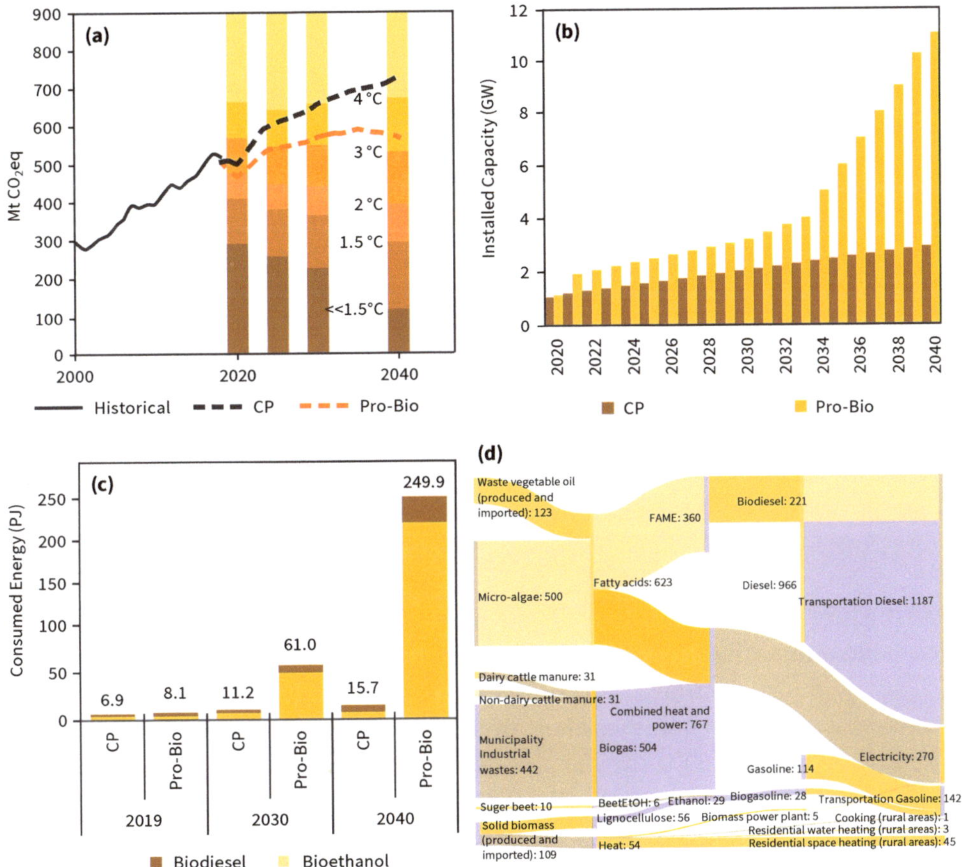

Figure 5. (**a**) The total GHG emissions under the CP and Pro-Bio scenarios; (**b**) the total installed capacity of power plants based on biogas, biomethane, biomass, and waste oil; (**c**) bioethanol and biodiesel consumption in the transportation sector; (**d**) the energy flow from biomass, wastes, and biofuels (PJ) in Turkey's energy system under the Pro-Bio scenario in 2040. Source: Graphic by authors.

4. Discussion

In this chapter, we assessed the effect of introducing biogas and biofuels more actively in Turkey's energy system till 2040. In the CP scenario, the contribution of biofuels to the transport sector is 275.4 PJ from 2014 till 2040, which displaces about 7.3 Mt of diesel and gasoline and avoids emitting 20.64 Mt CO_2-eq. into the atmosphere. However, by assuming higher biofuel blending ratios in the Pro-Bio scenario, an extra 153 Mt CO_2-eq. GHG emissions can be avoided within the same time frame. This is equivalent to not purchasing 57 Mt of petroleum products, which, by itself, can improve Turkey's trade deficit.

To provide the additional biofuels, a wider range of technical options should be considered. For this reason, preparing a strategic plan to develop bioenergy in each region respecting their ecological characteristics (ecotype) is crucial. In short-term solutions, agricultural residues produced on a provincial scale can play a significant role, as they entail fairly low costs and great accessibility. For instance, Turkey is one of the major hazelnut producers, and hazelnut is the most important tree nut crop for the country's economy; hence, the Turkish hazelnut cultivar's genome has recently been sequenced to unravel the critical properties for crop improvement (Avşar and Esmaeili Aliabadi 2017; Lucas et al. 2021). Most hazelnut production in Turkey occurs in Ordu, Samsun, and Giresun near the Black Sea Coast. The centralized production of hazelnut in these neighboring provinces provides a unique opportunity to exploit hazelnut wastes and shells as the main agricultural wastes for biogas production (Şenol 2019). Simultaneously, endemic terrestrial plants, such as Origanum Sp. Tekin-2017 (Cilgin 2020), should be evaluated to analyze their applications in biodiesel production.

Utilizing aquatic plant-like organisms, such as microalgae, can also be seen as a viable alternative to produce oil for biodiesel production. According to our calculations, the total surface area of all reservoirs behind hydroelectric power plants is above 4451 km^2. When the lakes are included, the total surface area of these waterbodies surpasses 8000 km^2. This potential can be utilized to cultivate microalgae for fuel production. Assuming a 45 tFM yield per hectare and year, Turkey can produce over 83.53 PJ of energy by exploiting 50% of this potential. Hence, achieving 500 PJ in 2040, which is required by the Pro-Bio scenario, might be infeasible without considering coastlines (e.g., the Black Sea and the Mediterranean Sea). To this end, investing in offshore technologies similar to the Offshore Membrane Enclosures for Growing Algae (OMEGA) project (Colen 2014; Wiley 2013) seems inevitable, which requires solving many techno-economic challenges to tap into this potential.

As the market diffusion of intermittent renewable resources (e.g., solar and wind) increases, and coal-burning power plants become less attractive for investors, independent system operators will have to be prepared to cope with the power balancing issue in real time. Flexible biogas-burning power plants can be viewed as a green alternative to gas-fired peaker plants to produce power on demand. Managing biogas optimally can complement the advantages of other renewable resources (Dotzauer et al. 2019; Lauer and Thrän 2018). Considering the technology mix, the produced power in the Pro-Bio scenario from biogas and vegetable cooking oil can prevent emitting 241.53 Mt CO_2-eq. until 2040.

5. Conclusions

In this chapter, a TIMES-based model was employed to compare the impact of higher bioenergy consumption in Turkey on greenhouse gas emissions via contrasting scenarios: the current policy and pro-bioenergy. According to the results, Turkey needs to invest in novel technologies to resolve the competition between biofuels and other critical supply chains (e.g., for food and medical applications), as the currently used waste cooking oil and sugar beet are hardly enough to support notable consumption levels. Furthermore, flexible biogas-based power plants should be supported using FITs as a green alternative to gas-fired peaker plants to hedge against volatilities caused by intermittent renewable sources. By so doing, power systems will be enabled to invest more in harnessing energy from solar and wind.

Author Contributions: Conceptualization, D.E.A., D.T. and A.B.; methodology and modeling, D.E.A.; visualization, D.E.A. and B.A.; formal analysis, D.E.A. and D.T.; investigation, D.E.A., D.T. and B.A.; data curation, D.E.A. and D.T.; writing–original draft preparation, D.E.A., D.T. and B.A.; writing–reviewing and editing, D.E.A. and A.B. All authors have read and agreed to the published version of the manuscript.

Funding: This research received no specific grant from any funding agency.

Acknowledgments: The authors would like to thank Sabanci University and Carmine Difiglio for their support.

Conflicts of Interest: The authors declare no conflict of interest.

Abbreviations

The following abbreviations are used in this manuscript:

ALBIYOBIR	The alternative energy and biodiesel manufacturers union
BENOPT	BioENergy OPTimization model
CAS9	CRISPR-Associated Protein 9
CRISPER	Clustered Regularly Interspaced Short Palindromic Repeats
CP	Current Policy scenario
EF	Emission Factor
EFOM	The Energy Flow Optimization Model
EU	The European Union
FIT	Feed-in tariff

GHG	Greenhouse gas
IEA	International Energy Agency
IPCC	Intergovernmental Panel on Climate Change
km	Kilometer
kWh	Kilowatt hour
MARKAL	Market allocation
Mt	Million tonne
OECD	Organization for Economic Co-operation and Development
OMEGA	Offshore membrane enclosures for growing algae
PJ	Petajoule
Pro-Bio	Pro-Bioenergy scenario
tFM	Tonne of Fresh Matter
TIMES	The Integrated MARKAL-EFOM System
US	The United States of America

References

AGWeek. 2011. Sugarbeets in Turkey. Available online: https://agweek.com/sugarbeet/sugarbeets-in-turkey (accessed on 1 November 2021).

Avşar, Bihter, and Danial Esmaeili Aliabadi. 2017. Identification of microRNA elements from genomic data of european hazelnut (*Corylus avellana* L.) and its close relatives. *Plant Omics* 10: 190–96. [CrossRef]

Bowden, Jeremy. 2019. EU Sees Dual Gas-Power Energy System as Best Bet to Reach Zero Carbon by 2050. Available online: https://informaconnect.com/eu-sees-dual-gas-power-energy-system-as-best-bet-to-reach-zero-carbon-by-2050 (accessed on 20 September 2021).

Brémond, Ulysse, Aude Bertrandias, Jean-Philippe Steyer, Nicolas Bernet, and Hélène Carrere. 2020. A vision of European biogas sector development towards 2030: Trends and challenges. *Journal of Cleaner Production* 287: 125065. [CrossRef]

Cilgin, Erdal. 2020. Investigation of biodiesel potential of new hybrid of *Origanum* Sp. Tekin-2017, native to Turkey. *Fuel* 277: 118180. [CrossRef]

Colen, Jerry. 2014. OMEGA Project (2009–2012). Available online: https://nasa.gov/centers/ames/research/OMEGA (accessed on 21 September 2021).

Daniel-Gromke, Jaqueline, Funda Cansu Ertem, Ronny Kittler, Fatih Gökgöz, Peter Neubauer, and Walter Stinner. 2016. Analyses of regional biogas potentials in Turkey. Paper presented at Eurasia 2016 Waste Management Symposium, Istanbul, Turkey, 2–4 May 2016.

Difiglio, Carmine, Bora Şekip Güray, Ersin Merdan, and Danial Esmaeili Aliabadi. 2020. *Turkey Energy Outlook 2020*. Istanbul: Sabancı University Istanbul International Center for Energy and Climate.

Dotzauer, Martin, Diana Pfeiffer, Markus Lauer, Marcel Pohl, Eric Mauky, Katharina Bär, Matthias Sonnleitner, Wilfried Zörner, Jessica Hudde, Björn Schwarz, and et al. 2019. How to measure flexibility–Performance indicators for demand driven power generation from biogas plants. *Renewable Energy* 134: 135–46. [CrossRef]

EMRA. 2005. Law on the Use of Renewable Energy Sources for Electric Energy Generation No. 5346. Available online: https://www.mevzuat.gov.tr/MevzuatMetin/1.5.5346.pdf (accessed on 13 October 2021).

Erkoyun, Ezgi, Daren Butler, and Jonathan Spicer. 2020. Turkey Trade Deficit Soars as Lira Drop Sets Off Gold Rush. Available online: https://www.reuters.com/article/turkey-economy-trade-int/turkey-trade-deficit-soars-as-lira-drop-sets-off-gold-rush-idUSKBN25T1CM (accessed on 21 September 2021).

Erkul, Nuran. 2020. Turkey Halts Ethanol-Mixed Fuel for More Disinfectants. Available online: https://www.aa.com.tr/en/energy/refining-petro-chemistry/turkey-halts-ethanol-mixed-fuel-for-more-disinfectants/28639 (accessed on 21 September 2021).

Esmaeili Aliabadi, Danial. 2019. *IICEC-Sabanci University TIMES Energy Model: Overview*. Istanbul: Sabancı University Istanbul International Center For Energy and Climate (IICEC), p. 90. [CrossRef]

Esmaeili Aliabadi, Esmaeili. 2020. Decarbonizing existing coal-fired power stations considering endogenous technology learning: A Turkish case study. *Journal of Cleaner Production* 261: 121100. [CrossRef]

Esmaeili Aliabadi, Danial, Emre Çelebi, Murat Elhüseyni, and Güvenç Şahin. 2021. Modeling, simulation, and decision support. In *Local Electricity Markets*. Edited by Tiago Pinto, Zita Vale and Steve Widergren. Amsterdam: Elsevier, pp. 177–97. [CrossRef]

European Commission. 2019. The European Green Deal. *Communication form the Commission to the European Parliament, the European Council, the Council, the European Economic and Social Committee and the Committee of the Regions*. Available online: https://eur-lex.europa.eu/legal-content/EN/TXT/?uri=COM (accessed on 21 September 2021).

Friedmann, Julio, Zhiyuan Fan, and Ke Tang. 2019. *Low-Carbon Heat Solutions for Heavy Industry: Sources, Options, and Costs Today.* New York: Columbia University Center on Global Energy Policy. Available online: https://www.energypolicy.columbia.edu/research/report/low-carbon-heat-solutions-heavy-industry-sources-options-and-costs-today (accessed on 21 September 2021).

FSR. 2019. Exclusive Interview with Borchardt (EC)! from Madrid Forum to the Gas Package. Available online: https://fsr.eui.eu/exclusive-interview-with-borchardt-ec-from-madrid-forum-to-the-gas-package (accessed on 20 September 2021).

Ghasemi Naghdi, Forough, Lina M. González González, William Chan, and Peer M. Schenk. 2016. Progress on lipid extraction from wet algal biomass for biodiesel production. *Microbial biotechnology* 9: 718–26. [CrossRef]

Gillenwater, Michael, Marian Martin Van Pelt, and Katrin Peterson. 2002. *Greenhouse Gases and Global Warming Potential Values.* Washington: Office of Atmospheric Programs of United States Environmental Protection Agency.

IEA. 2020a. Energy Efficiency Indicators: Highlights. Available online: https://iea.org/reports/energy-efficiency-indicators (accessed on 21 September 2021).

IEA. 2020b. World Energy Outlook 2020. Available online: https://www.oecd-ilibrary.org/energy/world-energy-outlook-2020_557a761b-en (accessed on 21 September 2021).

IEA. 2021. Turkey 2021: Energy Policy Review. Available online: https://www.oecd-ilibrary.org/energy/turkey-2021-energy-policy-review_0633467f-en (accessed on 21 September 2021).

IPCC. 2021. Climate Change 2021: The Physical Science Basis. Contribution of Working Group I to the Sixth Assessment Report of the Intergovernmental Panel on Climate Change. Available online: https://www.ipcc.ch/report/ar6/wg1/downloads/report/IPCC_AR6_WGI_Full_Report.pdf (accessed on 21 September 2021).

Lauer, Markus, and Daniela Thrän. 2018. Flexible biogas in future energy systems–sleeping beauty for a cheaper power generation. *Energies* 11: 761. [CrossRef]

Loulou, Richard, Uwe Remne, Amit Kanudia, Antti Lehtila, and Gary Goldstein. 2005. *Documentation for the TIMES Model—Part I: Energy Technology Systems Analysis Programme.* Paris: International Energy Agency.

Lü, Jing, Con Sheahan, and Pengcheng Fu. 2011. Metabolic engineering of algae for fourth generation biofuels production. *Energy & Environmental Science* 4: 2451–66. [CrossRef]

Lucas, Stuart J., Kadriye Kahraman, Bihter Avşar, Richard J. A. Buggs, and Ipek Bilge. 2021. A chromosome-scale genome assembly of European hazel (*Corylus avellana* L.) reveals targets for crop improvement. *The Plant Journal* 105: 1413–30. [CrossRef] [PubMed]

Maltsoglou, Irini, Luis Rincon, Ana Kojakovic, Evren Deniz Yaylaci, Manas Puri, and Olivier Dubios. 2016. *BEFS Assessment for Turkey: Sustainable Bioenergy Options from Crop and Livestock Residues.* Rome: Food and Agriculture Organization of the United Nations.

Matschoss, Patrick, Michael Steubing, Joachim Pertagnol, Yue Zheng, Bernhard Wern, Martin Dotzauer, and Daniela Thrän. 2020. A consolidated potential analysis of bio-methane and e-methane using two different methods for a medium-term renewable gas supply in Germany. *Energy, Sustainability and Society* 10: 1–17. [CrossRef]

Meisel, Kathleen, Markus Millinger, Karin Naumann, Franziska Müller-Langer, Stefan Majer, and Daniela Thrän. 2020. Future renewable fuel mixes in transport in Germany under RED II and climate protection targets. *Energies* 13: 1712. [CrossRef]

Millinger, Markus. 2020. BioENergy OPTimisation model–BENOPT. Available online: https://www.ufz.de/export/data/2/245341_Supplementary.pdf (accessed on 21 September 2021).

Millinger, Markus, Philip Tafarte, Matthias Jordan, Alena Hahn, Kathleen Meisel, and Daniela Thrän. 2021. Electrofuels from excess renewable electricity at high variable renewable shares: cost, greenhouse gas abatement, carbon use and competition. *Sustainable Energy & Fuels* 5: 828–43. [CrossRef]

Ng, I-Son, Shih-I Tan, Pei-Hsun Kao, Yu-Kaung Chang, and Jo-Shu Chang. 2017. Recent developments on genetic engineering of microalgae for biofuels and bio-based chemicals. *Biotechnology Journal* 2: 1600644. [CrossRef] [PubMed]

Ozdingis, Asiye Gul Bayrakci and Gunnur Kocar. 2018. Current and future aspects of bioethanol production and utilization in Turkey. *Renewable and Sustainable Energy Reviews* 81: 2196–203. [CrossRef]

Pauw, Pieter, Kennedy Mbeva, and Harro van Asselt. 2019. Subtle differentiation of countries' responsibilities under the paris agreement. *Palgrave Communications* 5: 1–7. [CrossRef]

Pedaliu, Effie G.H. 2020. The Biden Era: What Can Europe Expect from America's New President? Available online: https://blogs.lse.ac.uk/europpblog/2020/11/09/the-biden-era-what-can-europe-expect-from-americas-new-president/ (accessed on 21 September 2021).

Radakovits, Randor, Robert E. Jinkerson, Al Darzins, and Matthew C. Posewitz. 2010. Genetic engineering of algae for enhanced biofuel production. *Eukaryotic Cell* 9: 486–501. [CrossRef] [PubMed]

Republic of Turkey. 2015. Republic of Turkey Intended Nationally Determined Contribution. Available online: https://www4.unfccc.int/sites/submissions/INDC/PublishedDocuments/Turkey/1/The_INDC_of_TURKEY_v.15.19.30.pdf (accessed on 20 September 2021).

Schimschar, Sven, Thomas Boermans, David Kretschmer, Markus Offermann, and Ashok John. 2016. U-Value Maps Turkey: Applying the Comparative Methodology Framework for Cost-Optimality in the Context of the EPBD. Number BUIDE15722. Ecofys. Available online: https://izoder.org.tr/dosyalar/haberler/Turkiye-U-degerleri-haritasi-raporu-2016-Ingilizce.pdf (accessed on 21 September 2021).

Şenol, Halil. 2019. Biogas potential of hazelnut shells and hazelnut wastes in Giresun city. *Biotechnology Reports* 24: e00361. [CrossRef]

Tiryakioglu, Muhsin Baris. 2017. Turkey's EMRA Ups Max-Limit of Gas Distribution Regions. Available online: https://www.aa.com.tr/en/energy/regulation/turkeys-emra-ups-max-limit-of-gas-distribution-regions/1870 (accessed on 21 September 2021).

UNFCCC. 2018. Proposal from Turkey to Amend the List of Parties Included in Annex I to the Convention. Available online: https://unfccc.int/sites/default/files/resource/inf2.pdf (accessed on 21 September 2021).

United Nations. 2020. How Natural Gas Can Support the Uptake of Renewable Energy. Available online: https://www.un-ilibrary.org/content/books/9789210046589 (accessed on 21 September 2021).

Wiley, Patrick Edward. 2013. Microalgae Cultivation Using Offshore Membrane Enclosures for Growing Algae (OMEGA). Ph.D. thesis, University of California, Merced, CA, USA.

MDPI

St. Alban-Anlage 66

4052 Basel

Switzerland

Tel. +41 61 683 77 34

Fax +41 61 302 89 18

www.mdpi.com

MDPI Books Editorial Office

E-mail: books@mdpi.com

www.mdpi.com/books